EDIBLE COATINGS AND FILMS TO IMPROVE FOOD QUALITY

EDIBLE COATINGS and FILMS to IMPROVE FOOD QUALITY

EDITED BY

John M. Krochta
Department of Food Science and Technology
Department of Biological and Agricultural Engineering
University of California
Davis, California, U.S.A.

Elizabeth A. Baldwin
U.S. Department of Agriculture
Agricultural Research Service
Citrus and Subtropical Products Laboratory
Winter Haven, Florida, U.S.A.

Myrna O. Nisperos-Carriedo
Food Research Institute
Department of Agriculture
Werribee, Victoria, Australia

TECHNOMIC
PUBLISHING CO., INC.
LANCASTER · BASEL

Edible Coatings and Films to Improve Food Quality
a TECHNOMIC® publication

Published in the Western Hemisphere by
Technomic Publishing Company, Inc.
851 New Holland Avenue, Box 3535
Lancaster, Pennsylvania 17604 U.S.A.

Distributed in the Rest of the World by
Technomic Publishing AG
Missionsstrasse 44
CH-4055 Basel, Switzerland

Printed in the United States of America
10 9 8 7 6 5 4 3 2

Main entry under title:
    Edible Coatings and Films to Improve Food Quality

A Technomic Publishing Company book
Bibliography: p.
Includes index p. 357

Library of Congress Catalog Card No. 94-60493
ISBN No. 1-56676-113-1

Table of Contents

Preface

RESEARCH into edible coatings and films has been intense in recent years, with all indications that interest will continue. Motivations for developments in this field come from several directions. Consumer interests in health, food quality, convenience, and safety continue to increase, presenting food processors with new challenges to which edible coating and film concepts offer potential solutions. Consumers and processors alike have found new and renewed commitment to reducing the environmental consequences of packaging. By acting as barriers to moisture or oxygen, edible coatings can conceivably reduce the complexity and, therefore, improve recyclability of packaging. In some cases, edible films may be able to replace synthetic packaging films. Furthermore, food scientists and engineers have isolated new materials that present new opportunities in the formation and properties of edible coatings and films. In many cases, these materials are quite abundant in nature and have previously been regarded as surplus or waste. In some cases, these materials are being combined in new and creative ways to achieve heretofore unattainable coating and film properties.

With the explosion of interest and information in the area of edible coatings and films in recent years, a book that summarizes progress to this point was needed. Thus, *Edible Coatings and Films to Improve Food Quality* brings together some of the many researchers presently contributing to this field. The intent of the book is to achieve a number of goals, including introducing newcomers to this field, describing materials appropriate for use, summarizing properties, reviewing methods for application, describing approaches for mathematical modeling, and summarizing present and potential uses. Authors were asked to develop topics in a comprehensive manner, with the result that each chapter stands alone and as an important contributing element in the total book. Some overlap, therefore, occurs between chapters, but usually with the result of increased clarification due to the different contexts of each chapter. In addition to the review of the editors, each chapter was anonymously reviewed by experts in the chapter area to further insure accuracy and comprehensive presentation of subject matter.

The thanks of the editors go to the many researchers in this field who developed the information covered in the book, the authors who wrote the chapters, and Technomic Publishing Co., Inc. for their encouragement, patience, and support of this project.

CHAPTER 1

Edible Films and Coatings: Characteristics, Formation, Definitions, and Testing Methods

I. GREENER DONHOWE[1]
O. FENNEMA[2]

INTRODUCTION

ALTHOUGH the use of edible films in food products may seem new, food products were first enrobed in edible films and coatings many years ago. During the twelfth and thirteenth centuries, dipping of oranges and lemons in wax to retard water loss was practiced in China (Hardenberg, 1967). In England, during the sixteenth century, "larding," coating food products with fat, was used to prevent moisture loss in foods (Labuza and Contrereas-Medellin, 1981). Currently, edible films and coatings find use in a variety of applications, including casings for sausages, chocolate coatings for nuts and fruits, and wax coatings for fruits and vegetables. The technical challenges involved in producing stable foods suggest that edible films and coatings could be used to an even greater extent than they are currently. However, technical information concerning edible films is far from adequate, leaving the food scientist with the formidable task of developing a film for each food application.

The information presented in this chapter and subsequent ones will enable the food scientist to more readily develop edible films and to more thoroughly understand the fundamentals responsible for the physical properties of edible films.

RATIONALE FOR USING EDIBLE FILMS

Edible films can help meet the many challenges involved with the

[1]Griffith Laboratories North America, 1 Griffith Center, Alsip, IL 60658.
[2]Department of Food Science, University of Wisconsin, 1605 Linden Drive, Madison, WI 53706.

marketing of foods that are nutritious, safe, of high quality, stable, and economical by serving one or more of the functions listed in Table 1.1.

Edible films can be used as barriers to gases and water vapor. For this purpose they can be positioned either on the surface of the food, e.g., as a coating on fruits, or within the food, where they separate components of markedly different water activity.

In the first instance, the function is to restrict loss of moisture from the fruit to the environment or to lessen absorption of oxygen by the fruit and thereby slow respiration (Kester and Fennema, 1986). In the latter instance, the film may serve to stabilize water activity gradients and preserve different textural properties possessed by different food components. For example, an edible film could be used to separate the crisp component of a pizza from the moist semi-solid component.

Edible films can also be used as protective coatings for food ingredients that are susceptible to oxidation. Polysaccharide coatings on nutmeats serve this purpose (Murry and Luft, 1973).

Lessening migration of lipids is another potential application for edible films. Migration of fats and oils can be a problem in the confectionery industry, especially in products involving chocolate. For example, if oil from a peanut butter filling migrates into a chocolate coating, this will have an adverse effect on the properties of the chocolate (Nelson and Fennema, 1991; Paulinka, 1986).

Edible films and coatings can also improve the mechanical-handling properties, or structural integrity of a food product. In foods composed of many discrete particles, such as a pizza topping, an edible film could be used to secure these components in place during product distribution. Coatings can also provide some physical protection for food products

TABLE 1.1. **Possible Uses of Edible Films and Coatings.**

Use	Appropriate Types of Film
Retard moisture migration	Lipid, composite[a]
Retard gas migration	Hydrocolloid, lipid, or composite
Retard oil and fat migration	Hydrocolloid
Retard solute migration	Hydrocolloid, lipid, or composite
Improve structural integrity or handling properties	Hydrocolloid, lipid, or composite
Retain volatile flavor compounds	Hydrocolloid, lipid, or composite
Convey food additives	Hydrocolloid, lipid, or composite

[a]A composite film consists of lipid and hydrocolloid components combined to form a bilayer or conglomerate.

which are susceptible to injury during transport, such as fresh fruits and vegetables.

Food additives, such as flavors, antimicrobial agents, antioxidants, and colors can be incorporated into edible films and used to control location or rate of release of these additives in a food.

Additionally, edible films may offer an alternative to the commercial packaging materials used for food products. Using edible films for this purpose would likely reduce the packaging waste associated with processed foods.

Although it is clear that edible films and coatings have many potential applications in food, they have not, with few exceptions, been widely utilized by the food industry.

FILM COMPONENTS

Components of edible films and coatings can be divided into three categories: hydrocolloids, lipids, and composites. Suitable hydrocolloids include proteins, cellulose derivatives, alginates, pectins, starches, and other polysaccharides. Suitable lipids include waxes, acylglycerols, and fatty acids. Composites contain both lipid and hydrocolloid components. A composite film can exist as a bilayer, in which one layer is a hydrocolloid and the other a lipid, or as a conglomerate, where the lipid and hydrocolloid components are interspersed throughout the film. Specific types of film components will be discussed only briefly, since this information is covered in depth in Chapters 9–11.

HYDROCOLLOIDS

Hydrocolloid films can be used in applications where control of water vapor migration is not the objective. These films possess good barrier properties to oxygen, carbon dioxide, and lipids. Most of these films also have desirable mechanical properties, making them useful for improving the structural integrity of fragile products. Water solubility of polysaccharide films is advantageous in situations where the film will be consumed with a product that is heated prior to consumption. During heating, the hydrocolloid film or coating would dissolve, and ideally, would not alter the sensory properties of the food.

Hydrocolloids used for films and coatings can be classified according to their composition, molecular charge, and water solubility. In terms of composition, hydrocolloids can be either carbohydrates or proteins. Film-forming carbohydrates include starches, plant gums (for example alginates, pectins, and gum arabic), and chemically modified starches.

Film-forming proteins include gelatin, casein, soy protein, whey protein, wheat gluten, and zein. It should be noted, however, that great differences exist in how easily films of good integrity can be formed from these substances.

The charged state of a hydrocolloid can be useful for film formation. Alginates and pectins require the addition of a polyvalent ion, usually calcium, to facilitate film formation. They, as well as proteins, are susceptible to pH changes because of their charged state. For some applications, an advantage can be gained by combining hydrocolloids of opposite charge such as gelatin and gum arabic.

Although hydrocolloid films generally have poor resistance to water vapor because of their hydrophilic nature, those which are only moderately soluble in water, such as ethylcellulose, wheat gluten, and zein do provide somewhat greater resistance to the passage of water vapor than do the water-soluble hydrocolloids. The dependence of permeability on solubility is more thoroughly discussed in Chapter 7.

LIPIDS

Lipid films are often used as barriers to water vapor, or as coating agents for adding gloss to confectionery products. Their use in a pure form as free-standing films is limited, because most lack sufficient structural integrity and durability.

Waxes are commonly used for coating fruits and vegetables to retard respiration and lessen moisture loss. Formulations for wax coatings vary greatly and the compositions are often proprietary. Acetylated monoglycerides are frequently added to wax formulations to add pliability to the coating.

Shellac coatings, when formed on a supporting matrix, provide effective barrier properties to gases and water vapor (Hagenmaier and Shaw, 1991). Coatings to sucrose fatty acid esters reportedly are effective moisture barriers for maintaining the crispness of snack foods (Kester et al., 1990) and for extending the shelf-life of apples (Drake et al., 1987).

Although fatty acids and fatty alcohols are effective barriers to water vapor, their fragility requires that they be used in conjunction with a supporting matrix.

Many lipids exist in a crystalline form and their individual crystals are highly impervious to gases and water vapor (Fox, 1958). Since the permeate can pass between crystals, the barrier properties of crystalline lipids are highly dependent on the intercrystalline packing arrangement. Lipids consisting of tightly packed crystals offer greater resistance to diffusing gases than those consisting of loosely packed crystals. Also, crystals oriented with their major planes normal to permeate flow provide better barrier properties than crystals that are oriented differently (Fox, 1958).

Lipids existing in a liquid state or having a large proportion of liquid components offer less resistance to gas and vapor transmission than those in a solid state (Kamper and Fennema, 1984a; Kester and Fennema, 1989a, 1989b), indicating that molecular mobility of lipids detracts from their barrier properties.

The barrier properties of lipids having crystalline properties can be influenced both by tempering and by polymorphic form (Kester and Fennema, 1989e, 1989f).

COMPOSITES

Composite films can be formulated to combine the advantages of the lipid and hydrocolloid components and lessen the disadvantages of each. When a barrier to water vapor is desired, the lipid component can serve this function while the hydrocolloid component provides the necessary durability. Properties of lipid-hydrocolloid bilayer films have been studied extensively in the authors' laboratory (Greener and Fennema, 1989a, 1989b; Kester and Fennema, 1989a–1989f; Kamper and Fennema, 1984a, 1984b). Composite films consisting of a conglomerate of casein and acetylated monoglycerides have been studied by Krochta et al. (1990). These films can be used as coatings for processed fruits and vegetables. A composite film of gum acacia and glycerolmonostearate was reported to have good water vapor barrier properties at a 43.8–23.6% relative humidity gradient (Martin-Polo and Voilley, 1990).

FILM ADDITIVES

The functional, organoleptic, nutritional, and mechanical properties of an edible film can be altered by the addition of various chemicals in minor amounts. Plasticizers, such as glycerol, acetylated monoglyceride, polyethylene glycol, and sucrose are often used to modify the mechanical properties of a film. Incorporation of these additives may, however, cause significant changes in the barrier properties of the film. For example, the addition of hydrophilic plasticizers usually increases the water vapor permeability of the film.

Other types of film additives often found in edible film formulations are antimicrobial agents, vitamins, antioxidants, flavors, and pigments.

FILM FORMATION

Many techniques have been developed for forming films directly on food surfaces, or as separate, self-supporting films.

COACERVATION

Coacervation involves separation of a polymeric coating material from a solution by heating, altering pH, adding solvents, or altering the charge on the polymer involved (Deasy, 1984; Bankan, 1973). In simple coacervation, only one soluble polymer is involved (Deasy, 1984; Bankan, 1973). In complex coacervation, at least two oppositely charged macromolecules are combined to yield an insoluble mixed polymer by the mechanism of charge neutralization (Glicksman, 1982).

Coacervation may also be classified according to the type of phase separation: aqueous or nonaqueous. Aqueous phase separation requires a hydrophilic coating such as gelatin or gelatin–gum acacia that is deposited on a water-insoluble core particle. Nonaqueous phase separation usually involves a hydrophobic coating that is deposited on a core which may be either water-soluble or water-insoluble (Dziezak, 1988).

Although coacervation has been used extensively in the pharmaceutical industry, especially for encapsulation, few applications have developed in the food industry. This is probably the result of costly equipment; the lack of food-grade status for some encapsulating materials; and with flavor encapsulation, difficulty incorporating adequate flavor concentrations in microcapsules (Dziezak, 1988).

SOLVENT REMOVAL

For film-forming materials dispersed in aqueous solutions, solvent removal is a necessity for solid film formation. Rate and temperature of drying has been found to influence the resulting crystallinity and mechanical properties of cellulosic films (Greener, 1992; Reading and Spring, 1984).

SOLIDIFICATION OF MELT

Solidification of the melt by cooling is a common technique for preparing lipid films. Similar to the rate of solvent removal, the rate of cooling plays an important role in the overall physical properties of the resulting film. The rate of cooling influences the predominant polymorphic state, as well as degree of recrystallization in the solidified film. Kester and Fennema (1989e) reported that the water vapor and oxygen resistances of lipid films were dependent on the polymorphic state. After the initial solidification, the barrier properties of lipid films to water vapor and oxygen are further altered by tempering (Kester and Fennema, 1989f; Landmann et al., 1960).

FILM APPLICATION

Any of the above film-forming techniques can be utilized with any of the following application techniques.

DIPPING

This method lends itself to food products that require several applications of coating materials or require a uniform coating on an irregular surface. After dipping, excess coating material is allowed to drain from the product, and it is then dried or allowed to solidify. This method has been used to apply films of acetylated monoglycerides to meats, fish, and poultry, and to apply coatings of wax to fruits and vegetables.

SPRAYING

Films applied by spraying can be formed in a thinner, more uniform manner than those applied by dipping. Spraying, unlike dipping, is more suitable for applying a film to only one side of a food to be covered. This is desirable when protection is needed on only one surface, e.g., when a pizza crust is exposed to a moist sauce. Spraying can also be used to apply a thin second coating, such as the cation solution needed to cross-link alginate or pectin coatings.

CASTING

This technique, useful for forming free-standing films, is borrowed from methods developed for nonedible films. Coating is simple and allows film thickness to be controlled accurately on smooth, flat surfaces. Casting can be accomplished by controlled-thickness spreading or by pouring. Controlled-thickness spreading requires a spreader with a product reservoir and an adjustable gate, the height of which can be set accurately and with good reproducibility. The spreader is simply drawn over the receiving surface, depositing a layer of the film-forming solution of the desired thickness, which is subsequently dried. Alternatively, the film-forming solution can be poured into a confined area of a level receiving surface and subsequently dried.

OTHER METHODS

Coatings in liquid form can also be applied with brushes, by use of a falling-film enrobing technique, by panning, or with rollers (Guilbert, 1986).

Additional information on methods of applying edible coatings is presented in Chapter 8.

DEFINITIONS

When seeking information about the barrier properties of films and coatings, one is confronted with a wide variety of terms, such as water vapor permeability, water vapor transmission rate, permeability, permeability coefficient, permeance, and resistance. When investigators use different terms to describe the barrier properties of films, it is difficult to compare the results from different studies. It is appropriate, therefore, to briefly review permeation phenomenon and the various terms used to describe it. Additional information is provided in Chapter 7.

Gas transport can occur by two mechanisms: capillary diffusion and activated diffusion. Capillary diffusion dominates in materials which are porous or which have imperfections. Activated diffusion involves solubilization of the permeant gas in the film, diffusion through the film, and finally release at the opposite side of the film. In the absence of imperfections, gas permeation through a film is likely to occur by activated diffusion. If a gas is insoluble in a film and yet penetrates the film, the dominant mechanism is capillary flow.

The term "activated diffusion," used extensively in diffusion and permeability literature since the 1930s (Chao and Rizvi, 1988; Doty, 1946; Barrer, 1938), deserves further consideration. This term should not be confused with "facilitated diffusion," which applies to diffusion mechanisms in living cells. The term "activated diffusion" is applied to diffusion processes that are temperature-dependent and possess an activation energy.

Most of the mathematical expressions used to describe the barrier properties of edible films are identical to those used for synthetic films.

PERMEABILITY

A combination of Fick's first law of diffusion and Henry's law of solubility is used to express steady state permeability of a permeate through a nonporous barrier having no significant imperfections. Fick's first law states that the permeate flux, J, is dependent on permeate diffusivity (D), the differential of concentration (dC) in the film, and the differential of film thickness (dX). It can be expressed by the following equation:

$$J = -D\frac{dC}{dX} \tag{1.1}$$

The negative sign indicates that migration occurs in the direction of lower concentration.

If Henry's law of solubility is obeyed, the concentration (C) of permeate in the film is equal to the product of the solubility coefficient (S) and the permeate partial pressure in the adjacent air (P):

$$C = SP \tag{1.2}$$

Combining Equations (1.1) and (1.2) gives:

$$J = -DS\frac{dP}{dX} \tag{1.3}$$

in which dC is replaced by $S \cdot dP$, and where dP is the differential of the partial pressure of the permeate across the film. By rearrangement, the quantity DS, also known as permeability, can be expressed as:

$$DS = -\frac{JdX}{dP} = \text{permeability} \tag{1.4}$$

Experimentally, permeability can be determined according to the expression:

$$\text{permeability} = \frac{\text{permeate weight} \cdot \text{thickness}}{\text{area} \cdot \text{time} \cdot \text{partial pressure difference}} = \frac{\text{g} \cdot \text{cm}}{\text{m}^2 \cdot \text{day} \cdot \text{mmHg}}$$
$$\tag{1.5}$$

Some other commonly used units for permeability are listed in Table 1.2.

The terms "permeability" and "permeability coefficient" are often used interchangeably (ASTM, 1990). In some instances this practice is valid; in others, however, it is not. The term "permeability coefficient" should be used only when it has been established that permeability is constant, regardless of the pressure or concentration gradients (standardized on a basis of per unit of pressure or concentration). When this occurs, both Henry's law and Fick's law are obeyed. In contrast, the term "permeability" can be used either under conditions in which instance permeability and permeability coefficient are identical, or under circumstances where they are not. For example. "permeability" would be appropriate (and "permeability coefficient" would be inappropriate) for describing the rate at which water vapor permeates a hydrophilic film, since the permeability determined at 50% relative humidity (RH) differs from that at 95% RH.

Solute permeability of edible films can be expressed using Equation (1.4), if the partial pressure gradient term is replaced by a concentration gradient term (Vojdani and Torres, 1990).

TABLE 1.2. Common Terms Used to Describe Barrier Properties of Edible Films.

Term	Equation[a]	Common Units[b]	Acceptable SI Units[c]
Permeability	$M \cdot \Delta X \cdot (A \cdot t \cdot \Delta P)^{-1}$ or $M \cdot \Delta X \cdot (A \cdot t \cdot \Delta C)^{-1}$	g·mil·(100 in²·day·cmHg)⁻¹ g·mil·(m²·day·mmHg)⁻¹ cc·mil·(m²·day·atm)⁻¹ g·cm·(cm²·s·mmHg)⁻¹ cc·cm· (cm²·s·cmHg)⁻¹ mg·cm·[cm²·sec·(mg·ml⁻¹)]⁻¹	g·(Pa·s·m)⁻¹ kg·(Pa·s·m)⁻¹ Δmol·(Pa·s·m)⁻¹ m²·sec⁻¹
Permeability coefficient	$M \cdot \Delta X \cdot (A \cdot t \cdot \Delta P)^{-1}$	Same units as permeability	Same units as permeability
Permeance	$M \cdot (A \cdot t \cdot \Delta P)^{-1}$ or $M \cdot (A \cdot t \cdot \Delta C)^{-1d}$	g·(m²·day·mmHg)⁻¹ m·s⁻¹	g·(Pa·s·m²)⁻¹ kg·(Pa·s·m²)⁻¹ ϱmol·(Pa·s·m²)⁻¹
Transmission rate	$M \cdot (A \cdot t)^{-1}$ or $M \cdot \Delta X \cdot (A \cdot t)^{-1e}$	g·(m²·day)⁻¹ g·(m²·h)⁻¹ g ·mil·(m²·day)⁻¹ cc·mil·(100 in²·day)⁻¹ cc·cm· (cm²·day)⁻¹ g·cm·(cm²·h)⁻¹	g·(m²·s)⁻¹ ϱmol·(m²·s)⁻¹ g·mm·(m²·s)⁻¹
Resistance	$\Delta C \cdot A \cdot t \cdot M^{-1}$	s·m⁻¹	
Activation energy	$P_o \exp(-E_p \cdot R^{-1} \cdot T^{-1})$	kcal·mol⁻¹ kJ·mol⁻¹	kJ·mol⁻¹

[a] M = mass of permeate; ΔX = thickness of film; A = area; t = time; ΔP = pressure gradient; ΔC = concentration gradient; P_o = permeability constant; E_p = activation energy; R = gas content; T = temperature (kelvin).
[b] Units listed are found throughout the literature on packaging films.
[c] SI = Systéme International (SI) units. These are preferred.
[d] This expression for permeance is commonly used in describing barrier properties of plant lipids.
[e] Transmission rate is often reported with a thickness term in the expression.

Terms like "permeability," which take barrier thickness into account, are used with the assumption that a plot of permeation versus thickness yields a value of 1.00, or permeability is a function of thickness^{-x}, where $x = 1$. In actuality, x varies between 0.8 and 1.2 for most polymers (Salame and Steingiser, 1977). Therefore, a 2-mil film will exhibit a permeability that is 80–120% that of the value of a 1-mil film. Because of this inconsistency, film thickness should always be indicated, even when thickness is included in the expression being used. Readers interested in the theories that address this relationship are directed to Hwang and Kammermeyer (1974).

The terms "permeability" and "permeability coefficient" are most commonly used to describe films of homogeneous composition. The term "permeability" may be used for heterogeneous films; again, the partial pressure gradient and film thickness should also be reported.

PERMEANCE

Permeance is simply the permeability (or permeability coefficient) expression devoid of the term for film thickness. Permeance is a "performance evaluation," not an inherent property of the film (ASTM, 1990).

Units often used to express permeance are as follows:

$$\text{permeance} = \frac{\text{permeate weight}}{\text{area} \cdot \text{time} \cdot \text{partial pressure difference}} = \frac{\text{g}}{\text{m}^2 \cdot \text{day} \cdot \text{mmHg}}$$
(1.6)

or:

$$\text{permeance} = \frac{\text{permeate weight}}{\text{time} \cdot \text{area} \cdot \text{concentration gradient}} = \text{m} \cdot \text{sec}^{-1} \quad (1.7)$$

Other units sometimes used for permeance are listed in Table 1.2.

With this expression, area, time, and permeate partial pressure are controlled and the weight of the permeate passing through the film is measured. Even when thickness is known, it is not incorporated in the calculation, usually because the barrier is not homogeneous. Film thickness should always be reported when it is known.

The term "permeance" is often used when the film is heterogeneous, of unknown composition, and/or of unknown thickness. However, it is imperative that permeance values be accompanied by a statement clearly indicating that differences in permeance values may be caused by these differences in composition and/or thickness.

Equation (1.7) is commonly used to describe the barrier properties of plant cuticles.

TRANSMISSION RATE

Transmission rate, like permeance, is derived from the expression for permeability with the exception that film thickness and the partial pressure gradient of the permeate are disregarded.

Units used to express transmission rate are as follows:

$$\text{transmission rate} = \frac{\text{permeate weight}}{\text{area} \cdot \text{time}} = \frac{\text{g}}{\text{m}^2 \cdot \text{day}} \qquad (1.8)$$

Sometimes transmission rate is reported with a thickness term in the numerator of Equation (1.8). Other commonly used units are listed in Table 1.2.

When transmission rate is determined according to Equation (1.8), time and film area are controlled, and the weight of the permeate transmitted is measured. The partial pressure differential of the permeate and the thickness of the film are not considered, but should be reported when known. Transmission rate suffers from a deficiency similar to that of permeance, but worse, i.e., two variables are usually left unstandardized—partial pressure differential and film thickness. Transmission should not, therefore, be used to compare the barrier properties of films of different compositions or thicknesses except in circumstances where the pressure gradient and film thickness are unknown. In such a case, it is imperative that transmission values be accompanied by a statement indicating that differences in rates may be caused by difference in any or all of the nonstandardized variables, namely, film composition, thickness (both of the film and the critical barrier component), and/or the partial pressure difference across each barrier. Furthermore, when transmission values are reported, unknown variables should be clearly stated and the values of variables that are known, but not included in the expression for transmission, should be reported.

Use of the term "transmission rate" is considered appropriate for describing the permeabilities of films when they are tested at a single RH gradient that is considered normal. For example, TAPPI (Technical Association of Paper and Pulp Industries) uses transmission rate as the standard unit for determining the barrier properties of sheet materials, such as paper to water vapor (TAPPI, 1971). For this test, a 50–0% RH difference is used.

RESISTANCE

The expression "resistance" can also be used to describe the ability of a material to serve as a barrier to permeate. This expression is often used to describe the barrier properties of plant cuticular membranes (Holmgren et al., 1965), animal and insect lipids (Hadley, 1985), and occasionally other heterogeneous barriers (Kester and Fennema, 1989c–1989f). Resistance to water vapor transmission through a film is expressed as:

$$\text{resistance} = \frac{W_i - W_a}{J} = \text{sec} \cdot \text{m}^{-1} \tag{1.9}$$

where W_i is the saturated water vapor concentration (saturated vapor concentration at the surface temperature of the film in g water vapor·m^{-3}), W_a is the water vapor concentration at the test condition (g water vapor·m^{-3}), and J is the flux of water vapor (g·m^{-2}·sec^{-1}). Resistance is the inverse of permeance when expressed with concentration as the driving force [Equation (1.7)], and the stipulations applying to use of the permeance term apply equally to resistance.

Use of resistance (or permeance) is especially appropriate with films of heterogeneous composition. This expression disregards thickness, which is acceptable in this instance since total film thickness will not reflect the relative resistances provided by various film components. Despite this fact, an attempt should be made to keep film thickness reasonably uniform among samples being tested. Moreover, when reporting resistance values, film thickness should be reported if known, or estimated and reported if not known.

In summary, when working with homogeneous barrier materials, use of the terms "permeability" or "permeability coefficient" is most suitable. With heterogeneous materials, such as composites, the terms "permeance," or its inverse, "resistance," are suitable, provided one is keenly aware of the impropriety of comparing the barrier effectiveness of different films unless discussion is included indicating how differences in film thickness and/or composition might influence the results. Use of the term "transmission rate" is appropriate when films of standard thickness are tested at a single RH difference that is considered "normal." Transmission rate may also be used in circumstances where film thickness and partial pressure difference are unknown, provided appropriate accompanying statements are reported.

L NUMBER

Hong et al. (1990) used a dimensionless number, called the "L" number, to evaluate the moisture barrier properties of edible films and food products. The L number is the ratio of the moisture permeance of the food to the moisture permeance of the coating material. For coated food materials, L numbers less than 0.04 indicate that the main resistance to moisture migration is the food, and L numbers greater than 4 indicate that the main resistance is the coating. The authors also provide examples of how the L number can be used to select appropriate coatings. This approach appears to be successful, but totally ignores the role of coating thickness in achieving this success. For example, the authors report that paraffin and chocolate liquor have identical L numbers. But since thickness and composition of these coatings is not taken into account, the L number fails to inform the potential user that for identical barrier properties, the paraffin film need be only 3.9–5.7-mils (0.1–0.014-cm) thick whereas the chocolate liquor film must be 66.4–66.9-mils (0.167–0.170-cm) thick. As stressed earlier, permeance values, which are involved in determining the L number, are appropriate for the direct comparison of films of differing thickness and composition only when accompanied by statements regarding the roles of composition and thickness in the results. This stipulation was violated in this instance.

ACTIVATION ENERGY

For gases and vapors, the temperature dependences of the permeability, resistance, or permeance values of edible films often obey the Arrhenius relationship:

$$P = P_o e - (E_p/RT) \tag{1.10}$$

where P is permeability, P_o is a constant (a kinetic frequency factor), R is the gas constant, and E_p is the activation energy for permeation. The activation energy for permeation can be easily determined from the slope of plots of reciprocal temperature (kelvin) versus either the logarithm of permeability, the logarithm of permeance, or the logarithm of resistance. The activation energy determined by the above graphical procedure represents an inherent property of a material in the presence of a specific gas or vapor.

The activation energy, E_p, is the sum of the heat of solution, ΔH_{sol}, and the activation energy for diffusion, E_d (Myers et al., 1961):

$$E_p = \Delta H_{sol} + E_d \tag{1.11}$$

For most vapors and gases, ΔH_{sol} is close to zero in value, and E_d is large and positive, making E_p large and positive. E_p may become negative in value with water vapor permeating through a hydrophilic film. The change in value for E_p is due to a large negative value for ΔH_{sol} and a reduced E_d, both of which result in less water being absorbed in the film and diffusing through as the temperature is increased. Negative activation energies have been reported for water vapor migrating through lipid films cast on Whatman-50 filter paper (Kester and Fennema, 1989d).

Determination of an activation energy is only valid for films yielding a linear relationship between log permeability and temperature $(K)^{-1}$, i.e., in instances where the permeant-barrier interactions change linearly with change in temperature. This condition is violated with water vapor transmission through cellophane and a valid activation energy cannot be determined for this film-vapor combination. Also, a linear Arrhenius plot will not be obtained if structural changes, such as crystallization or polymorphic alterations, occur in the film over the temperature range being studied.

Gases migrating through polymeric films generally exhibit activation energies of 3–15 kcal/mole. Very small activation energies may be an indication of small cracks existing within the film, causing permeation to occur primarily by capillary diffusion.

Early work in the field of synthetic polymers established a linear positive correlation between E_p and the logarithm of the constant P_o [Equation (1.10)] for a given gas (Doty, 1946). Thus, materials with large activation energies were found to have large values of P_o, suggesting a relationship between the number of active sites and their probability for activation.

Summarized in Table 1.2 are the permeation expressions discussed here and the common units associated with these expressions. In Table 1.3, an attempt has been made to indicate appropriate uses for the various permeation expressions.

CHOICE OF UNITS

Confusion can and does arise from the large array of alternate combinations of units used to express the various permeation terms. Karel (1975) listed eleven combinations of units used for reporting permeability. Today the list is even longer, with the introduction of "perm" by the American Society for Testing and Materials (ASTM, 1990) with units of 5.7×10^{-11} $g \cdot Pa^{-1} \cdot s^{-1} \cdot m^{-2}$ for permeance and 1.45×10^{-9} $g \cdot Pa^{-1} \cdot s^{-1} \cdot m^{-1}$ for permeability. It is hoped that standardization will soon come to this field and it is recommended that SI units (Table 1.2) form the basis of this standardization.

TESTING METHODS

Many of the tests developed for determining permeability and other physical properties of edible films are based on standard test methods used

TABLE 1.3. Attributes of Equations Relating to Barrier Properties of Edible Films.

Term	Variable Measured[a]	Variables Accounted For[a]	Variables Unaccounted For[a]	Comments
Permeability	M	ΔX, A, t, ΔP or ΔC	None	Suitable for use with any homogeneous barrier. Identical to permeability coefficient only when permeability is constant with varying pressure or concentration gradient (standardized).
Permeability coefficient	M	ΔX, A, t, ΔP or ΔC	None	Suitable for use with any homogeneous barrier provided permeability is constant over varying pressure or concentration gradients (standardized).
Permeance	M	A, t, ΔP, or ΔC	ΔX	Suitable for barriers of heterogeneous composition such as conglomerate films where thickness of the film cannot be accurately related to a specific chemical component of the film that is responsible for the barrier properties. This expression can be used to compare barrier properties of films of differing compositions and thickness, provided values are accompanied by a statement indicating that differences in barrier properties may be caused by these differences in composition and/or thickness. When comparing barrier properties of films of the same composition, thickness should be similar and reported. When ΔC rather than ΔP is incorporated in the expression, then permeance is the inverse of resistance.

16

TABLE 1.3. (continued).

Term	Variable Measured[a]	Variables Accounted For[a]	Variables Unaccounted For[a]	Comments
Transmission rate	M	A, t	$\Delta X, \Delta P$	Since this expression does not include terms for thickness or pressure gradient, its use for comparing barrier properties of films with different compositions is limited. This term is suitable for reporting permeation data for films of standard thickness when results at a single RH gradient are especially relevant. This term may also be used for barriers of unknown thickness or composition with unknown pressure differentials occurring across the films. Comparisons of transmission rates for different films under these circumstances can be made, provided the precautionary statement mentioned for permeance is included. The pressure gradient and thickness, if known, should be reported.
Transmission rate	M	$A, t, \Delta X$	ΔP	This expression for transmission rate includes a thickness term, however the pressure gradient remains unaccounted for. The pressure gradient, if known, should be reported. The comments mentioned for the first transmission rate expression discussed in this table apply here as well.

(continued)

TABLE 1.3. (continued).

Term	Variable Measured[a]	Variables Accounted For[a]	Variables Unaccounted For[a]	Comments
Resistance	M	Q, t, M	ΔX	This expression is commonly used to describe barrier properties of plant membranes and insect lipids. Like permeance, a thickness variable is not included, making this term suitable for heterogeneous barriers. If resistance values are used to compare films of differing composition and thickness, the precautionary statement mentioned for permeance must be included with the reported data. Resistance is the inverse of permeance when the ΔC term is used.

[a]M = mass of permeate; ΔX = thickness of film; A = area; t = time; ΔP = pressure gradient; ΔC = concentration gradient.

for nonedible films. Permeability properties of films and coatings are discussed in greater detail in Chapter 7.

GAS PERMEABILITY

Karel (1975) reviewed several methods for measuring the gas permeability of packaging films. One method involves applying the permeate at elevated pressure to one side of the film and measuring the increase in pressure, with time, on the low pressure side of the film. The rate at which the test gas permeates the film can be quantified using mass spectrometry (Anonymous, 1975a).

Another method involves establishing a concentration gradient of the permeate across the film, while maintaining equal total pressures on both sides of the film. This is accomplished by pressurizing the permeate-poor side of the film with a nonpermeating gas, as needed. Gas that permeates can be determined with a gas chromatograph. Permeability can also be determined by measuring the volume of a gas that permeates a film when a pressure gradient is established across the film (Stern et al., 1964).

Since permeability of films to oxygen is often of importance, many analytical devices, some devised specifically for packaging films, have been used to measure concentrations of this gas. These include gas chromatographs, paramagnetic oxygen analyzers, galvanic cell electrodes and coulometric sensors (Anonymous, 1975b; Quast and Karel, 1972; Karel et al., 1963; Landrock and Proctor, 1952).

WATER VAPOR PERMEABILITY

Water vapor permeability of edible films, like that of flexible, nonedible films, is normally measured using a 100–0% RH gradient. In the past, results derived using a 100–0% RH gradient were generally regarded as "worse case" data, since exposure to a smaller RH gradient was assumed to be a less severe condition. However, recent studies have shown that for lipid-hydrocolloid bilayer films, a 97–65% RH gradient represents a more severe test condition (Greener and Fennema, 1989a; Kester and Fennema, 1989a; Kamper and Fennema, 1984b). These studies demonstrate the necessity of testing films under conditions of temperature and RH that will be encountered in practice. For example, if the film will be situated between a piece of luncheon meat and a cracker, the film should be tested at an RH gradient of 88–30%, since this properly represents conditions prevailing in the two components. In all instances, the experimenter should report the exact RH gradient and temperature conditions of the test.

Water vapor permeability is usually measured gravimetrically. A film is sealed over the opening of a special impervious cup containing a desiccant, and the cup is then placed in a chamber at constant temperature and RH (ASTM, 1990). The desiccant maintains a constant low RH environment (usually 0% RH) inside the cup. More rapid techniques, which are generally more expensive, have been developed, and these involve detection of water vapor by means of electrical resistance, electrolysis, or infrared absorption (ASTM, 1983; Anonymous, 1986).

SOLUTE PERMEABILITY

Detection of solute permeability can involve isotopes, or potentiometric methods (Torres and Karel, 1985; Schönerr et al., 1979). Regardless of the method chosen, all tests should be conducted, if possible, under conditions that closely resemble those in which the film will be used, and test conditions should be reported in detail.

LIPID PERMEABILITY

Permeability of edible films to lipid can be determined using techniques such as colormetric assays, isotopes, or incorporation of lipid probes (Nelson and Fennema, 1991; Tabouret, 1987).

OTHER TESTS

Tests to determine an edible film's ability to tolerate repeated surface impact and deformation were developed by Greener and Fennema (1989b). The surface impact test involved controlled impingement of glass beads against the lipid surface of a lipid-hydrocolloid edible film. The deformation test involved controlled stretching of the film over an uneven surface. Both tests were developed to simulate abusive conditions that might occur to an edible film during commercial use and were modified versions of standard tests (TAPPI, 1974, 1977).

Little research has been done on adhesion of edible films to a host substrate. An exception exists with respect to adhesion of edible coatings to pharmaceutical tablets (Johnson and Zografi, 1986; Fung and Parrot, 1980; Rowe, 1978).

The pharmaceutical literature offers a wealth of information on the physical and mechanical properties of edible films, and those interested in applications to foods should keep abreast of this literature. The interested reader can refer to two excellent reviews of pharmaceutical films for additional information (Banker, 1966; Munden et al., 1964).

SUMMARY

The use of edible films is attracting increasing interest in the food industry. This is because edible films are available with a wide range of properties that can help to alleviate many problems encountered with storing and transporting foods.

Numerous expressions have been used to quantify the barrier properties of edible films: permeability, permeability coefficient, resistance, permeance, and transmission rate. These expressions differ in appropriateness, and some are deceiving or inappropriate if not properly used. A concerted effort is needed to standardize combinations of units used for each expression and to develop guidelines for the proper use of these expressions.

Many methods exist for testing the permeability of edible films to various penetrants and for testing their durability. Some of these are standard methods and others are not. Standard methods should be used unless such tests are not available for a particular need, or their use is not feasible experimentally.

REFERENCES

Anonymous. 1975a. *Instruction for the Oxtran-100 Oxygen Permeability Tester.* Minneapolis, MN: Mocon Instruments, Inc.

Anonymous. 1975b. *MS Gas Transmission Apparatus: Model CS-180.* Cedar Knoll, NJ: Custom Scientific Instruments, Inc.

Anonymous. 1986. *Instruction Manual for the Permatran-W1A.* Minneapolis, MN: Mocon Instruments, Inc.

ASTM. 1983. "Standard Test Method for Water Vapor Transmission Rate of Sheet Materials Using a Rapid Technique for Dynamic Measurement," E398–83, *ASTM Book of Standards,* 4.06:712–716.

ASTM. 1990. "Standard Test Method for Water Vapor Transmission of Materials," E96–80, *ASTM Book of Standards,* 15.09:811–818.

Bankan, J. A. 1973. "Microencapsulation of Foods and Related Products," *Food Technol.,* 24(11):34–44.

Banker, G. S. 1966. "Film Coating Theory and Practice," *J. Pharm. Sci.,* 55:81–89.

Barrer, R. M. 1938. "Permeation, Diffusion and Solution of Gases in Organic Polymers," *Trans. Faraday Soc.,* 35:628–643.

Chao, R. R. and S. H. Rizvi. 1988. "Oxygen and Water Vapor Transport through Polymeric Film," in *Food and Packaging Interactions,* J. H. Hotchkiss, ed., Washington, DC: American Chemical Society.

Deasy, P. 1984. *Microencapsulation and Related Drug Processes.* New York: Marcel Dekker, Inc., pp. 21–60.

Doty, P. M. 1946. "On Diffusion of Vapors through Polymers," *J. Chem. Phys.,* 4(4):244–251.

Drake, S. R., J. K. Fellman and J. W. Nelson. 1987. "Postharvest Use of Sucrose Polyesters for Extending the Shelf-Life of Stored 'Golden Delicious' Apples," *J. of Food Sci.*, 52(5):1283–1285.

Dziezak, J. 1988. "Microencapsulation and Encapsulated Ingredients," *Food Technol.*, 42(4):136–148.

Fox, R. C. 1958. "The Relationship of Wax Crystal Structure to the Water Vapor Transmission Rate of Wax Films," *TAPPI*, 41(6):283–289.

Fung, R. and E. Parrot. 1980. "Measurement of Film-Coating Adhesiveness," *J. Pharm. Sci.*, 64(4):439–441.

Glicksman, M. 1982. *Food Hydrocolloids, Vol. 1.* Boca Raton, FL: CRC Press, Inc., pp. 86–90.

Greener, I. 1992. "Physical Properties of Edible Films and Their Components," Ph.D. dissertation, University of Wisconsin.

Greener, I. and O. Fennema. 1989a. "Barrier Properties and Surface Characteristics of Edible Bilayer Films," *J. Food Sci.*, 54(6)1393–1399.

Greener, I. and O. Fennema. 1989b. "Evaluation of Edible, Bilayer Films for Use as Moisture Barriers for Food," *J. Food Sci.*, 54(6):1400–1406.

Guilbert, S. 1986. "Technology and Application of Edible Protective Films," in *Food Packaging and Preservation: Theory and Practice*, M. Mathlouthi, ed., London, UK: Elsevier Applied Science Publishing Co., pp. 371–394.

Hadley, N. 1985. "Integumental Waterproofing," in *The Adaptive Role of Lipids in Biological Systems.* New York, NY: John Wiley & Sons, p. 129.

Hagenmaier, R. and P. Shaw. 1991. "Permeability of Shellac Coatings to Gases and Water Vapor," *J. Agric. Food Chem.*, 39(5):827–829.

Hardenberg, R. E. 1967. "Wax and Related Coatings for Horticultural Products–A Bibliography," *Agricultural Research Service Bulletin No. 965.* Ithaca, NY: Cornell University, p. 1.

Holmgren, P., P. G. Jarvis and M. Jarvis. 1965. "Resistances to Carbon Dioxide and Water Vapour Transfer in Leaves of Different Plant Species," *Physiol. Plant.*, 18:557–573.

Hong, Y. C., C. M. Koelsch and T. P. Labuza. 1990. "Using the L Number to Predict the Efficacy of Moisture Barrier Properties of Edible Food Coating Materials," *J. Food Proc. Pres.*, 15:45–62.

Hwang, S. T. and K. Kammermeyer. 1974. "The Effect of Thickness on Permeability," in *Polymer Science and Technology: Permeability of Plastic Films and Coatings to Gases, Vapors, and Liquids, Vol. 6*, H. B. Hopfenberg, ed., New York, NY: Plenum Press.

Johnson, B. A. and G. Zografi. 1986. "Adhesion of Hydroxypropyl Cellulose Films to Low Energy Solid Substrates," *J. Pharm. Sci.*, 75(6):529–533.

Kamper, S. L. and O. Fennema. 1984a. "Water Vapor Permeability of Edible Bilayer Films," *J. Food Sci.*, 49(6):1478–1481, 1485.

Kamper, S. L. and O. Fennema. 1984b. "Water Vapor Permeability of an Edible, Fatty Acid Bilayer Film," *J. Food Sci.*, 49(6):1482–1485.

Karel, M. 1975. "Protective Packaging of Foods," in *Principles of Food Science*, O. Fennema, ed., New York, NY: Marcel Dekker, pp. 399–464.

Karel, N., P. Issenberg and V. Jurin. 1963. "Application of Gas Chromatography to the Measurement of Gas Permeability of Packaging Materials," *Food Technol.*, 17:327–330.

Kester, J. J. and O. Fennema. 1986. "Edible Films and Coatings: A Review," *Food Technol.*, 40(12):47–59.

Kester, J. J. and O. Fennema. 1989a. "An Edible Film of Lipids and Cellulose Ethers: Barrier Properties to Moisture Vapor Transmission and Structural Evaluation," *J. Food Sci.*, 54:1383–1389.

Kester, J. J. and O. Fennema. 1989b. "An Edible Film of Lipids and Cellulose Ethers: Performance in a Model Frozen Food System," *J. Food Sci.*, 54(6):1390–1392, 1406.

Kester, J. J. and O. Fennema. 1989c. "Resistance of Lipid Films to Oxygen Transmission," *JAOCS*, 66(8):1129–1138.

Kester, J. J. and O. Fennema. 1989d. "Resistance of Lipid Films to Water Vapor Transmission," *JAOCS*, 66(8):1139–1146.

Kester, J. J. and O. Fennema. 1989e. "The Influence of Polymorphic Form on Oxygen and Water Vapor Transmission through Lipid Films," *JAOCS*, 66(8):1147–1153.

Kester, J. J. and O. Fennema. 1989f. "Tempering Influence on Oxygen and Water Vapor Transmission through a Stearyl Alcohol Film," *JAOCS*, 66(8):1154–1157.

Kester, J. J., C. Bernhardy, J. J. Elsen, J. Letton and M. Fox. 1990. U.S. Patent 4,960,600. October 2, 1990.

Krochta, J. M., A. E. Pavlath and N. Goodman. 1990. "Edible Films from Casein-Lipid Emulsions for Lightly-Processed Fruits and Vegetables," in *Engineering and Food, Vol. 2*, W. E. L. Spiess and H. Schubert, eds., New York, NY: Elsevier Applied Science.

Labuza, T. and R. Contrereas-Medellin. 1981. "Prediction of Moisture Protection Requirements for Foods," *Cereal Food World*, 26:335–343.

Landmann, W., N. V. Lovegren and R. O. Feuge. 1960. "Permeability of Some Fat Products to Moisture," *JAOCS*, 37:1–4.

Landrock, A. H. and B. E. Proctor. 1952. "Gas Permeability of Films," *Mod. Packag.*, 25(10):131–135, 199–201.

Martin-Polo, M. and A. Voilley. 1990. "Comparative Study of the Water Permeability of Edible Film Composed of Arabic Gum and Glycerolmonostearate," *Sci. Aliments*, 10:473–483.

Munden, H., G. Dekay and G. Banker. 1964. "Evaluation of Polymeric Materials. I. Screening of Selected Polymers as Film Coating Agents," *J. Pharm. Sci.*, 53(4):395–401.

Murry, D. G. and L. Luft. 1973. "Low-D.E. Corn Starch Hydrolysates," *Food Technol.*, 27(3):32–39.

Myers, A. W., J. A. Meyer, C. E. Rogers, V. Stannett and M. Szwarc. 1961. "Studies on the Gas and Vapor Permeability of Plastic Films and Coated Papers. Part IV. The Permeation of Water Vapor," *TAPPI*, 44(1):58–64.

Nelson, K. and O. Fennema. 1991. "Methylcellulose Films to Prevent Lipid Migration in Confectionery Products," *J. Food Sci.*, 56(2):504–508.

Paulicka, F. 1986. "Quality Consideration in the Selection of Confectionery Fats," *The Manufacturing Confectioner* (May):75–81.

Quast, D. G. and M. Karel. 1972. "Technique for Determining Oxygen Concentration within Packages," *J. Food Sci.*, 37:490–491.

Reading, S. and M. Spring. 1984. "The Effects of Binder Film Characteristics on Granule and Tablet Properties," *J. Pharm. Pharmacol.*, 36:421–426.

Rowe, R. C. 1978. "The Measurement of the Adhesion of Film Coatings to Tablet Surfaces: The Effect of Tablet Porosity, Surface Roughness, and Film Thickness," *J. Pharm. Pharmac.*, 30:343–346.

Salame, M. and S. Steingiser. 1977. "Barrier Polymers," *Polym. Plast. Technol. Eng.*, 8(2):155–175.

Schönerr, J., K. Eckl and H. Gruler. 1979. "Water Permeability of Plant Cuticles: The Effect of Temperature on Diffusion of Water," *Planta*, 147:21–26.

Stern, S. A., T. P. Sinclair and T. P. Gareis. 1964. "An Improved Permeability Apparatus of the Variable-Volume Type," *Mod. Plastics*, 10:50–53.

Tabouret, T. 1987. "Technical Note: Detection of Fat Migration in a Confectionery Product," *Int. J. of Food Sci. and Tech.*, 22:163–167.

TAPPI. 1971. "Water Vapor Transmission Rate of Sheet Materials at Normal Temperature," T448 su-71, *TAPPI Official Standards.*

TAPPI. 1974. "Abrasion Resistance of Wax Coatings," T685-PM-74, *TAPPI Technical Standards.*

TAPPI. 1977. "Static Creasing of Paper for Water Vapor Transmission Tests," T465 OS-77, *TAPPI Official Standards.*

Torres, J. A. and M. Karel. 1985. "Microbial Stabilization of Intermediate Moisture Food Surfaces. III. Effect of Surface Preservative Concentration and Surface pH Control on Microbial Stability of an Intermediate Moisture Cheese Analog," *J. Food Proc. and Pres.*, 9:107–119.

Vojdani, F. and J. A. Torres. 1990. "Potassium Sorbate Permeability of Methycellulose and Hydroxypropyl Methylcellulose Coatings: Effect of Fatty Acids," *J. Food Sci.*, 55(3):841–846.

CHAPTER 2

Edible Coatings for Fresh Fruits and Vegetables: Past, Present, and Future

ELIZABETH A. BALDWIN[1]

INTRODUCTION

MANY storage techniques have been developed to extend the useful marketing distances and holding periods for fresh horticultural commodities after harvest. Various authorities have estimated that, unfortunately, 25 to 80% of harvested fresh fruits and vegetables are lost due to spoilage. This results in much economic waste in developed countries and more devasting consequences in many tropical regions of the world (Wills et al., 1981). One method of extending postharvest shelf-life is the use of edible coatings. Such coatings are made of edible materials that are used to enrobe fresh produce, providing a semipermeable barrier to gases and water vapor. To understand the effect of edible coatings on harvested fruits and vegetables, a background knowledge of postharvest fruit physiology and storage techniques is necessary.

The problem of spoilage results when fresh fruits and vegetables are thought of as static materials when, in fact, they are living tissue. In the broadest of analogies it could be said that, like members of the animal kingdom, fruits and vegetables need to "breathe" or they will undergo cer-

[1]U.S. Citrus and Subtropical Products Laboratory, P.O. Box 1909, Winter Haven, FL 33880. South Atlantic Area, Agricultural Research Service, U.S. Department of Agriculture. Mention of a trademark or proprietary product is for identification only and does not imply a guarantee or warranty of the product by the U.S. Department of Agriculture. All programs and services of the U.S. Department of Agriculture are offered on a nondiscriminatory basis without regard to race, color, national origin, religion, sex, age, marital status, or handicap.

tain anaerobic reactions and, in effect, start to "suffocate." They use up oxygen (O_2) and release carbon dioxide (CO_2) as they respire. Also like animals, fruit and vegetable tissues "perspire," or lose moisture to the air, and are subject to attack by microorganisms. The rates at which fresh produce "breathes," called respiration, and loses water, called transpiration, are partially dependent upon the temperature, gaseous makeup, and humidity of the environment surrounding the commodity.

POSTHARVEST PHYSIOLOGY

EFFECT OF ATMOSPHERE AND TEMPERATURE

Controlled atmosphere (CA) environments can be established around commodities in storage rooms, and modified atmospheres (MA) can be created within selectively permeable plastic packaging, or in the internal atmosphere of fruit tissues in the case of edible coatings. During MA storage, fruit respiration itself causes a depletion of O_2 and a buildup of CO_2, while in CA storage, specific levels of gases are maintained and regulated by external sources. Care must be taken, however, that O_2 levels do not become so low that the fruits or vegetables start anaerobic reactions, which can result in off-flavors, abnormal ripening, and spoilage (Kader, 1986). The O_2 levels that cause anaerobic reactions vary among commodities according to the permeability of the commodity peel, respiration patterns, storage temperature, and type and thickness of applied coatings if present. Commodities of the climacteric type will "hyperventilate" as they ripen, i.e., they will exhibit an accelerated respiration pattern. In general, once ripening commences, it proceeds fairly rapidly in this type of produce, resulting in a short shelf-life. These commodities also increase their production of the volatile hormone ethylene in a process that requires O_2. Ethylene, in turn, stimulates ripening and senescence, thereby shortening shelf-life. Both controlled and modified atmosphere storage regimes, with low O_2, and high CO_2 concentrations, slow down respiration and retard ethylene production and therefore ripening (Kader, 1986).

Environmental levels of O_2 below 8% decrease ethylene production, and CO_2 levels above 5% prevent or delay many responses to ethylene by fruit tissue, including ripening. A minimum of 1 to 3% O_2 is required around a commodity, however, to avoid a shift from aerobic to anaerobic respiration. Under anaerobic conditions, the glycolytic pathway replaces the Krebs cycle, and pyruvic acid is decarboxylated to form acetaldehyde, CO_2, and ultimately ethanol, which leads to off-flavors (Kader, 1986). High storage temperatures cause an increase in fruit respiration and ripening, while low (above freezing) temperatures often slow down respiration, delay ripening, and usually retard microbial growth. Low-temperature storage (sometimes used as a quarantine treatment) can also result in chilling injury of some tropical and subtropical produce.

EFFECT OF HUMIDITY

Humid environments, often measured in terms of percent relative humidity (% RH), reduce fruit transpiration. Transpiration is the process of water movement from fruit cells ultimately to the surrounding atmosphere, following a gradient of high (100% RH in the fruit intercellular spaces) to low water concentration (the % RH of the air surrounding the fruit). Therefore, fruits and vegetables are often stored in highly humid environments (90–98% RH) to minimize water loss, subsequent weight loss, and shriveling (Woods, 1990). Humidity can alter coating permeability properties for both oxygen and water vapor (Hagenmaier and Shaw, 1990, 1992).

EFFECT OF FRUIT COATINGS

Many of the above techniques for preservation of fresh commodities can be enhanced or even replaced by the application of edible coatings that can provide partial barriers to moisture and gas (CO_2 and O_2) exchange. Coatings also improve the mechanical handling properties (Mellenthin et al., 1982), help maintain structural integrity, retain volatile flavor compounds (Nisperos-Carriedo et al., 1990), and carry food additives (antimicrobial agents, antioxidants, etc.) (Kester and Fennema, 1988). Use of coatings for fresh fruits and vegetables, however, is not really a new concept. Mother Nature provides fruits and vegetables with a natural waxy coating called a cuticle, consisting of a layer of cutin (fatty acid–related substances such as waxes and resins with low permeability to water). The pathway for water loss varies for different commodities, but for most fruits and vegetables, the majority of water is lost because of cuticular transpiration, while some also exits through stomates and lenticels (Woods, 1990). The purpose of edible coatings for fruits and vegetables is basically to mimic or enhance this natural barrier, if already present, or to replace it in cases where washing and handling has partially removed or altered it.

The development of so-called "wax" coatings (which may or may not actually include a wax) emphasized reduction of moisture loss due to hydrophobic components such as waxes, oils, and resins (similar to cutin) and added sheen due to wax or resin components such as shellac or carnauba wax, respectively. Waxing alone is sometimes reported to control decay (Waks et al., 1985), but often this occurs only when fungicides are incorporated into the wax (Miller et al., 1988; Radnia and Eckert, 1988; Brown, 1984). Other possible added ingredients include biological control agents (Wilson et al., 1991), growth regulators, anti-senescence substances (Rao and Chundawat, 1991; Kader et al., 1966; Lodh et al., 1963), chilling injury protectants (Nordby and McDonald, 1990), and even coloring agents that could affect fruit quality.

Recently, the emphasis in coating development for fresh fruits and vegetables has switched to the area of controlled gas exchange. Coatings are currently being developed from polysaccharides that attempt to create a modified atmosphere inside the fruit that will delay ripening and senescence in a manner similar to the more costly CA storage. In most cases, however, coating formulations that provide adequate gas exchange are often not good barriers to water, while those that are efficient at reducing water loss tend to create anaerobic conditions in the internal fruit atmosphere. The following will present a brief histroy of the use of edible coatings for fresh horticultural commodities, as well as a summary of what is being used today, and finally a discussion of problems, ideas, and new directions for the future.

HISTORY OF EDIBLE COATINGS

WAX AND OIL COATINGS

Wax has been used to retard desiccation of citrus fruits in China since the twelfth and thirteenth centuries. The Chinese noted that the waxes slowed water loss and caused fermentation (Hardenburg, 1967). Hot melt paraffin waxes have been used to coat citrus fruits in the United States since the 1930s, and carnauba wax and oil-in-water emulsions have been used for coating fresh fruits and vegetables since the 1950s (Kaplan, 1986). There have been reports over the last thirty years in the scientific and patent literature of edible films and coatings made from a variety of polysaccharides, proteins and lipids alone or in mixtures to produce composite films (Kester and Fennema, 1988; Ukai et al., 1975; Cothran, 1951). The most successful coatings were lipid films made of acetylated monoglycerides, waxes (beeswax, carnauba, candelilla, paraffin, and rice bran) (Warth, 1986; Lawrence and Iyengar, 1983; Paredes-Lopez et al., 1974), and surfactants. These coatings were primarily used to block transport of moisture, to reduce surface abrasion during handling of fruits (Hardenburg, 1967), and to control soft scald formation in apples (browning of the skin) (Kester and Fennema, 1988), coreflush in 'MacIntosh' apples, and Jonathan spot on 'Jonathan' apples (Fisher and Britton, 1940).

The early wax emulsions generally consisted of colloidal suspensions of one of the above oils or waxes dispersed in water by means of a soap. The acid portion of the soap was one of the higher molecular weight fatty acids (oleic or linoleic) and the basic portion was ammonia, sodium, or potassium hydroxide or, more commonly, triethanolamine. The dispersed phase of the emulsion was paraffin, carnuaba, or some wax to which portions of shellac, gum, resin, or mineral oil were sometimes added. The resulting particle sizes were less than 0.5 μm (Platenius, 1939). Carnauba was often used in combination with paraffin since the former wax imparted a hard, brittle, high-luster quality, while the latter provided an efficient moisture barrier (Claypool, 1940).

Many of the early waxes reduced permeability of O_2 and CO_2, resulting in increased internal CO_2 and decreased internal O_2 (Tables 2.1 and 2.2), thereby affecting fruit respiration (Trout et al., 1942, 1953; Smock, 1935). Early researchers noted the advantages and disadvantages of the restricted gas exchange caused by these early coating formulations. The restricted entrance of O_2 delayed ripening of apples and pears, and the coating reduced water loss in these commodities as well as for citrus, but also often resulted in off-flavors due to anaerobic conditions thus created (Tables 2.1 and 2.2) (Eaks and Ludi, 1960; Hitz and Haut, 1938; Smock, 1935; Magness and Diehl, 1924).

In early reports, fruit and vegetable response to coatings varied due to the type, and even the variety of fruit tested, as well as the thickness of the coating coverage. Claypool and King (1941) investigated several types of fruits and their responses to coatings. Surface characteristics of the peel influenced the effectiveness of the coating. For example, the presence of hairs, such as on apricots, the thickness and type of cuticle, the number of stomates, lenticels, and even cracks in the lenticels (corky, russeted areas) affected coating coverage, gas exchange, and water loss.

Pome Fruits

Many early coating studies were conducted on apples and pears. In the 1930s and 1940s, green apples and pears (Trout et al., 1953; Hitz and Haut, 1942; Fisher and Britton, 1940; Smock, 1935) were coated with paraffin, oil, or wax emulsions which resulted in delayed yellow color development, softening, and onset of mealiness. These observations were explained by the increased CO_2 and reduced O_2 levels in the internal atmosphere of the fruit (Table 2.1) (Magness and Diehl, 1924). Most of the wax and oil coatings were less permeable to CO_2 than to O_2, but Trout et al. (1942) found that the decrease in respiration and delayed color development in coated green apples correlated with a decrease in internal O_2 rather than an increase in internal CO_2. Delayed softening and retention of

TABLE 2.1. Effect of Wax, Oil, and Shellac Coatings on Internal Atmosphere, Weight Loss, and Ethanol Content in Apple Fruit.

TRT	T (°C)	STORAGE TIME (wks)	INT ATM % CO_2	INT ATM % O_2	% WT LOSS (% RH not given)	PPM ETOH
'Granny Smith' apples[1]						
Uncoated	4.5	1	2.3	16.8		
2 oil:1 shellac			3.7	11.5		
1 oil:1 shellac			4.1	8.3		
1 oil:2 shellac			4.1	8.6		
1 oil:4 shellac			4.7	11.9		
Shellac			6.3	10.3		
Uncoated	0/21	7/1.5				130
15% castor oil–shellac						1170
'Grimes Golden' and 'Golden Delicious' apples[2]						
Uncoated	16	12			4.9	
Brytene wax emulsion					3.0	
'Winesap' apples[3]						
Uncoated	0	20	2.3	18.4		
Paraffin			11.1	12.6		
Light mineral oil			10.1	10.4		
Heavy mineral oil			12.4	2.9		
Uncoated	0/27	16/3	9.2	14.2		
Oil			14.0	1.7		

[1]Trout et al., 1953.
[2]Hitz and Haut, 1938.
[3]Magness and Diehl, 1924.

30

TABLE 2.2. Effect of Coatings on Internal Atmosphere, Weight Loss, and Ethanol Content in Citrus Fruit.

TRT	T (°C)	STORAGE TIME (wks)	INT ATM % CO_2	INT ATM % O_2	% WT LOSS (50–60% RH)	PPM ETOH
'Valencia' oranges[1]						
Uncoated	10/20	2/6	8.5	15.2	12.6	
Brodex wax			11.2	10.6	7.0	
'Valencia' oranges[2]						
Uncoated	21	1	2.5	17.0		600
Veg. oil			5.5	4.0		1600
Beeswax			8.0	4.0		2350
2.0% TAL Pro-long			7.0	4.0		1400
'Murcott' tangerines[3]						
Uncoated at harvest	5/17	4/1				1090
Polyethylene wax						4680

[1]Eaks and Ludi, 1960.
[2]Nisperos-Carriedo et al., 1990.
[3]Cohen et al., 1990.

31

flavor, however, were related to the coating's resistance to CO_2. Different types of coatings were tested on apples (Table 2.1), including solutions of castor oil, shellac in alcohol, mixtures of castor oil and shellac, and aqueous emulsions of various waxes (paraffin, carnauba, beeswax) and oils (mineral and vegetable) sometimes combined with shellac (Trout et al., 1953). The shellac, used in these coatings, was dissolved in either alcohol or ammonia. All coatings resulted in decreased respiration for treated apples; 25 to 35% for oils and 18% for wax emulsions compared to uncoated fruit. In some formulations, such as emulsions of wax and oils in sodium oleate and shellac dissolved in ammonia, the diffusion of gases was dependent primarily on the continuous soap phase of the film rather than on the porosity of the entire membrane. Increasing the concentration of shellac in such coatings had no additional effect on respiration, but shellac itself caused greater resistance to CO_2 in 'Granny Smith' apples (Table 2.1), while castor oil, in particular, caused greater resistance to O_2 when applied to fruit. The effect of the coating treatments was dependent on storage temperature, thickness and type of coating, maturity at harvest, variety, and condition of fruit (Trout et al., 1953). Coated immature apples were more inclined to develop anaerobic respiration, resulting in increased ethanol (Table 2.1), acetaldehyde, and general off-flavor. Some coatings promoted an internal disorder called "brown heart" due to excess buildup of CO_2.

Most oil and wax coatings tested, with the exception of peanut, maize, and the lighter mineral oils, decreased weight loss. This effect was enhanced by the addition of shellac. It was found that paraffin wax emulsions were more effective than beeswax which, in turn, was more effective than carnauba in retarding moisture loss. The wax emulsions were more effective barriers to water than were the oils. Among the oils and resins tested, paraffin oil had the greatest resistance to water folowed by shellac, vegetable oil, and finally light mineral oil. This resistance was somewhat affected by the type of emulsifier used. The oils, however, were more effective than the waxes in varying internal atmosphere and decreasing respiration. The higher the alkalinity in these early coatings, the more weight loss was observed to occur. Fisher and Britton (1940) found that waxing apples with carnauba or paraffin emulsions reduced shrinkage at room temperature and low humidity, but showed no advantage over uncoated fruit at high humidity.

Commercially, in Australia, apples have been treated with vaseline, white oil, or similar compounds to improve appearances and keeping quality, while, in the United States, mineral oil and paraffin wax have been used (Trout et al., 1953; Mellor, 1928). In some cases, fermentation occurred at high storage temperatures and after long, cold storage periods.

Citrus Fruits

For citrus, paraffin and carnauba wax emulsions were found to increase internal CO_2, decrease internal O_2 (especially at higher storage temperatures), and reduce water loss (Table 2.2) (Eaks and Ludi, 1960). Citrus fruit is sensitive to anaerobic conditions for as the fruit matures, aerobic respiration decreases and the supplementary anaerobic pathway increases even without application of coatings or storage in a low O_2 atmosphere (Bruemmer, 1989). Ethanol and acetaldehyde levels (indicating anaerobic respiration) have also been found to increase in citrus over the harvest season (Davis, 1970). Application of waxes can exacerbate this anaerobic condition (Nisperos-Carriedo et al., 1990), especially in tangerines, resulting in off-flavors due to increased internal ethanol, acetaldehyde, and CO_2, as well as decreased internal O_2 (Table 2.2) (Cohen et al., 1990; Vines et al., 1968). A similar situation occurred when citrus was subjected to CA storage with low O_2 and high CO_2, which caused increases in ethanol and acetaldehyde in fruit tissue, resulting in reduced flavor scores (Ke and Kader, 1990; Davis et al., 1973). In some studies, growth regulators such as 2,4-dichlorophenoxyacetic acid (2,4-D) and 2,4,5-trichlorophenoxyacetic acid (2,4,5-T) were added to the wax as anti-senescent compounds to further prolong shelf-life for mandarin oranges (Lodh et al., 1963), lemons, and limes (Hatton, 1958). Coating grapefruit with a polyethylene wax emulsion reduced incidence of rind pitting when the fruit was stored at 0 to 10°C (Davis and Harding, 1960). Coatings have been shown to impart some cold protection to the fruit and thus prevent pitting, which is a symptom of chilling injury (McDonald, 1986; Aljuburi and Huff, 1984) or to promote such pitting in cucumber (Mack and Janer, 1942). A compound called Waxol 0-12 was used successfully in India to prevent water loss and decay in lemons (Sharma et al., 1989).

Banana Fruits

Wax-type coatings have been reported to prolong the shelf-life of tropical fruits. A finely divided crystalline–paraffin wax emulsion (Lawson, 1960), paraffin wax (Blake, 1966), a sisal-paraffin, sugarcane, and polyethylene wax emulsions (Dalal et al., 1970) all increased shelf-life, decreased bruising, and improved the appearance of bananas, without affecting flavor. The polytheylene emulsion was less effective than the others, but all extended the storage life of bananas by delaying ripening. These results, however, were affected by harvest maturity in that bananas harvested prior to 100 days from emergence of the inflorescence showed no delayed ripening due to coating treatments (Dalal et al., 1970). In another study, coating green bananas with polyethylene wax emulsion (15% solids) extended their storage life by one to two weeks, added glossy shine, decreased O_2 and increased CO_2 in the internal atmosphere of the fruit, and decreased shrinkage, respiration rate, and emergence of senescent spotting (Ben-Yehoshua, 1966). It was suggested that the decrease in spotting was due to the decreased internal O_2, resulting in decreased activity for the enzyme polyphenol oxidase which oxidizes peel compounds leading to production of brown pigments. A candelilla coating extended banana shelf-life and slowed ripening as evidenced by a retarded decrease in starch degradation (usually 20–25% in green bananas decreasing to 2% in ripe fruit) and slowed color change (Siade and Pedraza, 1977).

Mango and Other Tropical Fruits

Aqueous wax emulsions, consisting of vegetable (sisal, sugarcane, and carnauba) waxes and mineral petroleum (paraffin) with and without shellac and emulsifiers, increased storage life of mango, pineapple, banana, papaya, guava, and avocado (Dalal et al., 1971). Coating mango with a paraffin wax emulsion (7% solids) resulted in increased shelf-life (Mathur and Srivastava, 1955), especially at reduced storage temperature (12.8°C) (Bose and Basu, 1954), while coating with refined mineral oil resulted in fruit injury (Mathur and Srivastava, 1955). The oil coating appeared to decrease respiration more than did the wax coating, resulting in anaerobic conditions severe enough to injure the fruit. Both coatings, however, decreased weight loss. In another study, immature green mangoes dipped in 1.7 to 2.7% aqueous emulsions of a fungicidal wax containing 5% o-phenylphenol, microcrystalline petroleum wax, terpene resin, oleic acid, and triethanolamine increased retention of vitamin C and moisture while delaying ripening and extending shelf-life (Mathur and Subramanyam, 1956). A wax emulsion of 6% solids with 250 ppm maleic hydrazide or 20 ppm 2,4-D delayed ripening of mango fruits compared to fruits treated with the same wax with no growth regulators (Subramanyam et al., 1962).

Room temperature storage life of guavas was extended 80% by coating with a 3% carnauba-paraffin wax emulsion (Srivastava et al., 1962). No effect on ripening was evident, however. Coating pineapples in the "shipping green" stage with a 4% wax emulsion did not interfere with color development, but reduced acid concentration in the juice and fruit weight loss by 35 to 45%. Increasing the amount of wax, however, interfered with normal color development (Schappelle, 1941). Coating of fully husked coconuts with paraffin wax resulted in an increase of eight to ten months storage life before appreciable loss of nut water occurred. Shellac coatings, however, were not as successful (Muliyar and Marar, 1963).

Stone Fruits and Grapes

Waxing of peaches started on a commercial scale in the early 1960s. The principal benefit from waxing was to reduce water loss and to control decay through the use of fungicides carried in the wax. "Defuzzing" of peaches was necessary before waxing. Washed and "defuzzed" peaches lost weight (water) faster than unwashed controls, while waxed "defuzzed" fruit lost slightly less weight than controls and had a more attractive appearance (Kraght, 1966). When nectarines lose 4 to 5% of their fresh weight, shriveling becomes apparent. Waxing of fruit resulted in less weight loss and, therefore, reduced this defect (Mitchell et al., 1963). Postharvest application of 2,6-dichloro-4-nitroaniline (Botran) and Johnson's waxes reduced storage wastage in concord grapes (Blanpied and Hickey, 1963).

Vegetables and Melons

Coatings similar to those applied to fruits were also applied to vegetables as early as the 1930s. By the 1950s water waxes consisting of water, vegetable waxes (usually carnauba, sometimes candelilla), paraffin wax, and emulsifying agents (soaps, detergents) were usually formulated to contain 6 to 10% wax solids (Hartman and Isenberg, 1956). A typical formula of this type included carnauba wax, oleic acid, triethanolamine, water, and paraffin wax. Rutabagas were coated with melted paraffin, containing 2 to 5% mineral oil, vaseline, or beeswax, and 1 to 2% rosin (to reduce greasiness). Hydrocarbon waxes (solvent waxes) contained a highly purified liquid hydrocarbon solvent which evaporated from the surface of the vegetable, leaving a thin coating of wax. Other formulations contained melted wax or wax and nonvolatile mineral oil mixtures. Solid cake wax (5 to 15% carnauba wax in paraffin wax) was also used.

Wax emulsions decreased sugar and water loss in carrots, as well as water loss, shriveling, and shrinkage in cucumbers, most root crops, summer squash, pumpkins, sweet corn, eggplant, peppers, and tomatoes. Little benefit was achieved by coating leafy vegetables, (Platenius, 1939; Nelson, 1942). Wax coatings decreased O_2 consumption in carrots with no effect on CO_2 production, but since the respiratory quotient of carrots (and most root crops) is less than one, the shift in respiration was unlikely to have harmful effects. A water emulsion of a wax mixture (carnauba-paraffin) sharply reduced weight loss and spoilage in coated cucumber, but increased low-temperature injury symptoms (pitting) in cold storage (Mack and Janer, 1942). Paraffin-waxed potatoes showed reduced sprouting, weight loss, chlorophyll (greening), and solanine synthesis (Wu and Salunkhe, 1972) without adversely affecting respiration. Dalal et al. (1971) reported that treatment with vegetable waxes and paraffin prolonged the shelf-life of carrots, potatoes, tomatoes, muskemelon and cucumbers. Turnips were waxed to reduce weight loss and hot waxes (paraffin) worked better than cold (colloidal waxes in water suspension) (Franklin, 1961).

For a time the waxing of stem scars of tomatoes was attempted since the majority of tomato gas exchange occurs through this area (Emmert and Southwick, 1954; Brooks, 1938). Waxing the stem scars with mixtures of paraffin, beeswax, and mineral oil effectively delayed ripening of mature green tomatoes (Ayres et al., 1964; Hartman and Isenberg, 1956), but the process was labor intensive. Colored waxes were used on the red-skinned varieties of potatoes and sweet potatoes (Hartman and Isenberg, 1956). A carnauba–paraffin wax emulsion and a wax–sodium bisulfite mixture were used in an experiment to control decay of cantaloupes. Only the latter treatment was effective, however, but was unattractive due to a breakdown of the emulsion (Barger et al., 1948). Another advantage of wax coatings was the fact that coatings acted as lubricants to reduce surface scarring and chafing, especially for tomatoes and peppers. Coating of peppers with a carnauba wax emulsion with either 3 or 6% solids prolonged storage, reduced shriveling and decay, enhanced shine, reduced rates of respiration, and resulted in better retention of ascorbic acid (vitamin C) (Habeebunnisa et al., 1963).

POLYSACCHARIDE COATINGS

Polysaccharide-based coatings were also developed, but mostly for processed fruit and nut products. This is covered in more detail in Chapter 4. Pectin coatings were made from low methoxyl pectin (2–20% methylated), calcium chloride (as a cross-linker), a plasticizer, and in some cases organic acids (Miers et al., 1953; Schultz et al., 1948). The problem with pectin coatings was that they were quite permeable to water vapor. A typical low methoxyl pectin coating (3.8% methoxyl content, intrinsic viscosity of 3.2, and 40 microns thick) exhibited a water vapor transmission rate of 1400–4300 g water/m²/day at 25°C with an RH differential of 81 to 31%. This is a much higher water vapor transmission rate than for wax and oil coatings such as paraffin (170 g water/m²/day), beeswax (190–300 g water/m²/day), or vegetable oil (510–660 g water/m²/day) under similar conditions (Schultz et al., 1949). These early pectin coatings were tested on almonds, candied diced fruit, and dates (Swenson et al., 1953).

Derivatives of cellulose also make good film formers due to the linear structure of the polymer backbone. Nonionic cellulose ethers yield water-soluble films that are tough and flexible (Krumel and Lindsay, 1976). Hydroxypropyl and hydroxypropyl methyl cellulose form films that retarded moisture absorption and development of oxidative rancidity in nut meats (Ganz, 1969), and grease absorption in deep-fat frying of potatoes (Gold, 1969).

Amylaceous coatings were also tested on fruits. A solution of a hydroxy-propyl starch derivative that was high in amylose starch (70% w/v) with 12% glycerol as plasticizer, extended the shelf-life of pitted prunes, possibly due to O_2 barrier properties (Jokay et al., 1967). Starch films are reported to be semipermeable to CO_2, but have high resistance to passage of O_2 (Rankin et al., 1958).

SYNTHETIC POLYMERS AND DERIVATIVES OF NATURAL POLYMERS USED IN COATINGS

Several patents are on file from the late 1960s to early 1970s which describe edible coatings for fruits and vegetables made of synthetic and natural polymers (Danials, 1973). Raw fruits and vegetables were coated with an aqueous emulsion of lecithin–methyl anthranilate (Shillington and Liggett, 1970), a hydrolyzed lecithin coating (Mulder, 1969), acrylate and methacrylate polymers (DeLong and Shepherd, 1972), polymers of vinyl acetate (Danials, 1973), and organic latex and copolymers (vinyl acetate, acrylate, ethyl acrylate, and propyl acrylate) (Danials, 1973; Rosenfield, 1968). Such films claimed to have low vapor transmission, adequate permeability to O_2 and CO_2, and some microbial control which was often due to added fungicides (Danials, 1973). These polymers, with the exception of methyl anthranilate, are on the FDA list of substances that are acceptable for secondary direct food additives or indirect food additives (FDA, 1991).

COATINGS USED TODAY

WAX AND OIL COATINGS

Commercial use of wax preparations is rather extensive on citrus, apples, mature green tomatoes, rutabagas, and cucumbers. More limited use has been noted on other fruits including apricots, bananas, cherries, coconuts, dates, grapes, guavas, lychees, mangoes, peaches, nectarines, pears, persimmons, pineapples, peppers, cantaloupes, and honeydew melons. Coatings have also been applied to many other vegetables such as asparagus, beans, beets, carrots, celery, eggplant, kohlrabi, okra, parsnips, peppers, potatoes, radishes, squash, sweet potatoes, and turnips (Hardenburg, 1967).

Many coatings used today are similar to those used in the past. The wax-type coatings are made with natural or synthetic waxes and fatty acids (carnauba, polyethylene, oleic acid), oils (vegetable and mineral oil), wood rosin, shellac, and coumarone indene resin, emulsifiers, plasticizers, anti-foam agents, surfactants, and preservatives. Oxidized polyethylene is defined as the basic resin produced by the mild air oxidation of polyethylene, a petroleum by-product. Several grades of polyethylene wax are available to obtain desired emulsion properties (Eastman Chemical Inc., 1984, 1990). This wax, although less popular of late, is still used to make emulsion coatings for citrus. Generally, the amount of shellac and wood rosin (components of coatings that impart the most shine) vary based on the cost of the coating, i.e., those coatings with more shellac being higher priced. Coumarone indene resin is a petroleum and/or coal tar by-product. It is resistant to moisture and is often a component of the so-called solvent wax for citrus (Neville Chemical Co., 1988).

Most companies claim that their coatings reduce water loss, add various degrees of sheen, and provide adequate permeability to CO_2 and O_2. Coatings with shellac and rosin, however, generally have lower permeability to CO_2, O_2, and ethylene gases while coatings made with polyethylene wax have high permeability to such gases compared to those made with carnauba and other waxes. Permeability to O_2 is also affected by % RH and polar components used to raise pH and solubilize a polymer such as shellac (Hagenmaier and Shaw, 1992). Polymers that contain hydroxy, ester, and other polar groups tend to have lower O_2 permeability than those with hydrocarbon and other nonpolar groups (Ashley, 1985). Water vapor permeability of coatings is also sensitive to percent RH, especially for waxes that contain polar ingredients (Hagenmaier and Shaw, 1992). The characteristics of some commercial coatings are shown in Table 2.3, along with recommendations for their use on fresh produce.

In general, shellac waxes coat better, dry faster, and produce a higher shine, but may whiten in cases of extended transport where commodities go through several "sweats" due to humidity and temperature changes. New formulations of derivatized shellac, such as White French Varnish, an ester derivative from Mantrose Haeuser Co., Westport, CT, minimize this occurrence, but it is still enough of a problem that shellac coatings are usually not used on exported fruit. Carnauba wax, on the other hand, is less likely to whiten, adds some sheen, and is often substituted for shellac for exported produce.

Pome Fruits

A typical apple wax contains any of the following ingredients: water (solvent), isopropanol (cosolvent), shellac, carnauba, propylene glycol (plasticizer), morpholine, or ammonium oleates, casein, soy protein (binders), and methyl siloxanes (anti-foam agents). Use of wood rosin,

TABLE 2.3. Some Commercial Wax Coating Companies, Products, and Applications.

Company	Product	Major Ingredients	Application
American Machinery Corp. (AMC)	PacRite Apple Wax 61	water-based, carnauba-shellac emulsion	apples
	PacRite Apple Wax 55	aqueous, shellac-based emulsion compatible with thiabendazole	apples
	PacRite 425	Shellac and resin water emulsion	citrus (for prolonged export shipping)
	PacRite 585	Shellac and resin water emulsion	citrus (for domestic and export)
	PacRite Sunshine Wax	Water-based resin wax compatible with fungicides	citrus (for domestic shipments)
	Sealbrite Lustre Dry	Water-based, resin-shellac dispersion with or without 1000 ppm thiabendazole	citrus esp. limes (for domestic and European shipments)
	Stor-Rite 101	Oxidized homopolymer polyethylene shellac water emulsion	citrus esp. lemons and other fruits and vegetables
	PacRite Durafresh	mineral oil-based	tomatoes, cucumbers, green peppers, squash, peaches, plums, nectarines
	PacRite Stone Fruit 3	water-based, mineral oil fatty acid emulsion compatible with wettable powder fungicides	peaches, plums, nectarines (for domestic and export)
	PacRite Stone Fruit 35	water-based carnauba emulsion compatible with wettable powder fungicides	peaches, plums, nectarines (for domestic and export)

(continued)

39

TABLE 2.3. (continued).

Company	Product	Major Ingredients	Application
American Machinery Corp. (AMC) (continued)	PacRite Tropical Fruit Coating 213	water-based carnauba emulsion	mangoes, papayas, carambolas, melons, chayote, pineapples
Agri-Tech Inc.	Fresh-Cote 200, 207, 214, 241HL, 362	shellac-based	apples, pears
	Fresh-Cote CW1	carnauba-based	apples
	Fresh-Cote 711	oil emulsion	apples, eggplant, tomatoes, cucumbers
	Fresh-Cote 110	oil emulsion	peaches, apricots, nectarines
	Fresh-Cote 707	carnauba-based	peaches
Solutec Corp.	Vector 7	shellac-based with morpholine	apples (for domestic market)
	Apl-Brite 300C	carnauba-based	apples (for domestic and import)
	Citrus-Brite 300C	carnauba-based	citrus (for domestic and export)
S. C. Johnson	Primafresh Wax	carnauba-based wax emulsion	apple, citrus, and other firm-surfaced fruit
Pace Intl. Shield-Brite	Shield-Brite AP-40, AP-46, AP-700	shellac-based	apples
	Shield-Brite AP-50C	carnauba/shellac-based	apples (for extended tspt.)
	Shield-Brite AP-101	"natural waxes," "food grade refined gum lac (shellac?), phospho-protein	apples (for export.)

TABLE 2.3. (continued).

Company	Product	Major Ingredients	Application
Pace Intl. Shield-Brite (continued)	Shield-Brite C-50-C	carnauba-shellac	citrus
	Shield-Brite PR-150	shellac-based	pears
	Shield-Brite PR-160C	carnauba/shellac-based	pears
	Shield-Brite PR-190	"natural waxes"	pears
	Shield-Brite ST-200	vegetable oil/wax, xanthan gum	stone fruit
	Shield-Brite ST-400	vegetable oil, beeswax, xanthan gum	stone fruit
Food Machinery Corp. (FMC)	Sta-Fresh 360	natural resin	citrus, apples, stone fruits, pomegranates, tomatoes
	Sta-Fresh 223	natural and modified natural resins	citrus
	Sta-Fresh 320	natural and modified natural resins	citrus
	Sta-Fresh 705	synthetic resins	citrus, pineapple, canteloupes, sweet potatoes
	Flavorseal 93	synthetic resin	citrus
	Flavorseal 150	modified natural resin	citrus
Fresh Mark Corp.	Fresh Wax 3330	shellac and wood rosin	citrus
	Fresh Wax 625	oxidized polyethylene wax	cantaloupes, pineapples, apples, sweet potatoes, citrus
	Fresh Wax Veg. Wax 51 V	white oil, paraffin wax	cucumbers, tomatoes, other vegetables

(continued)

41

TABLE 2.3. (continued).

Company	Product	Major Ingredients	Application
Brogdex Co.	Apple Britex 559	carnauba wax emulsion	apple
	Melon Wax 551	carnauba wax with 1–2% Dowicide A as a fungicide	melon
	Banana Wax 509	emulsion wax	bananas
	Avocado, Chayote, Papaya, and Mango Wax 508	high shine wax	avocado, chayote, papaya and mango
	Pineapple Wax 510	water-based emulsion wax	pineapple
	Stone Fruit Wax 521	carnauba-based emulsion	stone fruits
	Stone Fruit Wax 522	vegetable oil available with 2.1% Botran (fungicide), but contains no animal fats or hydrocarbon	stone fruits
	Apple Wax 360	high shine resin-based (Drake and Nelson, 1990)	apples
	Citrus Wax 325	resin-based	citrus
	Britex 315, 551, 701	carnauba wax emulsion	citrus
	Storage Wax 505-05	concentrated polyethylene emulsion wax for short term storage and rapid degreening	lemons
	Storage Wax 505-25	concentrated polyethylene emulsion wax for long term storage and slow color development	lemons

42

TABLE 2.3. (continued).

Company	Product	Major Ingredients	Application
Surface Systems Int.	Semperfresh F	sucrose esters of fatty acids, sodium carboxymethyl cellulose, available in powder form to mix with water (little shine)	most fruits and vegetables
	Semperfresh G	same as Semperfresh F but in granular form	most fruits and vegetables
	Nu-Coat Flo	sucrose esters of fatty acids, sodium carboxymethyl cellulose, food approved anti-foaming agent, liquid concentrate	mango, melon, plum, pear, apple, cherry, guava, starfruit, durian, cucumber, aubergine, bean, courgette, capsicum
	Nu-Coat Flo C	same as Nu-Coat Flo	citrus, pineapples
	Ban-seel	same as Nu-Coat Flo liquid concentrate	bananas, plantains
	Brilloshine C	sucrose ester and wax (shine)	citrus
	Brilloshine L	same as Brilloshine C, liquid concentrate (shine)	most fruits and vegetables
	Snow-White	sucrose ester and ?, twin pack, liquid and powder (prevents oxidative browning)	processed potatoes
	White-Wash	sucrose ester and ?, liquid (prevents greening of potatoes exposed to light)	whole potatoes

which is approved for use in citrus coatings, has not yet been approved for use on apples. The apple industry is interested in clearing this compound for use in apple coatings. In a study with 'Delicious' and 'Golden Delicious' apples, no quality differences were evident among the apples waxed with commercial shellac, carnauba, or resin-based waxes. The waxed apples exhibited improved color, higher internal levels of CO_2 and ethylene, and reduced weight loss, but wax treatments had no effect on firmness (Drake and Nelson, 1990). A commercial carnauba-shellac coating (Shield-Brite PR 160C) was reported to delay ripening of 'd'Anjou' pears, resulting in lower external and higher internal concentrations of CO_2 than nonwaxed fruit, retained firmness, and delayed color changes (Drake et al., 1991).

Citrus Fruits

The so-called "water wax" that is most often used on citrus is composed primarily of water, shellac, and/or wood rosin with a small amount of morpholine (to raise pH in order to solubilize shellac), oleic acid, and frequently oxidized polyethylene (a petroleum product). The "solvent wax" for citrus is also commercially available, but is not used much by the industry because the solvent (petroleum ether) is expensive and requires special venting systems (Kaplan, 1986). In addition, solvent waxes are banned in Japan and have become unpopular with health-conscious consumers who prefer a more "natural" product. Lemons, designated for long-term storage, are coated with carnauba-paraffin-polyethylene wax (Kaplan, 1986), and limes are currently coated with mineral oil or carnauba wax. Coatings can also affect the degreening of citrus. 'Shmouti' and 'Washington Navel' oranges, harvested green, were dipped in a non-polyethylene water-emulsion wax to determine the effect on degreening. The amount of wax found on the fruits varied from 1.58 to 10.22/100 cm² of fruit surface. The amount of coating had no effect on weight loss, but increasing amounts of wax coating slowed down the rate of disappearance of the green color (Rajzman et al., 1973).

Other Fruits and Vegetables

Turnips are waxed to improve appearance, as well as to reduce water loss and therefore wilting (Franklin, 1961). Enrobing of turnip roots with a water-miscible, carnauba-based wax also temporarily darkened and intensified the purple color (Perkins-Veazie and Collins, 1991). Melons coated with Citruseal citrus wax had less weight loss, but also anaerobic tissue breakdown after six weeks at 8°C and five days at 15–25°C (Edwards and Blennerhassett, 1990). Cucumbers are commercially coated to reduce weight loss, but cucumbers coated with a commercial water wax showed increased levels of acetaldehyde, ethanol, and methanol, indicating anaerobic respiration after twenty-one days of storage at 7°C (Risse et al., 1987). The shelf-life of 'Yellow-heart' breadfruit (*Artocarpus altilis*) was extended by one week and chilling injury symptoms were reduced when the fruits were coated with Fresh Mark FM–5IV undiluted wax (Maharaj and Sankat, 1990). Waxing of papaya fruit with various resin-, polyethylene-, and carnauba-based waxes resulted in some delay in certain ripening parameters, a decrease in weight loss, and, in some cases, some off-flavor (Paull and Chen, 1989). In addition to carnauba-based waxes, mineral oil coatings are also used on stone fruits, tomatoes, and other fruits and vegetables. These coatings contain no water and some companies are either using or conducting research on vegetable oil as a replacement for mineral oil (personal communication, American Machinery Corp.). Early work with mineral oil on mango fruits proved unsuccessful because of subsequent anaerobic conditions due to reduced gas exchange (Mathur and Srivastava, 1955).

POLYSACCHARIDE COATINGS

Several types of composite coatings reported in the literature over the last ten years contain a polysaccharide base. For these coatings, the advantages are more in the area of gas exchange rather than retardation of water loss. As mentioned previously, polysaccharide films, because of their hydrophilic nature, provide only minimal moisture barriers (Kester and Fennema, 1988). The CO_2 and O_2 permeabilities of the polysaccharide-based coatings, however, result in retardation of ripening in many climacteric fruits, thereby increasing shelf-life, without creating severe anaerobic conditions. The idea is to mimic CA and or MA (plastic or shrink wrap) storage, which is labor-intensive and expensive, with an environmentally friendly biodegradable edible coating. Several water-soluble composite coatings became commercially available in the mid 1980s that are comprised of carboxymethyl cellulose (D-glucose units linked through B-1,4 glycosidic bonds), as the film former, with sucrose fatty acid ester emulsifiers (mono- and diglycerides, Mitsubishi Kasei Corp.) among other ingredients (Drake et al., 1987; Banks, 1984a, 1984b; Smith and Stow, 1984). The sucrose esters and polyesters are nonionic surfactants that serve as emulsifiers and are derived from sucrose stearate which, in turn, is derived from edible tallow or sucrose palmitate, myristate, or laurate from vegetable oils (Mitsubishi-Kasei, 1989). Those derived from tallow present problems for vegetarian or kosher-conscious consumers. A patent for this type of sucrose fatty acid polyester–cellulose coating was filed in the United States by Japanese scientists in 1975 (Ukai et al., 1975).

The first commercially available coating of this type was called TAL Pro-long (Courtaulds Group, London), and later Pro-long which was made up of sucrose polyesters of fatty acids and sodium salt of carboxymethyl cellulose. The dry ingredients were mixed with water to specific concentrations, depending on the commodity, and applied as a dip, drench, or spray to cover fruits and vegetables.

Another coating, Semperfresh (United Agriproducts, Greeley, Colorado), appeared on the market and is apparently of similar composition to TAL Pro-long. This coating claims to be an improved formulation of earlier sucrose polyester products with the major difference being the incorporation of a higher proportion of short-chain unsaturated fatty acid esters in its formulation (Drake et al., 1987). It is sold in powder or granular form which, when mixed with water to a concentration of 0.75 to 2.0%, is used to coat fruits and vegetables. Like TAL Pro-long, it is hygroscopic in nature, which suggests that it holds a thin film of water. CO_2 dissolves and diffuses through water twenty times faster than does O_2. Therefore, O_2 is limited in entering the fruit more than CO_2 is from escaping. As a result, coated fruit internal CO_2 levels, although they do increase, do not build up to harmful levels, as was observed with early wax coatings on 'Winesap' (Table 2.1) and other apple varieties (Hitz and Haut, 1942), while reduced O_2 is maintained, simulating CA storage. Semperfresh and similar coatings are not good barriers to movement of water, but have some effect on fruit transpiration due to blockage of stomatal, lenticular, and stem scar openings, which results in some decrease in water loss. These coatings, however, are not as effective in this respect as are conventional wax formulations. Mitsubishi International Corp., Fine Chemicals Department has announced two new polysaccharide films: Soageena (carrageenan-based line of coatings) and Soafil (edible polysaccharide) (Institute of Food Technology, 1991). No information is available, however, on the use of these films on fresh produce. Carrageenan is approved by the FDA as an emulsifier, stabilizer, and thickener for food systems (CFR# 172.620, Title 21, FDA, 1991), but not specifically as a component of coatings for the U.S.

Pome Fruits

Sucrose ester-carboxymethyl cellulose formulations retarded color development and retained acids and firmness compared to controls when tested on apples (Drake, et al., 1987; Banks, 1985a; Smith and Stow, 1984) and pears (Meheriuk and Lau, 1988). This was due to the effect of decreased ethylene evolution, increased internal fruit CO_2 and decreased

O_2. A 1.25% sucrose ester-carboxymethyl cellulose formulation (TAL Pro-long) failed to retard detrimental changes in fruit firmness, yellowing, and weight loss in 'Cox's Orange Pippin' apples during storage, but when a 1 to 4% coating formulation was applied after storage, it delayed yellowing and loss of firmness, increased internal CO_2, and reduced weight loss (Table 2.4) during a twenty-one day marketing period (Smith and Stow, 1984). Different storage temperatures seem to affect coating performance in terms of increasing internal CO_2 and, especially, decreasing internal O_2 concentrations. The lower storage temperatures (3–10°C) resulted in minor increases in internal CO_2 and negligible decreases in internal O_2 concentrations in the same variety of apples, whereas at the higher storage temperature (20°C), internal increases in CO_2 were greater while internal O_2 levels were drastically reduced (Table 2.4) (Banks, 1985a). These differences in internal gas changes due to storage temperature are probably related to the effect of temperature on fruit respiration.

When 'Cox's Orange Pippin,' 'Branley's seedling,' and 'Golden Delicious' apples were coated with TAL Pro-long, increased internal CO_2 and decreased O_2 were observed in fruit stored at 4 and 20°C, respectively (Table 2.4). Toxic levels of CO_2 accumulated in fruits coated with TAL Pro-long formulations of 3% or more (Banks, 1985b). The addition of a wetting agent increased permeability, but improved coverage. There was again variation in internal gas concentrations of coated fruit due to cultivar differences in peel permeability as was found in the early coating studies discussed in the previous section. TAL Pro-long coating increased resistance to some fungal rots including *Sclerotina* spp. and *Rhizopus nigricans* in apples, pears, and plums (Lowings and Cutts, 1982). A 3% TAL Pro-long coating reduced aminocyclopropanecarboxylic acid (ACC, ethylene precursor) and ethylene levels in 'Jonathan' apples stored at 8°C and at room temperature, and weight loss in fruit stored at room temperature. In 'Fuji' fruit stored at the low temperature, however, ACC content, ethylene-forming enzyme activity, and ethylene evolution were increased compared to controls (Lee and Kwon, 1990).

'McIntosh' and 'Golden Delicious' apples coated with Semperfresh exhibited reduced ripening rates as evidenced by delayed color development and increased firmness compared to uncoated controls during four months of cold storage (5°C). The coating had no effect on pH, titratable acidity, soluble solids, or flavor, however. These results are in contrast to those of Smith and Stow (1984) where TAL Pro-long was not found to affect the ripening rate of apples during cold storage, but only after removal from storage for marketing. The difference in results was attributed to the chemical differences between Semperfresh and TAL Pro-long (Santerre et al., 1989).

TABLE 2.4. Effect of Polysaccharide Coatings on Internal Atmosphere and Weight Loss in Apple Fruit.

TRT	T (°C)	STORAGE TIME (wks/hrs)	INT ATM % CO_2	INT ATM % O_2	% WT LOSS (% RH not given)
'Cox's Orange Pippin' apples[1]					
Uncoated	3.5	16	1.6	19.4	
1.25% sucrose ester–polysaccharide			3.2	17.7	
Uncoated	3.5/10	16/2	3.2	17.7	
1.25% sucrose ester–polysaccharide			5.1	16.0	
Uncoated	10	3			5.8
2.0% sucrose ester–polysaccharide					5.1
4.0% sucrose ester–polysaccharide					4.7
'Cox's Orange Pippin' apples[2]					
Uncoated	3–5	2	2.0	19.0	
1.0% TAL Pro-long			2.0	18.0	
3.0% TAL Pro-long			5.0	14.0	
Uncoated	4/20	2/48 hr	2.0	15.0	
1.0% TAL Pro-long			6.0	5.0	
3.0% TAL Pro-long			8.0	3.0	

[1]Smith and Stow, 1984.
[2]Banks, 1985a.

Citrus Fruits

Semperfresh extended the storage life of citrus, but storage rots were a limiting factor. The coating is, however, compatible with most fungicides used on citrus fruit (Curtis, 1988). An indirect and little understood effect of these coatings is that the coated fruits remain more turgid and juicy than uncoated fruits despite some loss in weight. Coated cirtus fruit also had better flavor, which appeared to be related to acidity measurements (Curtis, 1988). In another study TAL Pro-long coated 'Valencia' oranges resulted in higher flavor volatile levels in the juice and relatively low ethanol levels compared to oil- or wax-coated fruit (Nisperos-Carriedo et al., 1990). Since high shine is important for citrus, Semperfresh was applied in combination with a shellac-wax to citrus fruits. This treatment resulted in fruit with higher turgidity, less decay, and good flavor, although ethanol levels were increased. Whereas wax coatings have been reported to adversely affect flavor in citrus after long-term storage, fruit coated with Semperfresh exhibited enhanced flavor (Curtis, 1988).

'Persian' limes were treated with 1.5, 2.0, and 2.5% TAL Pro-long and stored at various temperature-RH combinations (8–25°C and 40–95% RH). The coating extended the shelf-life of coated limes by controlling weight loss and degreening when stored at high RH and low temperatures (Motlagh and Quantick, 1988).

Banana Fruits

Delayed ripening, along with similar changes in internal fruit CO_2 and O_2 levels, was also observed in coated bananas (Banks, 1983, 1984a, 1984b, 1985b). Sucrose esters of fatty acids were labeled with ^{14}C and included in fruit coatings which retarded ripening. The esters remained intact on surfaces of bananas, apples, and pears during storage at 17°C for thirty days. Only a small amount of migrated label was detected in pulpy tissue of the fruits. The mode of action of sucrose esters in retarding ripening was not dependent, therefore, on their migration into the pulpy tissue (Bhardwaj et al., 1984).

TAL Pro-long was also applied to banana fruit (Banks, 1985b). The fruit exhibited decreased O_2 levels and the coating prevented the climacteric rise in ethylene production, if applied before initiation of ripening. If ripening had already been initiated, the coating stopped climacteric ethylene in mid rise. Coated fruit also showed delayed chlorophyll loss (loss of green color) and climacteric rise in respiration, but these effects did not occur as rapidly as those described for ethylene, and declined relative to ripening initiation. The effect of the coating on color and respiration, therefore, appeared to be independent of its effect on ethylene production.

The fruit surface was analyzed using light microscopy, scanning electron microscopy, chlorine diffusion, and dispersive X-ray analysis (with coating labeled with aurothioglucose). It was observed that the coating caused stomatal blockage which physically impeded gaseous diffusion through stomates, the principal route of gas exhange across banana fruit skin (Banks, 1984b).

Mango and Cherry Fruits

Mango fruits also exhibited retarded ripening and therefore increased storage life when coated with 0.75 to 1.0% TAL Pro-long and stored at 25°C (Dhalla and Hanson, 1988). The authors also reported reduced weight loss in the coated fruit compared to uncoated controls, and increased ethanol formation in fruit pulp after thirteen days with 1% TAL Pro-long. No adverse effects on sensory quality was detected when a 0.75% formulation was used, however. Coated mangoes showed a slower decrease in titratable acidity and ascorbic acid as well as a retarded softening and carotenogenesis (loss of green color). Cherries coated with Semperfresh showed reduced moisture loss but not to the extent of those coated with Pennwal or Shield-Brite commerical waxes. Unfortunately, none of the coated fruit showed reduced stem discoloration, which greatly influences consumer perception of cherry quality (Drake et al., 1988).

Vegetables and Melons

Sucrose fatty acid esters (0.25%) were applied to shredded cabbage as a dip. After subsequent storage at 10°C the cabbage showed reduced browning, which was thought to be due to reduced O_2 at the cabbage surface, but also increased ethanol levels (Sakane et al., 1990). TAL Pro-long was not effective in decreasing respiration rate and water loss in fruits of Solanaceae such as tomato and sweet pepper (Nisperos-Carriedo and Baldwin, 1988; Lowings and Cutts, 1982), nor was Semperfresh effective in retarding water loss of melons (Edwards and Blennerhassett, 1990).

COATINGS FOR THE FUTURE

CHITOSAN COATINGS

Research on coating development from other polysaccharides has branched into the area of chitosan [(1,4)-linked 2-amino-2-deoxy-β-D-glucan], a deacetylated form of chitin which is derived from a naturally occurring cationic bipolymer (Hirano and Nagao, 1989; Davis et al.,

1988; Allan and Hadwiger, 1979). Chitin is the next most abundant polysaccharide after cellulose and is widely distributed as supporting material of crustacea (Postharvest News and Information, 1991). This interesting polysaccharide is also present in cell walls of phytopathogens and has been shown to inhibit growth of several fungi (Hirano and Nagao, 1989; Stossel and Leuba, 1984; Allan and Hadwiger, 1979). Chitosan also induced chitinase in plant tissues, a defense enzyme that degrades fungal cell walls (Hirano and Nagao, 1989), and elicited the anti-microbial phytoalexin, pisatin, in pea pods (*Pisum sativum*) (Hadwiger and Beckman, 1980). The chitosan oligomer that showed maximal anti-fungal activity to *Fusarium solani* and elicitation of pisatin formation in peas was the heptamer, and the functional groups on the polymer necessary for growth inhibition of the pathogen were discovered to be the cationized amino groups at pH 5.6 (Hirano and Nagao, 1989; Kendra and Hadwiger, 1984).

Using this polymer as a film former and natural preservative, a differentially permeable fruit coating called Nutri-Save (Nova Chem, Halifax, NS, Canada) was developed from N,O-carboxymethyl chitosan that also simulated CA storage for apples and pears in that it reduced rates of respiration (50%) and desiccation for these commodities (Elson et al., 1985). Methylation of the chitosan polymer in Nutri-Save resulted in a two-fold increase in the resistance to CO_2 compared to the unmethylated polymer. A 1 to 2% solution of this coating containing a surfactant provided a slight sheen to test commodities. The rate of water vapor transfer through the coating was determined to be 0.8 mg water/cm^2/hr, while O_2 and CO_2 permeability varied with % RH. Below 70% RH the coating is impermeable to both gases, but at 100% RH 44 μl O_2 and 3 μl CO_2 penetrates the film/cm^2/hr. The absorption of water opened the structure of the film and facilitated transport of gases that are soluble in water.

Pome Fruits

Certain formulations of Nutri-Save were reported to retain fruit firmness and titratable acids over nine months of storage for 'Golden' and 'Red Delicious' apples and 'Bartlett' and 'Clapps Favorite' pears, which is comparable to CA storage. After removal from cold storage, the fruits ripened normally (Elson et al., 1985). This is in contrast to the data presented by Meheriuk and Lau (1988) where 'Bartlett' pears ripened unevenly when coated with either TAL Pro-long or Nutri-Save. 'Newton' apples coated with Nutri-Save retained their green color longer than uncoated fruit (Meheriuk, 1990b).

A new method was recently developed for preparing the chitosan derivative, N,O-carboxymethyl chitosan from a reaction of chitosan and monochloroacetic acid under alkaline conditions. A coating made in this manner was used on apples which were then stored for six months at room temperature with no decrease in quality (Postharvest News and Information, 1991).

Strawberry Fruits

Coating strawberries with Nutri-Save produced encouraging results (Elson et al., 1985). Another chitosan coating was formulated from 1.0 to 1.5% chitosan in water using hydrogen chloride to dissolve the chitosan (later adjusted with sodium hydroxide) and Tween 80 at pH 5.6 as a wetting agent. This formulation controlled decay of inoculated strawberries at 13°C. Normally, strawberries are stored at 1°C with high RH and elevated CO_2 to deter infection by *Botrytis cinerea* and *Rhizopus* sp. Chitosan-coated berries were firmer and higher in titratable acidity and synthesized anthocyanin at a slower rate than those treated with the fungicide iprodione (Rovral) or untreated fruit (El Ghaouth et al., 1991).

Vegetables

Encouraging results were reported when Nutri-Save was applied to tomatoes (El Ghaouth et al., 1992), peppers, squash, cauliflower, brussel sprouts, and broccoli (Elson et al., 1985). Banaras et al. (1989), however, reported that neither Nutri-Save nor Semperfresh were effective in extending the postharvest life of bell or long green peppers (0.5, 1.0, or 1.5%, w/v) when the fruit were stored at 21°C due to wilting and color development. At the present time, chitosan polymers are not on the FDA list of approved food additives although Nutri-Save may be available in Canada.

COMPOSITE COATINGS

So far, a water-soluble coating with good water vapor barrier properties has not been developed. This is largely due to the increased solubility of water vapor in water-soluble films, resulting in high water vapor permeability. One possible solution is to formulate composite coatings that are heterogeneous films of hydrophobic particles within a hydrophilic matrix. Several such coatings were formulated from paraffin or beeswax, emulsifiers, triethanolamine, or coconut oil and carboxymethyl cellulose, oleic acid or sodium oleate. These coatings were applied to peaches which were stored at room temperature with 57–63% RH. The films with cellulose, however, resulted in higher fruit weight loss than uncoated fruit, while the hydrophobic films resulted in low O_2 permeability and fermentation (Erbil and Mnftngil, 1986).

An experimental coating called Nature-Seal™ was recently formulated at the U.S. Department of Agriculture, Agricultural Research Service, and was patented by the U.S. Department of Agriculture (Nisperos-Carriedo and Baldwin, 1993). Nature-Seal is currently in the process of commercial testing through the U.S. Government Technology Transfer Program. This composite polysaccharide-based coating also uses cellulose derivatives as film formers, but does not contain sucrose fatty acid esters as does TAL Pro-long and Semperfresh. In laboratory tests, Nature-Seal retarded ripening of climacteric fruits including tomatoes and mangoes, and delayed browning of carambolas (Nisperos-Carriedo et al., 1992; Sanchez, 1990), but, as with other water-soluble coatings, it does not have the moisture barrier properties of the more hydrophobic wax coatings. An experimental film consisting of chitosan and lauric acid was less permeable to water vapor, but more permeable to gases than chitosan alone (Wong et al., 1992). When lauric acid was exchanged for palmitic, octanoic, or butyric acids, methyl laurate, acetylated monoglyceride, or propylene glycol stearate, the water vapor transfer rate increased. Only lauric acid containing formulations resulted in a microstructure that consisted of a sheet-like structure that was stacked in layers, presenting an efficient barrier to moisture transfer.

BILAYER COATINGS

The advantage of the good water barrier properties of lipid coatings and the good gas permeability properties and nongreasy texture of polysaccharide coatings can also be combined to form edible bilayer films. One such film, consisting of a layer of stearic and palmitic acids and a layer of hydroxypropyl cellulose, substantially reduced transfer of water (Kamper and Fennema, 1985). Another film comprised of 53% hydroxypropyl methylcellulose and 45% stearic acid had a permeability of 0.17 g water mil/(m²/day/mmHg) at 85% RH compared to 48 g water mil/(m₂/day/mmHg) (Hagenmaier and Shaw, 1990) for hydroxypropyl methylcellulose alone. There are no data, however, on the effect of bilayer films on whole fruits or vegetables.

PROTEIN COATINGS

Proteins are also under investigation for use in coatings for fruits and vegetables. Proteins for edible coatings may be derived from maize, wheat, soybeans, and collagen (gelatin); these provide good barriers to O_2 and CO_2, but not to water (Park and Chinnan, 1990; Postharvest News and Information, 1990b). Protein coatings are produced exclusively from renewable edible ingredients and degrade more readily than polymeric material. Such films could supplement the nutritional value of coated foods, and would be useful for individual packaging of small portions of food, particularly products that are not currently individually packaged such as peas, beans, nuts, and berries (Gennadios and Weller, 1990). A zein composite coating, containing vegetable oils, glycerin, citric acid, and anti-oxidants, prevented rancidity in products such as nuts by acting as a moisture barrier (Andres, 1984) but also probably restricted O_2 transport in addition to serving as a carrier for anti-oxidants. Zein has been used in formulation of edible coatings for pharmaceutical tablets and confectionery products including nuts and dried fruits, often as a substitute for shellac. A zein film was reported to retard ripening of tomatoes (Park et al., 1990). Wheat gluten films are reported to be good barriers to O_2 and CO_2 and their mechanical properties are comparable to polymeric films, but these films have high water permeability due to their hydrophilic nature (Gennadios and Weller, 1990).

NEW DEVELOPMENTS AND GOALS FOR EDIBLE COATINGS

USE OF EDIBLE COATINGS IN QUARANTINE TREATMENTS

Edible coatings are currently being tested for use in quarantine treatments. A technique for fruit fly (*Anastrepha suspensa*) disinfestation in fruits is under investigation by USDA researchers. Apparently some coatings cause larval suffocation in infested fruit perhaps due to reduced O_2 levels. Use of coatings in combination with other quarantine treatments such as hot water dips, fumigation, and extended cold storage was investigated. There is a possibility that the use of coatings could replace or reduce the severity of these established procedures. All of the above quarantine treatments result in some damage to fruit peel and possibly flavor, thereby reducing fruit quality and value (Hallman et al., 1992).

LABELING REGULATIONS

The latest regulations from the Food and Drug Administration (FDA, 1991) require labels on food, including the list of ingredients used, but the law permits exemptions where compliance is "impractical." Individual labeling of fresh fruits and vegetables falls into the exempt category. Foods received in bulk containers at retail establishments should display labeling information on the bulk container, which must be in plain view, or through the use of a counter card, sign, or other appropriate device that prominently bears the required ingredient information (Lecos, 1982). Identification of components in coatings is not clearly defined, however. On June 26, 1991, the FDA published a proposal for ingredient listing that required specification of the subingredients of the wax, or, if the packers/repackers do not have a "consistent pattern" of wax use, the origin of the wax (e.g., shellac-based or vegetable-based, etc.) should be printed instead (Terpstra, 1991). The International Apple Institute joined other commodity organizations in commenting on the FDA's proposal. These industries would rather list only the origin of the wax rather than the subingredients, and only label the shipping container and not the retail package. The January 6, 1993 issue of the *Federal Register* contained the FDA final rule for labeling all produce treated with postharvest coatings. Shippers and retailers have until May 8, 1994 to comply with the new regulations. All coated products must have one of the following minimum statements on the shipping container: coated with (animal-, petroleum-, lac-resin or shellac-, vegetable-, or beeswax-based) wax. The terms "food-grade" and "to maintain freshness" are optional (Citrograph, 1993).

Restrictions on coating formulations present a confusing picture, both for domestic and exported produce. For example, the apple coating industry and wood rosin producers are interested in verifying the legality of using wood rosin (approved for citrus) in apple waxes. In some countries overseas, certain types of waxes used here are banned. Norway recently banned all imports of waxed fruit effective November 1, 1990 (Postharvest News and Information, 1990a). Petroleum waxes, morpholine, and carnauba wax are prohibited in the United Kingdom and petroleum waxes in Japan (personal communication, Pace International/Shield-Brite).

GOALS FOR FUTURE RESEARCH

Permeability Properties of Coatings

A major goal for future research is to reduce the hydrophilic nature of polysaccharide and protein coatings while maintaining the desirable gas permeability properties of these types of films. Polysaccharide polymers, such as cellulose, must first undergo hydrolysis or derivitization to become water-soluble. The size of the polymer, or type of salts and functional groups added result in different permeability properties (microcrystalline cellulose, sodium carboxymethyl cellulose, methyl cellulose, hydroxypropyl cellulose, hydroxypropylmethyl cellulose, etc.). These different types of cellulose polymers can also be mixed to achieve a desired level of permeability. The solubility properties of proteins can also be manipulated by the addition of various functional groups. Soy protein, for example, showed significantly improved solubility and emulsifying acitvity over a lower pH range upon phosphorylation by a protein kinase (Campbell et al., 1992).

Use of Additives

Growth regulators such as auxin derivatives and analogues, polyamines, and gibberellic acid must be tested and approved by the FDA before they can be incorporated into edible coatings for fruits and vegetables. The potential benefit of these additives is great, since these compounds delay ripening, senescence, and color changes and, in the case of gibberellic acid, serve to strengthen fruit peel (Evans and Malmberg, 1989; Sharma et al., 1989; Kader et al., 1966; Lodh et al., 1963). The plant growth regulator, GA (150 ppm), was added to a 6% Waxol coating and significantly delayed the ripening of 'Lacatan' bananas and enhanced marketability (Rao and Chundawat, 1991). Public pressure to reduce use of fungicides emphasizes the need to find alternative means of controlling postharvest diseases of fruits and vegetables. Natural salts, such as calcium chloride (Meheriuk, 1990a), and biological control agents, including yeasts, bacteria, and fungi (Jeffries and Jeger, 1990), could be incorporated into edible coatings for disease control.

In general, edible coatings made from natural waxes, resins, polysaccharides, and proteins, represent an environmentally ideal package since they are biodegradable and can be consumed with the packaged product. The main ingredients are produced from renewable resources, in contrast to paraffin, mineral oil, oxidized polyethylene, and plastics, which are manufactured from a limited supply of fossil fuels. In the future, coatings may reduce the need for energy-intensive refrigeration and costly CA storage.

REFERENCES

Aljuburi, H. J. and A. Huff. 1984. "Reduction in Chilling Injury to Stored Grapefruit (*Citrus paradisi* Macf.) by Vegetable Oils," *Sci. Hort.*, 24:53–58.

Allan, C. R. and L. A. Hadwiger. 1979. "The Fungicidal Effect of Chitosan on Fungi of Varying Cell Wall Composition," *Exp. Mycology*, 3:285–287.

Andres, C. 1984. "Natural Edible Coating Has Excellent Moisture and Grease Properties," *Food Proc.* (Dec.):48–49.

Ashley, R. J. 1985. "Permeability and Plastics Packaging," in *Polymer Permeability*, J. Comyn, ed., New York, NY: Elsevier, pp. 269–308.

Ayres, J. C., A. A. Kraft and L. C. Peirce. 1964. "Delaying Spoilage of Tomatoes," *Food Technology*, 9:100–103.

Banaras, M., N. K. Lownds and P. W. Bosland. 1989. "Evaluation of Skin Coatings and Modified Atmosphere Packaging for Storage of Pepper Fruits," *Program and Abstracts for the 86th Annual Meeting of Amer. Soc. Hort. Sci.*, p. 97.

Banks, N. H. 1983. "Evaluation of Methods for Determining Internal Gases in Banana Fruit," *J. Exp. Bot.*, 44:871:879.

Banks, N. H. 1984a. Some Effects of TAL Pro-long Coating on Ripening Bananas," *J. Exp. Bot.*, 35:127–137.

Banks, N. H. 1984b. "Studies of the Banana Fruit Surface in Relation to the Effects of TAL Pro-long Coating on Gaseous Exchange," *Sci. Hort.*, 24:279–286.

Banks, N. H. 1985a. "Internal Atmosphere Modification in Pro-long Coated Apples," *Acta Hort.*, 157:105–112.

Banks, N. H. 1985b. "Responses of Banana Fruit to Pro-long Coating at Different Times Relative to the Initiation of Ripening," *Sci. Hort.*, 26:149–157.

Barger, W. R., J. S. Wiant, W. T. Pentzer, A. L. Ryall and D. H. Dewey. 1948. "A Comparison of Fungicidal Treatments for the Control of Decay in California Cantaloupes," *Phytopath.*, 38:1019–1024.

Ben-Yehoshua, S. 1966. "Some Effects of Plastic Skin Coating on Banana Fruit," *Trop. Agric. Trin.*, 43:219–232.

Bhardwaj, C. L., H. F. Jones and I. H. Smith. 1984. "A Study of the Migration of Externally Applied Sucrose Esters of Fatty Acids through the Skins of Banana, Apple and Pear Fruits," *J. Sci. Food Agric.*, 35:322–331.

Blake, J. R. 1966. "Some Effects of Paraffin Wax Emulsions of Bananas," *Queensland J. Agric. Animal Sci.*, 23:49–56.

Blanpied, G. D. and K. D. Hickey. 1963. "Concord Grape Storage Trials for Control of *Botrytis cinera* and *Penecillium* sp.," *Plant Dis. Rep.*, 47:986–989.

Bose, A. N. and G. Basu. 1954. "Studies on the Use of Coating for Extension of Storage Life of Fresh Fajli Mango," *Food Res.*, 19:424–428.

Brooks, C. 1938. "Some Effects of Waxing Tomatoes," *Amer. Soc. Hort. Sci. Proc.*, 35:720.

Brown, E. 1984. "Efficacy of Citrus Postharvest Fungicides Applied in Water or Resin Solution Water Wax," *Plant Dis.*, 68:415–418.

Bruemmer, J. H. 1989. "Terminal Oxidase Activity during Ripening of Hamlin Orange," *Phytochem.*, 28:2901–2902.

Campbell, N. F., F. F. Shih and W. E. Marshall. 1992. "Enzymatic Phosphorylation of Soy Protein Isolate for Improved Functional Properties," *J. Agric. Food Chem.*, 40:403–406.

Citrograph. 1993. "Waxes: A Common Sense Rule," *Citrograph* (Feb.)

Claypool, L. L. 1940. "The Waxing of Deciduous Fruits," *Amer. Soc. Hort. Sci. Proc.*, 37:443–447.

Claypool, L. L. and J. R. King. 1941. "Fruit Waxing in Relation to Character of Cover," *Proc. Amer. Soc. Hort. Sci.*, 38:261–265.

Cohen, E., Y. Shalom and I. Rosenberger. 1990. "Postharvest Ethanol Buildup and Off-Flavor in 'Murcott' Tangerine Fruits," *J. Amer. Soc. Hort. Sci.*, 115:775–778.

Cothran, C. D. March 27, 1951. U.S. Patent 2,700,025.

Curtis, G. J. 1988. "Some Experiments with Edible Coatings on the Long-Term Storage of Citrus Fruits," *Proc. 6th Int. Citrus Congress*, 3:1514–1520.

Dalal, V. B., W. E. Eipeson and N. S. Singh. 1971. "Wax Emulsion for Fresh Fruits and Vegetables to Extend Their Storage Life," *Indian Food Packer*, 25:9–15.

Dalal, V. B., P. Thomas, N. Nagaraja, G. R. Shah and B. C. Amla. 1970. "Effect of Wax Coating on Bananas of Varying Maturity," *Indian Food Packer*, 24:36–39.

Danials, R. 1973. *Edible Coatings and Soluble Packaging.* Park Ridge, NJ: Noyes Data Corp., p. 360.

Davis, D. G., C. M. Elson and E. R. Hayes. 1988. "*N,O*-Carboxymethyl Chitosan, a New Water Soluble Chitin Derivative," *4th Int. Conf. on Chitin and Chitosan,* Trondhein, Norway, August 22–24.

Davis, P. L. 1970. "Relation of Ethanol Content of Citrus Fruits to Maturity and to Storage Conditions," *Proc. Fla. State Hort. Soc.,* 83:294–298.

Davis, P. L. and P. C. Harding. 1960. "The Reduction of Rind Breakdown of Marsh Grapefruit by Polyethylene Emulsion Treatments," *J. Amer. Soc. Hort. Sci.,* 75:271–274.

Davis, P. L., B. Roe and J. H. Bruemmer. 1973. "Biochemical Changes in Citrus Fruits during Controlled-Atmosphere Storage," *J. Food Sci.,* 38:225–229.

DeLong, C. F. and T. H. Shepherd. June 13, 1972. U.S. Patent 3,669,691.

Dhalla, R. and S. W. Hanson. 1988. "Effect of Permeable Coatings on the Storage Life of Fruits. II. Pro-long Treatment of Mangoes (*Mangifera indica* L. cv. Julie)," *Int. J. of Food Sci. and Technol.,* 23:107–112.

Drake, S. R. and J. W. Nelson. 1990. "Storage Quality of Waxed and Nonwaxed 'Delicious' and 'Golden Delicious' Apples," *J. Food Qual.,* 13:331–341.

Drake, S. R., R. Caverlieri and E. M. Kupferman. 1991. "Quality Attributes of d'Anjou Pears after Different Wax Drying Temperatures and Refrigerated Storage," *J. Food Qual.,* 14:455–465.

Drake, S. R., J. K. Fellman and J. W. Nelson. 1987. "Postharvest Use of Sucrose Polyesters for Extending the Shelf-Life of Stored 'Golden Delicious' Apples," *J. Food Sci.,* 52:1283–1285.

Drake, S. R., E. M. Kupferman and J. K. Fellman. 1988. " 'Bing' Sweet Cherry (*Prunus avium*) Quality as Influenced by Wax Coatings and Storage Temperature," *J. Food Sci.,* 53:124–126, 156.

Eaks, I. L. and W. A. Ludi. 1960. "Effects of Temperature, Washing and Waxing on the Composition of the Internal Atmosphere of Orange Fruits," *Proc. Amer. Hort. Soc.,* 76:220–228.

Eastman Chemical Inc. 1984. "A Guide to the Use of Epolene Waxes under United States FDA Food Additive Regulations," Publication No. F-243D, pp. 1–6.

Eastman Chemical Inc. 1990. "Emulsification of Epolene Waxes," Publication No. F-302, pp. 1–17.

Edwards, M. E. and R. W. Blennerhassett. 1990. "The Use of Postharvest Treatments to Extend Storage Life and to Control Postharvest Wastage of Honeydew Melons (*Cucumis melo* L. var. inodorus Naud.) in Cool Storage," *Aust. J. of Exp. Agric.,* 30:693–697.

El Ghaouth, A., J. Arul, R. Ponnampalam and M. Boulet. 1991. "Chitosan Coating Effect on Storability and Quality of Fresh Strawberries," *J. Food Sci.,* 56:1618–1631.

El Ghaouth, A., R. Ponnompalam, F. Castaigne and J. Arul. 1992. "Chitosan Coating to Extend the Storage Life of Tomatoes," *HortSci.,* 27:1016–1018.

Elson, C. M., E. R. Hayes and P. K. Lidster. 1985. "Development of the Differentially Permeable Fruit Coating 'Nutri-Save' for the Modified Atmosphere Storage of Fruit," in *Controlled Atmosphere for Storage and Transport of Perishable Agricultural Commodities,* M. Blankenship, ed., Raleigh, NC: North Carolina State Univ., pp. 248–262.

Emmert, F. H. and F. W. Southwick. 1954. "The Effect of Maturity, Apple Emanations, Waxing, and Growth Regulators on the Respiration and Red Color Development of Tomato Fruits," *Amer. Soc. Hort. Sci. Proc.*, 63:393–401.

Erbil, H. Y. and N. Mnftngil. 1986. "Lengthening the Postharvest Life of Peaches by Coating with Hydrophobic Emulsions," *J. Food Proc. and Pres.*, 10:269–279.

Evans, P. T. and R. L. Malmberg. 1989. "Do Polyamines Have Roles in Plant Development," *Annu. Rev. Plant. Physiol. Plant Mol. Biol.*, 40:235–269.

FDA. 1991. Title 21–Food and Drugs. Washington, DC: Federal Register, U.S. Food and Drug Administration, Code of Federal Regulations.

Fisher, D. V. and J. E. Britton. 1940. "Apple Waxing Experiments," *Sci. Agric.*, 21:70–79.

Franklin, E. W. 1961. "The Waxing of Turnips for the Retail Market," *Canada Dept. of Agric.*, Publication No. 1120, 3 pp.

Ganz, A. J. 1969. "CMC and Hydroxypropylcellulose–Versatile Gums for Food Use," *Food Prod. Dev.*, 3:65.

Gennadios, A. and C. L. Weller. 1990. "Edible Films and Coatings from Wheat and Corn Proteins," *Food Technol.*, 44:63–69.

Gold, W. C. 1969. U.S. Patent 3,424,591.

Habeebunnisa, M., C. Pushpa and J. C. Srivastava. 1963. "Studies on the Effect of Protective Coating on the Refrigerated and Common Storage of Bell Peppers (*Capsicum frutescense*)," *Food Sci.* (Mysores), 12:192–196.

Hadwiger, L. A. and J. M. Beckman. 1980. "Chitosan as a Component of Pea (*Fusarum solani*) Interactions," *Plant Physiol.*, 66:205–211.

Hagenmaier, R. D. and P. E. Shaw. 1990. "Moisture Permeability of Edible Films Made with Fatty Acid and (Hydroxypropyl) Methylcellulose," *J. Agric. Food Chem.*, 38:1799–1803.

Hagenmaier, R. D. and P. E. Shaw. 1992. "Gas Permeability of Fruit Coating Waxes," *J. Amer. Soc. Hort. Sci.*, 117:105–109.

Hallman, G. J., M. O. Nisperos-Carriedo, E. A. Baldwin and C. A. Campbell. 1994. "Mortality of Caribbean Fruit Fly (Diptera: Tephritidae) Immatures in Coated Fruits," *J. Econ. Ent.* (in press).

Hardenburg, R. E. 1967. "Wax and Related Coatings for Horticultural Products. A Bibliography," *Agr. Res. Bull.*, 51–15, U.S. Dept. of Agric., Washington, DC.

Hartman, J. and F. M. Isenberg. 1956. "Waxing Vegetables," *New York Agric. Exten. Ser. Bull. No. 965*, pp. 3–14.

Hatton, T. T. 1958. "Some Effects of Waxes and 2,4,5-Trichlorophenoxyacetic Acid as Postharvest Treatments on Persian Limes," *Fla. State Hort. Soc.*, 71:312–315.

Hirano, S. and N. Nagao. 1989. "Effects of Chitosan, Pectic Acid, Lysozyme, and Chitinase on the Growth of Several Phytopathogens," *Agric. Biol. Chem.*, 53:3065–3066.

Hitz, C. W. and I. C. Haut. 1938. "Effect of Certain Waxing Treatments at Time of Harvest upon the Subsequent Storage Quality of 'Grimes Golden' and 'Golden Delicious' Apples," *Proc. Amer. Soc. Hort. Sci.*, 36:440–447.

Hitz, C. W. and I. C. Haut. 1942. "Effects of Waxing and Pre-Storage Treatments upon Prolonging the Edible and Storage Qualities of Apples," *Univ. of MD Agric. Exp. Sta. Tach. Bull.*, No. A14, 44 pp.

Institute of Food Technology. 1991. "New from Mitsubishi," *Annual Meeting and Food Expl. Program and Exhibit Directory*, Dallas Convention Center, June 1–5, 1991, Chicago, IL.

Jeffries, P. and M. J. Jeger. 1990. "The Biological Control of Postharvest Diseases of Fruit," *Postharvest News and Information,* 1:365–368.

Jokay, L., G. E. Nelson and E. C. Powell. 1967. "Amylaceous Coatings for Foods," *Food Technol.,* 21:1064–1066.

Kader, A. A. 1986. "Biochemical and Physiological Basis for Effects of Controlled and Modified Atmospheres on Fruits and Vegetables," *Food Technol.,* 40:99–104.

Kader, A. A., L. L. Morris and E. C. Maxie. 1966. "Effect of Growth-Regulating Substances on the Ripening and Shelf-Life of Tomatoes," *HortSci.,* 1:90–91.

Kamper, S. L. and O. R. Fennema. 1985. "Use of an Edible Film to Maintain Water Vapor Gradients in Foods," *J. Food Sci.,* 50:382–384.

Kaplan, H. J. 1986. "Washing, Waxing and Color Adding," in *Fresh Citrus Fruits,* W. F. Wardowdki, S. Nagy and W. Grierson, eds., Westport, CT: AVI Publishing Co., p. 379.

Ke, E. and A. A. Kader. 1990. "Tolerance of 'Valencia' Oranges to Controlled Atmospheres as Determined by Physiological Responses and Quality Attributes," *J. Amer. Soc. Hort. Sci.,* 115:779–783.

Kendra, D. and L. A. Hadwiger. 1984. "Characterization of the Smallest Chitosan Oligomer that is Maximally Antifungal to *Fusarium solani* and Elicits Pisatin Formation in *Pisum sativum* (Peas)," *Exp. Mycology,* 8:276–281.

Kester, J. J. and O. R. Fennema. 1988. "Edible Films and Coatings: A Review," *Food Technol.,* 42:47–59.

Kraght, A. J. 1966. "Waxing Peaches with the Consumer in Mind," *Produce Marketing,* 9:20–21.

Krumel, K. L. and T. A. Lindsay. 1976. "Nonionic Cellulose Ethers," *Food Tech.,* 30:36–43.

Lawrence, J. F. and J. R. Iyengar. 1983. "Determination of Paraffin Wax and Mineral Oil on Fresh Fruits and Vegetables by High Temerature Gas Chromatography," *J. Food Safety,* 5:119.

Lawson, J. A. 1960. "Banana Packing and Waxing," *West Austral. Dept. Agr. Jour.* (Ser. 4), 1:41–45.

Lecos, C. 1982. "How to Shine an Apple," *FDA Consumer* (February):8–11.

Lee, J. C. and O. W. Kwon. 1990. "Effects of Pro-long on Storage Quality and Ethylene Evolution in 'Jonathan' and 'Fuji' Apple Fruits," *J. Kor. Soc. Hort. Sci.,* 31:247–254.

Lodh, B. B., S. De, S. K. Mukherjee and A. W. Bose. 1963. "Storage of Mandarin Oranges. II. Effects of Hormones and Wax Coatings," *J. Food Sci.,* 28:519–524.

Lowings, P. H. and D. G. Cutts. 1982. "The Preservation of Fresh Fruits and Vegetables," *Proc. Inst. Food Sci. Tech. Ann. Symp.,* July, 1981, Nottingham, UK, p. 52.

Mack, W. B. and J. R. Janer. 1942. "Effects of Waxing on Certain Physiological Processes of Cucumbers under Different Storage Conditions," *Food Res.,* 7:38–47.

Magness, J. R. and H. C. Diehl. 1924. "Physiological Studies on Apples in Storage," *J. Agr. Res.,* 27:1–38.

Maharaj, R. and C. K. Sankat. 1990. "Refrigerated Storage of the Breadfruit and the Effect of Washing, Packaging, and Storage in Water," *American Soc. Agric. Engin.,* Paper No. 90–6509, 31 pp.

Mathur, P. B. and H. C. Srivastava. 1955. "Effect of Skin Coatings on the Storage Behavior of Mangoes," *Food Res.,* 20:559–566.

Mathur, P. B. and H. Subramanyam. 1956. "Effect of a Fungicidal Wax Coating on the Storage Behavior of Mangoes," *J. Sci. Food Agric.*, 7:673–676.

McDonald, R. E. 1986. "Effects of Vegetable Oils, CO_2, and Film Wrapping on Chilling Injury and Decay of Lemons," *HortSci.*, 21:476–477.

Meheriuk, M. 1990a. "Effects of Diphenylamine, Gibberellic Acid, Daminozide, Calcium, High CO_2 and Elevated Temperatures on Quality of Stored 'Bartlett' Pears," *Can. J. Plant Sci.*, 70:887–892.

Meheriuk, M. 1990b. "Skin Color in 'Newton' Apples Treated with Calcium Nitrate, Urea, Diphenylamine, and a Film Coating," *HortSci.*, 25:775–776.

Meheriuk, M. and O. L. Lau. 1988. "Effect of Two Polymeric Coatings on Fruit Quality of 'Bartlett' and 'd'Anjou' Pears," *J. Amer. Soc. Hort. Sci.*, 113:222–226.

Mellenthin, W. M., P. M. Chen and D. M. Borgic. 1982. "In-Line Application of Porous Wax Coating Materials to Reduce Friction Discoloration of 'Bartlett' and 'd'Anjou' Pears," *Hort. Sci.*, 17:215–217.

Mellor, J. R. 1928. "Changes Produced in Apples by the Use of Cleaning and Oil-Coating Process," *J. Agric. Res.*, 36:429–436.

Miers, J. C., H. A. Swenson, T. H. Schultz and H. S. Owens. 1953. "Pectinate and Pectate Coatings. I. General Requirements and Procedures," *Food Technol.*, 7:229–231.

Miller, W. R., D. Chun, L. A. Risse, T. T. Hatton and T. Hinsch. 1988. "Influence of Selected Fungicide Treatments to Control the Development of Decay in Waxed or Film Wrapped Florida Grapefruit," *Proc. 6th Int. Cit. Cong.*, 3:1471–1477.

Mitchell, F. G., J. H. Larue, J. P. Gentry and M. H. Gerdts. 1963. "Packing Nectarines to Reduce Shrivel," *California Agric.*, 17:10–11.

Mitsubishi-Kasei. 1989. "Ryoto Sugar Ester, Technical Information," Mitsubishi-Kasei, pp. 1–20.

Motlagh, F. H. and P. C. Quantick. 1988. "Effects of Permeable Coatings on the Storage Life of Fruits. I. Pro-long Treatment of Limes," *Int. J. Food Sci. and Technol.*, 23:99–105.

Mulder, T. H. June 24, 1969. U.S. Patent 3,451,826.

Muliyar, M. K. and M. M. K. Marar. 1963. "Studies on the Keeping Quality of Ripe Coconuts in Storage," *Indian Coconut J.*, 17:13–18.

Nelson, R. C. 1942. "A Method of Studying the Movement of Respiratory Gases through Waxy Coatings," *Plant Physiol.*, 17:509–514.

Neville Chemical Co. 1988. "Production Information, Resins, Plasticizers, Nonstaining Antioxidants and Chlorinated Paraffins," Brochure No. NCCFAL88, pp. 1–15.

Nisperos-Carriedo, M. O. and E. A. Baldwin. 1988. "Effect of Two Types of Edible Films on Tomato Fruit Ripening," *Proc. Fla. State Hort. Soc.*, 101:217–220.

Nisperos-Carriedo, M. O. and E. A. Baldwin. March 30, 1993. U.S. Patent 5,198,254.

Nisperos-Carriedo, M. O., E. A. Baldwin and P. E. Shaw. 1992. "Development of Edible Coating for Extending Postharvest Life of Selected Fruits and Vegetables," *Proc. Fla. State Hort. Soc.*, 104:122–125.

Nisperos-Carriedo, M. O., P. E. Shaw and E. A. Baldwin. 1990. "Changes in Volatile Flavor Components of Pineapple Orange Juice as Influenced by the Application of Lipid and Composite Film," *J. Agric. and Food Chem.*, 38:1382–1387.

Nordby, H. E. and R. E. McDonald. 1990. "Squalene in Grapefruit Wax as a Possible Natural Protectant against Chilling Injury," *Lipids*, 25:807–810.

Paredes-Lopez, O., E. Camargo-Rubio and Y. Gallardo-Navarro. 1974. "Use of Coatings of Candelilla Wax for the Preservation of Limes," *J. Sci. Food Agric.*, 25:1207–1210.

Park, H. J. and M. S. Chinnan. 1990. "Properties of Edible Coatings for Fruits and Vegetables," *Amer. Soc. Agric. Engin.*, Paper No. 90-6510, 19 pp.

Park, H. J., M. S. Chinnan, and L. Shewfelt. 1990. "Effect of Corn-Zein Film Coating on the Storage Life of Tomatoes," Paper No. 291, *51st Annual Meeting of Inst. Food Technol.*, Anaheim, CA, June 16–20.

Paull, R. E. and N. J. Chen. 1989. "Waxing and Plastic Wraps Influence Water Loss from Papaya Fruit during Storage and Ripening," *J. Amer. Soc. Hort. Sci.*, 114:937–942.

Perkins-Veazie, P. M. and J. K. Collins. 1991. "Color Changes in Waxed Turnips during Storage," *J. Food Qual.*, 14:313–319.

Platenius, H. 1939. "Wax Emulsions for Vegetables," *Cornell Univ. Agric. Exp. Sta. Bull. No. 723.*, 43 pp.

Postharvest News and Information. 1990a. "General News: Waxed Fruit Banned in Norway," *Postharvest News and Information*, 1:433.

Postharvest News and Information. 1990b. "General News: Protein Films from Maize, Wheat and Soybeans—Will They Extend Fruit and Vegetable Shelf Life," *Postharvest News and Information*, 1:435.

Postharvest News and Information. 1991. "General News: Chitosan-Derivative Keeps Apples Fresh," *Postharvest News and Information*, 2:75.

Radnia, P. M. and J. W. Eckert. 1988. "Evaluation of Imazalil Efficacy in Relation to Fungicide Formulation and Wax Formulation," *Proc. 6th Int. Cit. Cong.*, 3:1427–1434.

Rajzman, A., H. Heller and S. Abramovicz. 1973. "Effect of a Wax Coating Applied to Oranges on the Weight Loss and Disappearance of Green Color in the Stored Fruit," *Rehavot Nat. Univ. Inst. Agr.*, Prelim. Rep. No. 729, 12 pp.

Rankin, J. C., I. A. Wolff, H. A. Davis and C. E. Rist. 1958. "Permeability of Amylose Film to Moisture Vapor, Selected Organic Vapor, and the Common Gases," *Ind. Eng. Chem., Chem. Eng. Data Ser.*, 3:120–123.

Rao, D. V. R. and B. S. Chundawat. 1991. "Chemical Regulation of Ripening in Banana Bunches cv. 'Lacatan' at Non-Refrigerated Temperatures," *Haryana J. Hort. Sci.*, 20:6–11.

Risse, L. A., D. Chun, R. E. McDonald and W. R. Miller. 1987. "Volatile Production and Decay during Storage of Cucumbers Waxed, Imazalil-Treated, and Film-Wrapped," *Hort Sci.*, 22:274–276.

Rosenfield, D. November 12, 1968. U.S. Patent 3,410,696.

Sakane, Y., N. Arita, S. Shimokana, H. Ito and Y. Osajima. 1990. "Storage of Shredded Cabbage in Plastic Films Using Ethylene-Acetaldehyde or Sucrose Fatty Acid Esters," *Nippon Shokuhin Kogyo Gakkaishi*, 37:281–286.

Sanchez, D. 1990. "Keep It under an Edible Coat," *Agr. Res.* (March):4–5.

Santerre, C. R., T. F. Leach, J. N. Cash. 1989. "The Influence of the Sucrose Polyester Semperfresh on the Storage of Michigan Grown 'McIntosh' and 'Golden Delicious' Apples," *J. Food Process. Preserv.*, 13:293–305.

Schappelle, N. A. 1941. "A Physiological Study on the Effects of Waxing Pineapples of Different Stages of Maturity," *Puerto Rico Univ. Sta. Res. Bul.*, 3:30–32.

Schultz, T. H., H. S. Owens and W. D. Maclay. 1948. "Pectinate Films," *J. Colloid Sci.*, 3:53–62.

Schultz, T. H., J. C. Miers, H. S. Owens and W. D. Maclay. 1949. "Permeability of Pectinate Films to Water Vapor," *J. Phys. Colloid. Chem.*, 53:1320–1330.

Sharma, R. K., R. Singh, J. Kumar and S. S. Sharma. 1989. "Shelf Life of Baramasi Lemon as Affected by Some Chemicals," *Res. and Dev. Reporter*, 6:78–82.

Shillington, W. L. and J. J. Liggett. October 13, 1970. U.S. Patent 3,533,810.

Siade, G. and E. Pedraza. 1977. "Extension of Storage Life of Banana (Giant Cavendish) Using Natural Wax Candelilla," *Acta. Hort.*, 62:327–335.

Smith, S. M. and J. R. Stow. 1984. "The Potential of a Sucrose Ester Coating Material for Improving the Storage and Shelf-Life Qualities of 'Cox's Orange Pippin' Apples," *Ann. Appl. Biol.*, 104:383–391.

Smock, R. M. 1935. "Certain Effects of Wax Treatments on Various Varieties of Apples and Pears," *Proc. Amer. Soc. Hort. Sci.*, 33:284–289.

Srivastava, H. C., N. S. Kapur, V. B. Dalal, H. Subramanyam, S. D'Souza and K. S. Rao. 1962. "Storage Behavior of Skin Coated Guavas (*Psidium guajava*) under Modified Atmosphere," *Food Sci.* (Mysore), 11:224–248.

Stossel, P. and J. C. Leuba. 1984. "Effect of Chitosan, Chitin and Some Aminosugars on Growth of Various Soilborne Phytopathogenic Fungi," *Phytopath. Z.*, 111:82–90.

Subramanyam, H., N. V. Moorthy, V. B. Dalal and H. C. Srivastava. 1962. "Effect of a Fungicidal Wax Coating with or without Growth Regulator on the Storage Behavior of Mangoes," *Food Sci.*, (Mysores), 11:236–239.

Swenson, H. A., J. C. Miers, T. H. Schultz and H. S. Owens. 1953. "Pectinate and Pectate Coatings. II. Application to Nut and Fruit Products," *Food Technol.*, 7:232–235.

Terpstra, E. 1991. "Right to the Core: Identifying Waxes," *Fruit Grower* (October):10.

Trout, S. A., E. G. Hall and S. M. Sykes. 1953. "Effects of Skin Coatings on the Behavior of Apples in Storage," *Aust. J. Agr. Res.*, 4:57–81.

Trout, S. A., E. G. Hall, R. N. Robertson, F. M. V. Hackney and S. M. Sykes. 1942. "Studies in the Metabolism of Apples. I. Preliminary Investigations on Internal Gas Composition and Its Relation to Changes in Stored 'Granny Smith' Apples," *Aust. J. Exp. Biol. Med. Sci.*, 20:219–231.

Ukai, N. T. Tsutsumi and K. Marakami. February 25, 1975. U.S. Patent 3,997,674.

Vines, H. M., W. Grierson and G. J. Edwards. 1968. "Respiration, Internal Atmosphere, and Ethylene Evolution of Citrus Fruit," *J. Amer. Soc. Hort. Sci.*, 92:227–234.

Waks, J., M. Schiffmann-Nadel, E. Lomaniec and E. Chalutz. 1985. "Relation between Fruit Waxing and Development of Rots in Citrus Fruit during Storage," *Plant Disease*, 69:869–870.

Warth, A. H. 1986. *The Chemistry and Technology of Waxes*. New York, NY: Reinhold Pub. Co., pp. 37–192.

Wills, R. H., T. H. Lee, D. Graham, W. B. McGlasson and E. G. Hall. 1981. *Postharvest, an Introduction to the Physiology and Handling of Fruit and Vegetables*. Westport, CN: AVI Publishing Co. Inc., pp. 1–2.

Wilson, C. L., M. E. Wisniewski, C. L. Biles, R. McLaughlin, E. Chalutz and S. Droby. 1991. "Biological Control of Post-Harvest Diseases of Fruits and Vegetables: Alternatives to Synthetic Fungicides," *Crop Protection*, 10:172–177.

Wong, W. W., F. A. Gastineau, K. S. Gregorski, A. J. Tillin and A. E. Pavlath. 1992. "Chitosan-Lipid Films: Microstructure and Surface Energy," *J. Agric. Food Chem.*, 40:540–544.

Woods, J. L. 1990. "Moisture Loss from Fruits and Vegetables," *Postharvest News and Information*, 1:195–199.

Wu, M. T. and D. K. Salunkhe. 1972. "Control of Chlorophyll and Solanine Synthesis and Sprouting of Potato Tubers by Hot Paraffin Wax," *J. Food Sci.*, 37:629–630.

CHAPTER 3

Development of Edible Coatings for Minimally Processed Fruits and Vegetables

DOMINIC W. S. WONG[1]
WAYNE M. CAMIRAND[1]
ATTILA E. PAVLATH[1]

INTRODUCTION

IN recent years, the market demand for minimally processed fruits and vegetables has witnessed a rapid expansion. "Minimal processing" is defined to include all operations such as washing, sorting, trimming, peeling, slicing, coring, etc., that would not extensively affect the fresh-like quality of the produce (Shewfelt, 1987). The product remains biologically and physiologically active, in that the tissues are living and respiring, with a shifting of cellular processes and interactions in response to the tissue damage inflicted by the operations. Minimal processing usually increases the degree of perishability of the processed materials. This is in contrast to all other conventional processing systems in which the products are stabilized and their shelf-life is extended. An increased demand for minimally processed produce has led to a need for more knowledge and new techniques to address the problems not usually encountered in conventional processing.

UNDESIRABLE CHANGES CAUSED BY MINIMAL PROCESSING

One of the most critical problems in minimal processing is the occur-

[1]U.S. Dept. of Agriculture, Agricultural Research Service, Western Regional Research Center, 800 Buchanan Street, Albany, CA 94710. Reference to a company and/or product is only for the purposes of information and does not imply approval or recommendation of the product to the exclusion of others which may also be suitable. All programs and services of the U.S. Department of Agriculture are offered on a nondiscriminatory basis without regard to race, color, national origin, religion, sex, age, marital status, or handicap.

rence of undesirable physiological changes. Minimal processing causes disruption of cell tissue and breakdown of cell membranes, leading to a myriad of immediate membrane-related events: leakage of ions, loss of components, alteration in flux potential, and loss of tugor (Davies, 1987).

BREAKDOWN CAUSED BY ENZYME ACTIONS

Cutting a fruit or vegetable destroys the integrity of the cell tissues and the compartmentation of endogenous enzymes and substrates. Some of these enzyme-catalyzed changes are well known. Mechanical damage allows the enzyme polyphenol oxidase to come into contact with its substrate, phenolic compounds and O_2, leading to a rapid formation of melanin. The molecular basis of this mechanism is a subject of extensive investigation by scientists in the fields of both agriculture and medicine (Hearing and Tsukamoto, 1991). Hydrolytic depolymerization of pectic substrates by polygalacturonase and pectic lyase causes cell wall degradation. Both enzymes are targets for genetic engineering in an effort to control the textural quality of tomato fruits and other commodities (Kramer et al., 1990; Smith et al., 1990).

Degradation of membrane lipids by lipoxygenase and lipase has been implicated as one of the major causes of tissue breakdown in a number of vegetables (Wardale and Galliard, 1977; Galliard et al., 1976). Reaction between lipoxygenase and its endogenous lipid substrates generates hydroperoxides that are related to senescence and scald induction (Feys et al., 1980; Lynch and Thompson, 1984; Altschuler et al., 1989). Breakdown of these hydroperoxides causes the formation of carbonyl compounds in damaged tissues (Klein et al., 1984; Hatanaka et al., 1992). Lipoxygenase-catalyzed formation of fatty acid radicals can react with cell components, leading to further breakdown. Bleaching of plant pigments, β-carotene and chlorphyll a, has been shown to occur during lipoxygenase-mediated reactions (Klein et al., 1984; Schierberle et al., 1981).

In addition to the above well-known examples, numerous less visible reactions occur unnoticed, but can cause drastic changes to the flavor, texture and palatability of the product. Enzyme reactions initiate and catalyze biological pathways which may otherwise be inactive, generating undesirable products. These reactions frequently catalyze and activate further enzyme reactions, resulting in a cascade of events (Schwimmer, 1983). Enzyme activity is known to be strongly related to high moisture content. Cut tissue surfaces therefore are highly suitable for the enhancement of enzyme activity. Maximum stability can be attained in dehydrated foods by controlling the rate of chemical reactions with a_w between 0.2 and 0.3 (Labuza, 1980; Eichner, 1986).

FORMATION OF SECONDARY METABOLITES

In addition to increased enzyme actions, stresses such as cut injury or infection often induce plant tissues to accumulate unusual metabolites (Haard and Cody, 1978; Grisebach, 1987). One of the better known examples is the accumulation of glycoalkaloids in damaged potato tubers. Polyphenolic compounds are produced in damaged tissues of sweet potatoes. Chlorogenic acid, isochlorogenic acid, caffeic acid, and methyl caffeate are among the common products formed via the phenylpropanoid pathway. Uritani and Asahi (1980) reported that the increase in polyphenols started after a lag period of twelve hours and reached a maximum after three days. Increased production of polyphenols was also accompanied by lignin formation and an increase of coumarins. In tomato, potato, and bean pod, resynthesis of a protective layer involves suberization. The materials formed in wound healing are suberin polymers consisting of ω-hydroxy and dicarboxylic acids of long chain fatty acids and alcohols as major components (Dean and Kolattukudy, 1976; Sukumaran et al., 1990). Suberin also contains a high percentage of phenolic compounds. Kolattukudy (1980) showed that depolymerization of suberin preparations yielded a large amount of insoluble material as colored residues.

INCREASED PRODUCTION OF ETHYLENE

Cut injury also stimulates the biogenesis of ethylene, which affects cell expansion, leaf abscission and fruit ripening. Wound-induced ethylene is caused by the increased formation of 1-aminocyclopropane-1-carboxylic acid (ACC) and the subsequent conversion of ACC to ethylene (Boller and Kende, 1980; Yu and Yang, 1980). The increased rate of ethylene production in response to mechanical damage has been reported for a variety of produce, including cantaloupe (McGlasson and Pratt, 1964; Hoffman and Yang, 1982), banana, potato tuber (McGlasson, 1969), tomato (MacLeod et al., 1976), sweet potato (Saltveit and Locy, 1982), winter squash (Hyodo et al., 1983), carrot (Kahl and Laties, 1989), and avocado (Starrett and Laties, 1991). Increased ethylene formation causes physiological disorders and consequently affects the quality attributes of the product. Physiological changes induced by elevated ethylene concentration include: (1) increased cell permeability, (2) loss of compartmentation, (3) increased senescence and respiratory activity, and (4) increased activity of enzymes (Hyodo et al., 1978). Some specific examples of quality effects include accumulation of phenolic compounds in carrots, sprouting in potatoes, lignification in asparagus, brown spot (russet spotting) in lettuce, and general softening of texture (Reid, 1992).

The rate of ethylene production stimulated by stress typically occurs with an initial short lag period, followed by a progressive increase, reaching a peak before subsiding to a stable level. Hyodo et al. (1983) reported that cut winter squash showed increased rate of ethylene production after a three-hour lag, and the rate reached its peak in thirty hours, followed by a decline to a low level forty hours after incubation at 24°C. The rate of increase varies depending on the type of commodity, cultivar, state of ripeness, temperature of incubation, method of preparation, as well as the impact of damage. Although the rate increase is usually in the range of 5–20-fold, more than 100-fold increase in response to wounding has been reported in some cases (Hyodo and Nishino, 1981; Hoffman and Yang, 1982; Hyodo et al., 1983).

INCREASED RESPIRATION

Physical damage also causes an increase in respiration. Several commodities have been studied, including banana (Palmer and McGlasson, 1969), tomato (MacLeod et al., 1976), cantaloupe (McGlasson and Pratt, 1964), salad mixture (lettuce, celery, carrot, radish, onion, endive) (Priepke et al., 1976), pear and strawberry (Rosen and Kader, 1989), and lettuce (Cantwell, 1992). The increase in the rate of respiration in these studies is typically in the range of 2–10-fold. The increase of respiration, as measured by the CO_2 evolved, is the result of many biochemical events that subsequently cause changes in color, flavor, texture, and nutritional quality. Biochemical changes triggered by cellular response include: (1) stimulation of degradation of carbohydrates, (2) activation of glycolysis and the pentose phosphate pathway, (3) activation in mitochondrial activity, and (4) increase in the synthesis of proteins and enzyme activities (Uritani and Asahi, 1980). These activations and accelerations in the cellular processes serve to provide energy and precursors for the biosynthesis of secondary metabolites that are important to wound healing.

CHANGES IN MICROBIAL FLORA

Minimally processed products have cut surfaces with tissues exposed to contaminants. The exudate (juice) from cut tissues constitutes a favorable growth medium for microorganisms (Maxcy, 1982). Various processing steps also increase the opportunities for contamination and growth. Storage conditions and postharvest handling may also modify the environment, causing a shift in the microflora that may result either in decreasing or increasing health risk (Brackett, 1987). Fungal pathogens are the major cause of postharvest spoilage of fruits and vegetables (Sommer et al., 1992). The number of molds present on a wide variety of vegetables at harvest has been reported to be in the range of 4.2 to 6.7 × 10^3 cfu/g. *Aureobasidium pullulans, Fusarium* species, *Alternaria tenuis, Epicoccum nigrum, Mucor* species, *Chaetomium fimeti,* and *Rhizopus nigricans* were among the most frequently isolated in high numbers (Webb and Mundt, 1978). However, the major flora in vegetables consists of bacterial contaminants. Brocklehurst et al. (1987) reported high concentrations of bacteria in retail packs of prepared salad vegetables stored at 7°C at the end of the storage life, with *Pseudomonas* and *Enterobacter* species being predominant. Pectolytic fluorescent *Pseudomonas* species have been identified to be the cause of soft rots of fresh vegetables (Liao and Wells, 1987). Low temperature storage may shift the conditions for the growth of psychrotrophic microorganisms and foodborne pathogens capable of growth at 5°C. These include *Clostridium botulinum* type E, *Yersinia enterocolitica, Escherichia coli, Listeria monocytogenes,* and *Aeromonas hydrophilia* (Palumbo, 1986). Plastic film packaging that alters the atmosphere and humidity may also affect the microfloral pattern. Vacuum packaging of carrot slices resulted in changing microbial composition, with *Leuconostop* species being predominant, as compared to air packs which contained the initially predominant *Erwinia* species as the main spoilage microorganism (Buick and Damoglou, 1987).

TECHNIQUES/APPROACHES FOR PRESERVATION OF MINIMALLY PROCESSED PRODUCTS

A number of potential treatments have been studied to minimize deteriorative effects caused by the application of minimal processing that adversely affect quality characteristics in an effort to extend the shelf-life of the product. Shelf-life is defined as "the time period that a product can be expected to maintain an acceptable quality under specified storage condition" (Shewfelt, 1987). Obviously, no single treatment can adequately be claimed as the solution. In fact, many of these treatments address only a specific aspect of the numerous factors involved in deterioration.

LOW TEMPERATURE STORAGE

Low temperature is effective in lowering the metabolic breakdown of cell tissues. Many biochemical reactions can be controlled to some extent by a decrease of enzyme activity. Most fruits of tropical and subtropical climates, however, are subject to chilling injury when stored at nonfreezing temperatures below 10–12°C. The primary mechanism of chill injury seems to involve a physical change of cell membrane, resulting in symptoms such as discoloration, pitting, necrosis, and growth inhibition (Wang, 1982). In general, low temperature alone is inadequate for retaining acceptable quality in minimally processed products. In addition to membrane lipid phase transition, other biochemical mechanisms have also been postulated to be involved in chilling injury. These include: (1) protein and enzyme dysfunction, (2) disruption of cellular calcium, and (3) lipid peroxidation (Parkin et al., 1989). Eaks (1980) showed that the respiration rate, ethylene production, and the content of ethanol and acetaldehyde of the internal atmosphere of cirtus fruit increased when exposed to a chilling temperature of 0–5°C.

SPECIAL PREPARATION PROCEDURES

Damage caused by peeling, slicing, coring, etc., can be minimized by studying the effects of the different techniques employed for each of the processing steps. For example, the stability of shredded lettuce was extended when cutting was done with a sharp blade using a slicing action (Bolin et al., 1977). Abrasive peeling of carrots causes the formation of a whitish translucent lignin-type surface, similar to a suberin layer in wound healing in potato. The extent of its formation is related to the degree of cel-- lular damage. Bolin and Huxsoll (1991b) reported that peeling with a sharp blade prevented whitening even in prolonged storage. Studies using scanning electron microscopy confirmed that the formation of white blush in carrot sticks was caused by sheer and disruption of surface cell tissues and subsequent dehydration of the damaged cells (Tatsumi et al., 1991).

It has also been demonstrated that removal of cellular debris and fluid generated by cutting can eliminate some of the adverse effects. Bolin and Huxsoll (1991a) showed that rinsing cut pieces with water, followed by centrifugation to remove the added moisture, increased the storage life of salad-cut lettuce.

USE OF ADDITIVES

The addition of chemicals is by far the most extensively investigated method for controlling deteriorative reactions (Borenstein, 1987). For the inhibition of browning, ascorbic acid was used in combination with citric acid (Santerre et al., 1988; Lattanzio and Linsalata, 1989), or calcium chloride (Ponting et al., 1972). Ascorbic acid-2-phosphate has been shown to act synergistically with polyphosphate in preventing browning in cut apple (Sapers and Hicks, 1989). Diamine derivatives of coumarin and 4-substituted resorcinol are effective inhibitors of shrimp melanosis and have potential application in apple, potato, avocado, and grape juice (McEvily et al., 1991). The inhibition constant, K_i is 0.1 μm for 4-hexylresorcinol and mushroom polyphenol oxidase using o-dihydroxy-phenylalanine at pH 6.5 and 2.5°C (Osuga et al., 1994).

Acidification can be employed to preserve tomato, snap bean and pea in bulk storage for subsequent handling and processing (Basel, 1982; Basel and Gould, 1983). Juliot et al. (1989) showed that blanched carrot slices acidified with a combination of acetic acid/citric acid/fumaric acid retained fresh-like qualities when stored in a CO_2 atmosphere. Alternatively, *lactobacillus* bacteria can be added to generate enough lactic acid to lower the pH, providing safe and stable products. Lactic acid bacteria also produce substances that are inhibitory to the growth of other microorganisms (Daeschel, 1989).

Control of microbial contamination or spoilage was achieved by including sorbic acid in dipping formulations for fruit salad and peach slices (Rushing and Senn, 1962; Heaton et al., 1969). In addition to organic acids and other traditional antimicrobial agents (Wagner and Moberg, 1989), a number of naturally occurring antimicrobial additives have potential application in food systems (Banks et al., 1986).

Lidster et al. (1986) showed that chlorogenic acid, catechins, and quercetin glycosides present in apple extracts inhibited tissue softening of whole apples caused by β-galactosidase–catalyzed cell wall degradation. Divalent ions usually prevent loss of firmness, while monovalent ions in general cause softening of tissues (McFeeters et al., 1989, McFeeters and Fleming, 1990). This metal ion effect is usually attributed to the complexation between calcium and pectic acid, although possible additional mechanisms may be required to account for texture regulation. Calcium serves as a binding point for protein-protein complexes, which is important for maintaining cell-to-cell adhesion. Calcium may inhibit the activity of cell wall–degrading enzymes. Cross-linking interaction with pectic polymers may also render the substrate inaccessible to enzyme action. Calcium may increase the transition temperature (rigid $<->$ liquid crystalline) of cell membrane lipid structures. Calcium has been found to suppress the respiratory activity and ethylene production. Calcium may also be involved in the regulatory functions of many enzymes (Poovaiah et al., 1988).

MODIFIED/CONTROLLED ATMOSPHERE

The use of elevated CO_2/reduced O_2 concentrations in packaging is known to retard senescence. The postharvest storage life of apple, pear, kiwifruit, cabbage, tomato, strawberry, banana, and other commodities can be extended by the use of this technique (Kadar et al., 1989). The tolerance limit to elevated CO_2/reduced O_2 concentrations varies depending on the types of commodities. Minimum O_2 concentration and maximum CO_2 concentration range between 0.5–5.0% and 2–15% respectively (Kader, 1992). Although the concept of modified atmosphere has found considerable application on raw produce, minimally processed fruits and vegetables are receiving very limited attention in this regard. Broccoli florets stored under 2–3% O_2, and 2–3% CO_2 modified atmosphere did not exhibit a better storage quality than broccoli pieces stored in air (Ballantyne et al., 1988b). In sealed packages of shredded lettuce, initial flushing with 5% O_2, 5% CO_2 followed with an equilibrated modified atmosphere of 1–3% O_2, 5–6% CO_2 doubled the shelf-life as compared to the controls (Ballantyne et al., 1988a). Dipping slice pear and strawberry in 1% $CaCl_2$ in combination with 0.5% O_2 or 12% CO_2 storage atmosphere has been shown to reduce respiration and ethylene production and retain a greater degree of firmness (Rosen and Kader, 1989).

THE USE OF EDIBLE COATINGS

The concept of using edible films as protective coatings for foods has been in practice for a long time. The most well-known example is the coating of fresh fruits and vegetables with waxes or waxy-water emulsions. The idea of coating cut pieces of fruits and vegetables sounds attractive. Theoretically at least, one could construct a coating that (1) forms an efficient barrier to moisture loss, (2) has a selective permeability to gases, (3) controls migration of water-soluble solutes to retain the natural color pigments and nutrients, and (4) incorporates additives such as coloring, flavor, or preservatives that impart specific functions and properties. In practice, the application of such an ideal coating is a formidable task.

Basically, there are three groups of materials commonly used in preparing films or coatings: proteins, polysaccharides, and lipids, including waxes, emulsifiers, and their derivatives. The structural, functional, physical, and morphological properties of film ingredients have been extensively reviewed (Kester and Fennema 1986) and are discussed in Chapters 9, 10, and 11 in this book.

FILM VERSUS EMULSION COATING

It is important to distinguish films in general from emulsion coatings in particular. Films may consist of single or multiple components, may be dry or moist, single or bilayer, and may not necessarily be an emulsion (Greener and Fennema, 1989; Hagenmaier and Shaw, 1990; Kamper and Fennema, 1984; Kester and Fennema, 1989; Vojdani and Torres, 1989). Many of the films studied are actually dry films of layered structures. Most films are not suitable for foods with high surface water activity because they swell, disolve and/or disintegrate on contact with water (Guilbert, 1986). In contrast, an emulsion coating contains immiscible components with one component dispersed as fine droplets (dispersed phase) in the other (continuous phase). Emulsion coating requires the consideration of stability, which is related to a balance between attractive and repulsive forces—including van der Waals, electrostatic, steric, and hydration forces. The size of the dispersed component in an emulsion is one of the critical elements contributing to the stability. In a mixture of oil-water-emulsifier, oil droplet sizes can range from 0.2–50 μm in a macroemulsion and from 10–100 nm in a microemulsion (Das and Kinsella, 1990). The former is thermodynamically unstable, while the latter is very stable. Microemulsions are oil-surfactant-water systems existing as liquid crystalline phases (Sjoblom and Friberg, 1978; Hernqvist, 1987).

While is is feasible to create simple microemulsions in an ideal system, most foods have more complex compositions. Furthermore, an emulsion in this case exists not as an isolated system, but as a coating that is in contact with a high moisture cut surface. Thus, research is needed for developing stable emulsions in food systems for these particular applications.

ADHESION OF COATING

Binding a lipid material onto a cut surface covered with juice presents a considerable technical problem. One solution is to employ some means of setting the coating material by forming a tight matrix. For example, an emulsion mixture of casein and acetylated monoglyceride will form a coagulum by adjusting the pH to the isolectric point of 4.6 (Krochta et al., 1988). The lipid molecules are presumably trapped within the matrix of the casein coagulum.

An alternative employed in some of our work involves the use of polymers with functional groups for ionic cross-linking. In a caseinate/acetylated monoglyceride/alginate emulsion, the association of polyglucuronic acid and calcium ions provides junction zones which cross-link the alginate polymers into a three-dimensional network containing casein, with the acetylated monoglyceride dispersed in the interstices. This type of binding technique using alginate has the additional advantage of forming cross-links between the coating and the endemic pectin on the cut surface, if the cut piece is first prepared by dipping into a solution containing calcium. In most experiments, ascorbate solutions containing calcium are used, both for the purpose of cross-linking and for preventing the cut surface from browning. In theory, low methoxyl pectin should also provide the same type of cross-linking, although preliminary experiments substituting alginate with pectin have resulted in lower moisture retention, perhaps due to the difference in viscosity of the two solutions used.

MONOLAYER EMULSION VERSUS MULTILAYER COATING

Krochta et al. (1988) initiated the use of emulsion coating consisting of sodium caseinate and acetylated monoglyceride using isoelectric pH for coagulation. Recently, the same group has conducted a series of investigations on the use of milk protein–based coatings for minimally processed vegetables. These are primarily caseinate-lipid coatings. Sodium caseinate/stearic acid mixtures tested on the coating of peeled carrots in storage at 2.5°C, 70% RH, and an air flow of 20 cc/min, were shown to eliminate the formation of white blush and decrease the rate of dehydration. Optimization of the formulations estimated by response surface methodology suggested that a coating composition of 1.4–1.6% sodium caseinate/0.1–0.2% stearic acid increased water vapor resistance by 84%, and retarded the formation of white blush that usually formed at cut carrot surfaces (Avena-Bustillos et al., 1993). Coating with caseinate/acetylated monoglyceride resulted in 75% reduction of moisture loss in uncut celery sticks stored at 2.5°C and 70% RH. Respiration and ethylene production rates showed no changes compared with the uncoated samples (Avena-Bustillos et al., 1994). The same coating applied to whole apple gave results that are less conclusive.

An emulsion of mixed sodium caseinate, acetylated monoglyceride and alginate has been used for coating apple pieces and other cut fruits (Figure 3.1). Coatings containing only one component did not give satisfactory results, while emulsions of mixed components seemed to perform best. Results indicate that the optimal composition ratio of caseinate:acetylated monoglyceride:alginate was 10:15:0.5%. The emulsified product was viscous. Lowering the concentration of any one component caused a drastic change in moisture retention. In general, proteins or polysaccharides are mostly hydrophilic and cannot be used as moisture barriers on cut surfaces with high surface water activity. Water from the cell tissues is simply drawn through the coating where evaporation readily occurs. All of our tests indicate that coatings containing protein or polysaccharide alone provide no detectable protection against water loss. The function of these two polymers is to provide a newtork formation where lipid molecules are dispersed. The composition and density of the final coating formed by an emulsion can significantly alter its effectiveness. Poor performance is expected from a coating with a coarse and loose network of polymers. There are at least two major drawbacks to using an emulsion coating. First of all, the products are difficult to handle since the coating remains wet. Secondly, such a coating may act as a sacrificial layer rather than as a true barrier to moisture. Since the two different properties cannot be distinguished, it becomes difficult to correlate the effectiveness of an emulsion coating to the nature of its components.

Thus far, we have described only emulsion coatings applied as a single layer. From a theoretical point of view, a layer of pure waxy material alone forms a much better moisture barrier than an emulsion of the waxy material in an aqueous phase of the same thickness. Development of a method to adhere a hydrophobic waxy coating on the surface may result in improvement. Bilayer coatings of cellulose derivatives and lipid reduced

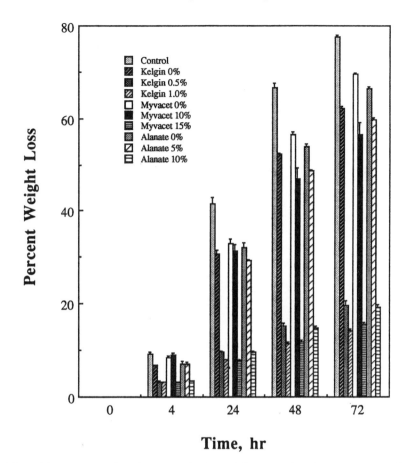

Figure 3.1 Percent water loss of coated apple pieces. The concentration of one component varied while the concentrations of the other two components were kept constant at their optimum ratios. The optimum composition ratio for caseinate:acetylated monoglyceride:alginate was 10:15:0.5. All noncoated (control) and coated samples were kept at ambient temperature (23 ± 1°C) in desiccator with anhydrous $CaSO_4$. Kelgin MV = tradename of Kelco for sodium alginate, Myvacet 5-07 = tradename of Eastman Kodak for acetylated monoglyceride, Alanate 110 = tradename of New Zealand Milk Products Inc. for sodium caseinate.

moisture loss of apple pieces by about 75% after seventy-two hours of storage at ambient temperature (23.0 ± 1.0°C) and 50% RH. Respiration was decreased by 70.0 ± 7.22% and ethylene diffusion was reduced by 93.4 ± 3.42% in coated apples as compared to the control without coatings (Table 3.1) (Wong et al., 1992). Similar types of coatings applied to a number of other commodities also showed protection from both water loss and gas diffusion.

TABLE 3.1. Gas Diffusion Rates of Cut Apples with and without Coating.

Sample	Respiration rate (ml CO_2/kg/hr)			Ethylene Production (ml C_2H_4/kg/hr) x 10^{-4}		
	Raw	Coated	%	Raw	Coated	%
1	15.33 (0.997)	3.94) (0.994)	74.3	285.44 (1.000)	20.86 (0.998)	92.7
2	19.4 (0.998)	5.85 (0.880)	69.9	553.47 (0.996)	36.02 (0.998)	93.5
3	28.10 (0.999)	9.60 (0.985)	65.8	843.74 (0.990)	36.90 (0.999)	95.6
4	26.90 (0.980)	8.65 (0.992)	67.8	480.99 (1.000)	39.19 (0.998)	91.9
5	22.31 (0.987)	6.20 (0.911)	72.2	787.55 (0.989)	36.08 (0.998)	95.4
6	16.58 (0.970)	3.70 (0.986)	77.7	87.81 (1.000)	12.59 (0.996)	85.7
7	30.01 (0.896)	6.50 (0.996)	78.3	332.63 (0.991)	46.52 (0.998)	86.0
8	23.44 (0.950)	10.68 (0.994)	54.5	276.62 (0.999)	17.61 (0.997)	93.6

Coating ingredients are: cellulose derivatives and acetylated monoglyceride.
Headspace analyses were conducted by GC at several time intervals. Calculation of the rates was based on the initial slope of the curve before saturation.
Number in parentheses = r^2 of the regression line from which the slope was determined.

ADDITIONAL CHALLENGES

There are a number of questions that need to be addressed when considering the coating of minimally processed produce.

The degradation initiated by cutting is basically a wound healing process initiated by the plant. Applying an edible coating is, in effect, substituting an alternative protection for the natural process. It also invariably alters the many reactions occurring in the natural processes, the results of which have not been adequately assessed. Research in this area is crucial for any success in coating freshly cut fruits and vegetables. In fact, it is uncertain whether the currently developed coatings actually have a direct inhibitory action on the physiological changes induced by minimal processing.

In contrast to the large amount of data available on the respiration rates and ethylene production rates of whole fruits in air and in different storage atmospheres, temperatures, and other conditions, there is no comparable information on minimally processed products. More studies are needed, especially on the internal atmosphere modification of coated products, to determine the limits of beneficial or detrimental effects in the use of edible coating in controlling gas diffusion.

Lipid materials usually impart a waxy or gummy mouthfeel. This undesirable organoleptic property is especially critical when one considers the subtle taste and texture of many fruits and vegetables, and may become not only undesirable but objectionable. This creates a dilemma, because the lipid component is indispensable for the water-barrier property of the coating. One may consider decreasing the thickness of the coating; but again, the density and thus the thickness determine to a certain extent the effectiveness of the coating. Another solution is to mask the undesirable attributes by incorporating flavoring compounds. The use of films as carriers of functional additives such as flavors, colors, antioxidants, or antimicrobial agents has received much attention (Karel and Langer, 1988; Rico-Peña and Torres, 1991; Vojdani and Torres, 1989). The use of lipids in the formulation of coatings may not be critical if other storage conditions are considered, such as lowering the storage temperature and/or maintaining a high-humidity atmosphere in a sealed bag.

It is unlikely that application of an edible coating alone can be the solution for minimal processing. The use of an edible coating should be considered as a viable approach that can be used in combination with other methods to enhance stability of minimally processed products.

THE OSMOTIC MEMBRANE (OSMEMB) PROCESS

Reduction of a_w usually increases product stability. For example. Gallander and Kretchman (1976) impregnated apple slices with fruit juices and sucrose syrups, and the a_w reduction was achieved by the infusion of solids. Another process known as "dehydrofreezing" involves osmotic dehydration followed by low temperature freezing (Lazar, 1968; Torreggiani et al., 1991). Dehydrofrozen products can be rehydrated to a fresh-like quality, and hence can be regarded as minimally processed products. Recently, we have considered the potential of coating cut pieces of fruit with an edible film, followed by limited osmotic dehydration, in an attempt to improve the stability and also to maintain the product's soft texture and fresh character. The product could subsequently be reconstituted to produce a fresh-like quality.

Most past research on osmotic dehydration focused on using it as an intermediate process prior to thermal dehydration, freezing, or freeze-drying (Dixon and Jen, 1977; Farkas and Lazar, 1969; Hawkes and Flink, 1978; Ponting et al., 1966). More recently an osmotically dehydrated shelf product with a high water content was produced using vacuum-packaging and pasteurization (Torreggiani et al., 1987). More efficient osmotic dehydration, however, could be obtained by coating the food pieces with a water-permeable polymeric material, prior to osmotic treatment (Camirand et al., 1968; Camirand and Forrey, 1969). Advantages of the OSMEMB process include the following:

(1) Greater degree of osmotic dehydration than can be obtained without coating

(2) Dehydration to a final moisture level without supplemental thermal air-drying or vacuum-drying

(3) Reduction in the loss of desired constituents, e.g., colorants, flavor compounds, and nutrients

(4) Allowing the use of lower molecular weight osmotic agents with higher osmotic pressures (e.g., glycerol)

(5) Reduction in the amount of osmotic agent (solute) that enters the food

(6) Coatings providing greater integrity or physical restraint to food pieces which can withstand mixing during osmotic dehydration and subsequent handling during storage and transportation

(7) Less microbial contamination during storage (Schultz et al., 1948)

(8) Greater esthetic appeal for the products, particularly those with clear polysaccharide coatings

(9) Less oxygen diffused into the food

OSMEMB COATINGS

Edible coatings for the OSMEMB process should have the following basic properties:

(1) Good mechanical strength

(2) Good sensory properties

(3) Easy and rapid film formation

(4) High water diffusivity and low solute diffusivity

(5) Intact coatings forming on the wet surface without dissolving into the osmotic agent

Thus far our work on OSMEMB coatings have been limited to polysaccharides, since they would appear to fit the requirements in this research (Camirand et al., 1992; Lewicki et al., 1984). We minimize heat treatment in order to produce coated products that will remain considerably "fresh" in character by limiting the operating temperatures to 30°C or below. Ethanol (95%) was the only nonaqueous solvent used (Camirand et al., 1992). Some polysaccharides which do not form a stable coating without an evaporation step were fixed using ionotropic gels formed by using calcium salts. Fruit pieces were submerged in solutions of sodium alginates or sodium low-methoxyl pectin (LMP) and then dipped into a calcium salt solution to form an insoluble gel. Polysaccharides that do not cross-link easily were incorporated into mixtures containing alkali-alginates or alkali-LMP solutions after cross-linking. The non–cross-linked polymer will be entangled within the cross-linked polymer, forming a semi-interpenetrating polymer network (semi-IPN) (Sperling, 1988). The semi-IPN strategy allows the successful production of coatings containing linear or soluble polymers, thus expanding the number of polymers that might be used.

MATHEMATICAL AND PHYSICAL MODELS

Important parameters determined from osmotically-dehydrated food research should include the % water loss, the % solid gained, and the % weight loss. In fact, we found a correlation between product quality, water loss, and the inverse of solutes entered. Thus, the performance ratio (P_r) is defined as the ratio of "water out" to "solute in" for any food piece at a given time during the osmotic dehydration process.

We generated a mathematical model based on Fick's law for diffusion for an idealized mass transfer between a membrane-coated food piece and an osmotic agent or between two reservoirs separated by a mass-selective membrane. We found that a P_r for the coating can be reduced to $P_r = (D_w K_w)/(D_s K_s)$ where D_w = the diffusion coefficient of water through the membrane, K_w = the distribution coefficient between the solution and the membrane for water, D_s = the diffusion coefficient of solute through the membrane, and K_s = the distribution coefficient between the solution and the membrane for the solute, all constants. Thus the P_r derived here is independent of membrane thickness, time, interfacial area, and the geometry of the system. These results give us confidence in using a bank of osmotic diffusion cells for evaluating OSMEMB coatings (Camirand et al., 1992).

OSMEMB FOOD PRODUCTS COMPARED

When we compared P_r and "% water loss" for coated and noncoated foods using the same osmotic agent and process time, an olive OSMEMB product had a 186% increase in P_r and a 6.2% increase in water loss, and a prawn OSMEMB product had a 58% increase in P_r and a 10% increase in water loss. In all cases, the coated products had a much improved appearance. Since the P_r of a coating can make a great difference in the final result, a model system consisting of diffusion cells was used to evaluate and compare the P_r values for a variety of coatings. The results of such experiments are presented in Table 3.2. Of all the coatings containing ionotropic gel substances, the highest P_r result was obtained using a 69% sucrose solution as the osmotic agent and a 3% LMP solution for coating no. 11. In general, semi-IPN coatings (nos. 5, 6, 8, 9, 10) exhibit higher P_r than coatings of a single polymer (Ca-LMP) solution at the same or lower concentrations (no. 4).

In order to better understand the practicality and perhaps limits of increasing the P_r values on a particular food, we applied a simple algebraic mass balance model presented in Figure 3.2. For this particular sample and osmotic agent, it would appear that a P_r of 20 would be sufficient, and special efforts to increase P_r values greater than this may be unwarranted. It should be realized that the "% solids gained" curve will be below these values at any particular P_r value before equilibrium. If an OSMEMB

TABLE 3.2. Carbohydrate Polymer Coatings[a, b]—Mean
P_r Values and Solids Compositions.

Coating No.	P_r Values[c]			Solids in Polymer Solutions (%)		
	Sucrose	Glycerol	Dextrose	Total Solids	LMP	Other Solids
1	—	—	—	3.60	—	0.90 SA + 2.7 PFCS
2[d]	—	—	—	10.0	—	10.0 EC
3	2.13	1.63	1.40	2.5	2.5	—
4	3.66	3.02	1.49	2.5	2.5	—
5	4.27	2.82	1.56	17.5	2.5	15.0 MD
6[e]	—	3.47	—	5.33	2.5	2.83 PFCS
7	4.66	4.03	—	3.6	—	0.90 SA + 2.7 PFCS
8	4.66	4.37	2.04	3.47	2.5	0.975 PS
9	7.31	3.70	—	7.0	1.5	5.8 MC
10	8.06	5.08	—	5.33	2.5	2.83 PFCS
11	8.81	—	—	3.0	3.0	—
12[d]	—	12.76	—	10.0	—	10.0 EC

[a]Adapted with permission from *Carbohydrate Polymers*, 17:39. Copyright © 1992 Elsevier Applied Science Publisher.
[b]LMP = low methoxyl pectin; PFCS = pure food corn starch; SA = sodium alginate; EC = ethyl cellulose; MD = maltodextrin; PS = potato starch; MC = methyl cellulose. All coatings used 5M CaCl$_2$ solution dipped for 30 seconds followed by a 60 second water dip.
[c]P_r definition: see text for details.
[d]Ethanol (95%) was used as solvent.
[e]Cross-linked coating was dipped into 95% ethanol for 30 seconds.

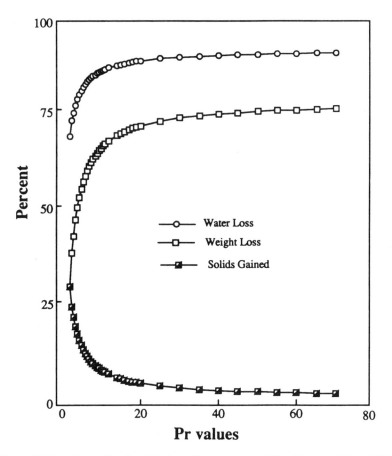

Figure 3.2 Projection of the % weight loss, % water loss and % solids gained that might be achieved for apple pieces using coatings with P_r values that assumed between 2 and 70. The model assumes results achieved after osmotic dehydration reaches equilibrium where $R = 5$, $D = 25$, $C = 75$, $A = 12$ and $B = 85$ where R = ratio of weight of syrup to weight of food; D = % initial water in syrup; C = % of initial solids in syrup; A = % initial soluble solids in the food; B = % initial water in the food; P_r = (g water out/g solids in) or (X/W); X = water out (g/100 g) or $X = (B*R*C - A*R*D)/[A+(R*D/P_r) + (B/P_r) (R*C)]$; W = % solids gained on original weight (g/100 g) = X/P; E = % wt loss = $X - W$; F = % water loss on original water in the food = $(X/B) * 100$.

coated product remains intact during storage and handling and has a reasonable P_r value (above that of the uncoated food) then the degree of rehydration will also be correspondingly high (unpublished data by the authors).

CONCLUSION

There has been increasing interest in edible coatings for lightly processed fruits and vegetables for improving the shelf-life, adding unique sensorial properties, or even producing products that would be recognized as having a high-quality fresh character by the public. There has been considerable success in achieving these objectives. However, fundamental as well as practical questions including physiological changes of cut tissues, microbial growth during storage, organoleptic problems relating to the coating, etc., must be addressed. The challenges remain great and interesting.

REFERENCES

Altschuler, M., W. S. Grayburn, G. B. Collins and D. F. Hildebrand. 1989. "Developmental Expression of Lipoxygenases in Soybeans, *Plant Sci.*, 63:151–158.

Avena-Bustillos, R. J., J. M. Krochta and M. E. Saltveit. 1994. "Optimization of Caseinate-Based Edible Coatings on 'Red Delicious' Apples and Celery Sticks," submitted to *J. Amer. Hort. Sci.*

Avena-Bustillos, R. J., L. A. Cisneros-Zevallos, J. M. Krochta and M. E. Saltveit, Jr. 1993. "Optimization of Edible Coatings on Minimally Processed Carrot to Reduce White Blush Using Response Surface Methodology," *Trans. ASAE*, 36(3):801–805.

Ballantyne, A., R. Stark and J. D. Selman. 1988a. "Modified Atmosphere Packaging of Shredded Lettuce," *Int. J. Food Sci. & Technol.*, 23:267–274.

Ballantyne, A., R. Stark and J. D. Selman. 1988b. "Modified Atmosphere Packaging of Broccoli Florets," *Int. J. Food Sci. & Technol.*, 23:353–360.

Banks, J. G., R. G. Board and N. H. C. Sparks. 1986. "Natural Antimicrobial Systems and Their Potential in Food Preservation of the Future," *Biotechnol. Biochem.*, 8:103–147.

Basel, R. M. 1982. "Acidified Bulk Storage of Snap Beans and Peas," *J. Food Sci.*, 47:2082–2083.

Basel, R. M. and W. A. Gould. 1983. "An Investigation of Some Important Storage Parameters in Acidified Bulk Storage of Tomatoes," *J. Food Sci.*, 48:932–934.

Bolin, H. R. and C. C. Huxsoll. 1991a. "Effect of Preparation Procedures and Storage Parameters on Quality Retention of Salad-Cut Lettuce," *J. Food Sci.*, 56:60–67.

Bolin, H. R. and C. C. Huxsoll. 1991b. "Control of Minimally Processed Carrot (*Daucus carota*) Surface Discoloration Caused by Abrasion Peeling," *J. Food Sci.*, 56:416–418.

Bolin, H. R., A. E. Stafford, A. D. King, Jr. and C. C. Huxoll. 1977. "Factors Affecting the Storage Stability of Shredded Lettuce," *J. Food Sci.*, 42:1319–1321.

Boller, T. and H. Kende. 1980. "Regulation of Wound Ethylene Synthesis in Plants," *Nature*, 286:259–260.

Borenstein, B. 1987. "The Role of Ascorbic Acid in Foods," *Food Technol.*, 41:98–99.

Brackett, R. E. 1987. "Microbiological Consequences of Minimally Processed Fruits and Vegetables," *J. Food Quality*, 10:195–206.

Brocklehurst, T. F., C. M. Zaman-Wong and B. M. Lund. 1987. "A Note on the Microbiology of Retail Packs of Prepared Salad Vegetables," *J. Appl. Bacterial.*, 63:409–415.

Buick, R. K. and A. P. Damoglou. 1987. "The Effect of Vacuum Packaging on the Microbial Spoilage and Shelf-Life of 'Ready-To-Use' Sliced Carrots," *J. Sci. Food Agric.*, 38:167–175.

Camirand, W. M. and R. R. Forrey. 1969. "Osmotic Dehydration of Coated Foods." U.S. Patent 3,425,848.

Camirand, W. M., R. R. Forrey, K. Popper, F. P. Boyle and W. L. Stanley. 1968. "Dehydration of Membrane-Coated Foods by Osmosis," *J. Sci. Food Agric.*, 19:472–474.

Camirand, W. M., J. M. Krochta, A. G. Pavlath, D. Wong and M. E. Cole. 1992. Properties of Some Edible Carbohydrate Polymer Coatings for Potential Use in Osmotic Dehydration," *Carbohydrate Polymers*, 17:39–49.

Cantwell, M. 1992. "Postharvest Handling Systems: Minimally Processed Fruits and Vegetables," in *Postharvest Technology of Horticultural Crops, Second Edition*, A. A. Kader, ed., Universtiy of California, Division of Agriculture and Natural Resources, Publ. 3311, Chapter 32, pp. 277–281.

Daeschel, M. A. 1989. "Antimicrobial Substances from Lactic Acid Bacteria for Use as Food Preservatives," *Food Technol.*, 43:164–167.

Das, K. P. and J. E. Kinsella. 1990. "Stability of Food Emulsions: Physicochemical Role of Protein and Nonprotein Emulsifiers," *Adv. Food Nutr. Res.*, 34:81–201.

Davies, E. 1987. "Plant Responses to Wounding," in *The Biochemistry of Plants, Vol. 12*, D. D. Davies, ed., P. K. Stumpf and E. E. Conn, ed.-in-chief, Academic Press, Chapter 7, pp. 243–264.

Dean, B. B. and P. E. Kolattukudy. 1976. "Synthesis of Suberin during Wound-Healing in Jade Leaves, Tomato Fruits, and Bean Pods," *Plant Physiol.*, 58:411–416.

Dixon, G. M. and J. J. Jen. 1977. "Changes of Sugars and Acids of Osmovac-Dried Apple Slices," *J. Sci. Food Agric.*, 42:1126–1127.

Eaks, I. L. 1980. "Effect of Chilling on Respiration and Volatiles of California Lemon Fruit," *J. Amer. Soc. Hort. Sci.*, 105:865–869.

Eichner, K. 1986. "The Influence of Water Content and Water Activity on Chemical Changes in Foods of Low-Moisture Content under Packaging Aspects," in *Food Packaging and Preservation: Theory and Practice*, M. Mathlouthi, ed., London: Elsevier Appl. Sci. Publ.

Farkas, D. F. and M. E. Lazar. 1969. "Osmotic Dehydration of Apple Pieces: Effects of Temperature and Syrup Concentration of Rates," *Food Technol.*, 23:688–690.

Feys, M., W. Naesens, P. Tobback and E. Maes. 1980. "Lipoxygenase Activity in Apples in Relation to Storage and Physiological Disorders," *Phytochemistry*, 19:1009–1011.

Gallander, J. F. and D. W. Kretchman. 1976. "Apple Slice Impregnation with Fruit Juices," *Hort. Sci.*, 11:395–397.

Galliard, T., J. A. Matthew, M. J. Fishwick and A. J. Wright. 1976. "The Enzymic Degradation of Lipids Resulting from Physical Disruption of Cucumber (*Cucumis sativis*) Fruit," *Phytochem.*, 15:1647–1650.

Greener, I. K. and O. Fennema. 1989. "Barrier Properties and Surface Characteristics of Edible Bilayer Films," *J. Food Sci.*, 54:1393–1399.

Grisebach, H. 1987. "New Insights into the Ecological Role of Secondary Plant Metabolites," *Comments Agric. Food Chem.*, 1:27–45.

Guilbert, S. 1986. "Technology and Application of Edible Protective Films," in *Food Packaging and Preservation: Theory and Practice*, M. Mathlouthi, ed., London: Elsevier Appl. Sci. Publ.

Haard, N. F. and M. Cody. 1978. "Stress Metabolites in Postharvest Fruits and Vegetables—Role of Ethylene," in *Postharvest Biology and Biotechnology*, H. O. Hultin and M. Milner, eds., Westport CN: Food Nutr. Press, Inc.

Hagenmaier, R. D. and P. E. Shaw. 1990. "Moisture Permeability of Edible Films Made with Fatty Acid and (Hydroxypropyl)methylcellulose," *J. Agric. Food Chem.*, 38:1799–1803.

Hatanaka, A., T. Kajiwara, K. Matsui and A. Kitamura. 1992. "Expression of Lipoxygenase and Hydroperoxide Lyase Activities in Tomato Fruits," *Z. Naturforsch*, 47e:369–374.

Hawkes, J. and J. M. Flink. 1978. "Osmotic Concentration of Fruit Slices Prior to Freeze Dehydration," *J. Food Proc. Preserv.*, 2:265–284.

Hearing, V. J. and K. Tsukamoto. 1991. "Enzymatic Control of Pigmentation in Mammals," *FASEB J.*, 5:2902–2909.

Heaton, E. K., T. S. Boggess, Jr. and K. C. Li. 1969. "Processing Refrigerated Fresh Peach Slices," *Food Technol.*, 23:96–100.

Hernqvist, L. 1987. "Polar Lipids in Emulsions and Microemulsions," in *Food Emulsions and Foams*, E. Dickerson, ed., London: The Royal Society of Chemistry.

Hoffman, N. E. and S. F. Yang. 1982. "Enhancement of Wound-Induced Ethylene Synthesis by Ethylene in Preclimacteric Cantaloupe," *Plant Physiol.*, 69:317–322.

Hyodo, H. and T. Nishino. 1981. "Wound-Induced Ethylene Formation in Albedo Tissue of Citrus Fruit," *Plant Physiol.*, 67:421–423.

Hyodo, H., H. Kuroda and S. F. Yang. 1978. "Induction of Phenylalanine Ammonia-Lyase and Increase in Phenolics in Lettuce Leaves in Relation to the Development of Russet Spotting Caused by Ethylene," *Plant Physiol.*, 62:31–35.

Hyodo, H., K. Tanaka and K. Watanabe. 1983. "Wound-Induced Ethylene Production and 1-Aminocyclopropane-1-carboxylic Acid Synthase in Mesocarp Tissue of Winter Squash Fruit," *Plant Cell Physiol.*, 24:963–969.

Juliot, K. N., R. C. Lindsay and S. C. Ridley. 1989. "Directly-Acidified Carrot Slices for Ingredients in Refrigerated Vegetable Salads," *J. Food Sci.*, 54:90–95.

Kader, A. A. 1992. "Modified Atmospheres during Transport and Storage," in *Postharvest Technology of Horticultural Crops, Second Edition*, A. A. Kader, ed.,University of California, Division of Agriculture and Natural Resources, Publ. 3311, Chapter 11, pp. 85–92.

Kader, A. A., D. Zagory and E. L. Kerbel. 1989. "Modified Atmosphere Packaging of Fruits and Vegetables," *Crit. Rev. Food Sci. & Technol.*, 28:1–30.

Kahl, G. and G. G. Laties. 1989. "Ethylene-Induced Respiration in Thin Slices of Carrot Root," *J. Plant Phylsiol.*, 134:496.

Kamper, S. L. and O. Fennema. 1984. "Water Vapor Permeability of an Edible, Fatty Acid, Bilayer Film," *J. Food Sci.*, 49:1482–1485.

Karel, M. and R. Langer. 1988. "Controlled Release of Food Additives," in *Flavor Encapsulation*, ACS Sym. Ser. 370, S. J. Risch and G. A. Reineccius, eds., Washington, DC: American Chemical Society.

Kester, J. J. and O. R. Fennema. 1986. "Edible Films and Coatings: A Review," *Food Technol.*, 40:47–59.

Kester, J. J. and O. Fennema. 1989. "An Edible Film of Lipids and Cellulose Ethers: Barrier Properties to Moisture Vapor Transmission and Structural Evaluation," *J. Food. Sci.*, 54:1383–1389.

Klein, B. P., S. Grossman, D. King, B.-S. Cohen and A. Pinsky. 1984. "Pigment Bleaching, Carbonyl Production and Antioxidant Effects during the Anaerobic Lipoxygenase Reaction," *Biochim. Biophys. Acta*, 793:72–79.

Kolattukudy, P. E. 1980. "Cutin, Suberin, and Waxes," in *The Biochemistry of Plants, Vol. 4*, P. K. Stumpf and E. E. Conn, eds., Academic Press, Chapter 18, pp. 571–645.

Kramer, M., R. Sanders, R. Sheehy, M. Melis, M. Kuehm and W. R. Hiatt. 1990. "Field Evaluation of Tomatoes with Reduced Polygalacturonase by Antisense RNA," *Horti. Biotechnol.*, 1990:347–355.

Krochta, J. M., J. S. Hudson, W. M. Camirand and A. E. Pavlath. 1988. "Edible Films for Light-Processed Fruits and Vegetables," paper presented at *International Winter Meeting of the American Society of Agricultural Engineers*, Chicago, IL.

Labuza, T. P. 1980. "The Effect of Water Activity on Reaction Kinetics of Food Deterioration," *Food Technol.*, 34(4):36–41, 59.

Lattanzio, V. and V. Linsalata. 1989. "The Beneficial Effect of Citric and Ascorbic Acid on the Phenolic Browning Reaction in Stored Artichoke (*Cynara scolymus* L.) Heads, *Food Chem.*, 33:93–106.

Lazar, M. E. 1968. "Dehydrofreezing of Fruits and Vegetables," in *The Freezing Preservation of Foods*, D. K. Tressler, W. B. van Arsdel and M. J. Copley, eds., Westport, CN: AVI, Chapter 12, pp. 347–376.

Lewicki, P. P., A. Lenart and W. Pakula. 1984. "Influence of Artificial Semi-Permeable Membranes on the Process of Osmotic Dehydration of Apples," *Food Technol. and Nutr.*, 16:17–24.

Liao, C. H. and J. M. Wells. 1987. "Diversity of Pectolytic, Fluorescent *Pseudomonads* Causing Soft Rots of Fresh Vegetables at Produce Markets," *Postharvest Pathology & Mycotoxins*, 77:673–677.

Lidster, P. D., A. J. Dick, A. DeMarco and K. B. McRae. 1986. "Application of Flavonoid Glycosides and Phenolic Acids to Suppress Firmness Loss in Apples," *J. Amer. Soc. Hort. Sci.*, 111:892–896.

Lynch, D. V. and J. E. Thompson. 1984. "Lipoxygenase-Mediated Production of Superoxide Anion in Senescing Plant Tissue," *FEBS Lett.*, 173:251–254.

MacLeod, R. F., A. A. Kader and L. L. Morris. 1976. "Stimulation of Ethylene and CO_2 Production of Mature-Green Tomatoes by Impact Bruising," *Hort. Sci.*, 11:604–606.

Maxcy, R. B. 1982. "Fate of Microbial Contaminants in Lettuce Juice," *J. Food Protection*, 45:335–339.

McFeeters, R. F. and H. P. Fleming. 1990. "Effect of Calcium Ions on the Thermodynamics of Cucumber Tissue Softening," *J. Food Sci.*, 55:446–449.

McFeeters, R. F., M. M. Senter and H. P. Fleming. 1989. "Softening Effects of Monovalent Cations in Acidified Cucumber Mesocarp Tissue," *J. Food Sci.*, 54:366–370.

McGlasson, W. B. 1969. "Ethylene Production by Slices of Green Banana Fruit and Potato Tuber Tissue during the Development of Induced Respiration," *Aust. J. Biol. Sci.*, 22:489–491.

McGlasson, W. B. and H. K. Pratt. 1964. "Effects of Wounding on Respiration and Ethylene Production by Cantaloupe Fruit Tissue," *Plant Physiol.*, 39:128–132.

McEvily, A. J., R. Iyengar and S. Otwell. 1991. "Sulfite Alternative Prevents Shrimp Melanosis," *Food Technol.*, 44:80–86.

Osuga, D., A. van der Schaff and J. R. Whitaker. 1994. "Control of Polyphenol Oxidase Activity Using a Catalytic Mechanism," in *Protein Structure-Function Relationship in Foods*, R. Y. Yada, ed., New York: Chapman & Hall.

Palmer, J. K. and W. B. McGlasson. 1969. "Respiration and Ripening of Banana Fruit Slices," *Aust. J. Biol. Sci.*, 22:87–99.

Palumbo, S. A. 1986. "Is Refrigeration Enough to Restrain Foodborne Pathogens," *J. Food Protection*, 49:1003–1009.

Parkin, K. L., A. Marangoni, R. L. Jackman, R. Y. Yada and D. W. Stanley. 1989. "Chilling Injury. A Review of Possible Mechanisms," *J. Food Biochem.*, 13:127–153.

Ponting, J. D., R. Jackson and G. Watters. 1972. "Refrigerated Apple Slices: Preservative Effects of Ascorbic Acid, Calcium and Sulfites," *J. Food Sci.*, 37:434–436.

Ponting, J. D., G. G. Watters, R. R. Forrey, R. Jackson and W. L. Stanley. 1966. "Osmotic Dehydration of Fruits," *Food Technol.*, 20(10):125.

Poovaiah, B. W., G. M. Glenn and A. S. N. Reddy. 1988. "Calcium and Fruit Softening: Physiology and Biochemistry," *Hort. Rev.*, 10:107–151.

Priepke, P. E., L. S. Wei and A. I. Nelson. 1976. "Refrigerated Storage of Prepackaged Salad Vegetables," *J. Food Sci.*, 41:379–382.

Reid, M. S. 1992. "Ethylene in Postharvest Technology," in *Postharvest Technology of Horticultural Crops*, A. A. Kader, ed., University of California, Division of Agriculture and Natural Resources, Publ. 3311, Chapter 13, pp. 97–108.

Rico-Peña, D. C. and J. A. Torres. 1991. "Sorbic Acid and Potassium Sorbate Permeability of an Edible Methylcellulose-Palmitic Acid Film: Water Activity and pH Effect," *J. Food Sci.*, 56:497–499.

Rosen, J. C. and A. A. Kader. 1989. "Postharvest Physiology and Quality Maintenance of Sliced Pear and Strawberry Fruits," *J. Food Sci.*, 54:656–659.

Rushing, N. B. and V. J. Senn. 1962. "Effect of Preservatives and Storage Temperatures on Shelf Life of Chilled Citrus Salad," *Food Technol.*, 16:77–79.

Saltveit, M. E. and R. D. Locy. 1982. "Cultivar Differences in Ethylene Production by Wounded Sweet Potato Roots," *J. Amer. Soc. Hort. Sci.*, 107:1114–1117.

Santerre, C. R., J. N. Cash and D. J. Vannorman. 1988. "Ascorbic Acid/Citric Acid Combinations in the Processing of Frozen Apple Slices," *J. Food Sci.*, 53:1713–1736.

Sapers, G. M. and K. B. Hicks. 1989. "Inhibition of Enzymatic Browning in Fruits and Vegetables," in *Quality Factors of Fruits and Vegetables, Chemistry and Technology*, J. J. Jen, ed., Washington, DC: ACS, ACS Sym. Ser. 405, Chapter 3, pp. 29–43.

Schieberle, P., W. Grosch, H. Kexel and H. L. Schmidt. 1981. "A Study of Oxygen Isotope Scrambling in the Enzymic and Non-Enzymic Oxidation of Linoleic Acid," *Biochim. Biophys. Acta,* 666:322–326.

Schultz, T. H., H. S. Owens and W. D. Maclay. 1948. "Pectinate Films," *J. Colloid Sci.,* 3:53–62.

Schwimmer, S. 1983. "Biochemistry in the Food Industry: Molecular Pabulistics and Thanatobolism," *TIBS,* 8:306–310.

Shewfelt, R. L. 1987. "Quality of Minimally Processed Fruits and Vegetables," *J. Food Quality,* 10:143–156.

Sjoblom, E., and S. Friberg. 1978. "Light-Scattering and Electron Microscopy Determinations of Association Structures in W/O Microemulsions," *J. Colloid Interface Sci.,* 67:16–30.

Smith, C. J. S., C. F. Watson, C. R. Bird, J. Ray, W. Schuch and D. Grierson. 1990. "Expression of a Truncated Tomato Polygalacturonase Gene Inhibits Expression of the Endogenous Gene in Transgenic Plants," *Mol. Gen. Genet.,* 224:477–481.

Sommer, N. F., R. J. Fortlage and D. C. Edwards. 1992. "Postharvest Diseases of Selected Commodities," in *Postharvest Technology of Horticultural Crops,* A. A. Kader. ed., University of California, Division of Agriculture and Natural Resources, Publ. 3311, Chapter 11, pp. 117–160.

Sperling, L. H. 1988. "Interpenetrating Polymer Networks," *Chemtech,* 18(2):104–109.

Starrett, D. A. and G. G. Laties. 1991. "Involvement of Wound and Climacteric Ethylene in Ripening Avocado Discs," *Plant Physiol.,* 97:720–729.

Sukumaran, N. P., J. S. Jassal and S. C. Verma. 1990. "Quantative Determination of Suberin Deposition during Wound Healing in Potatoes (*Solanum tuberosum* L.)," *J. Sci. Food Agric.,* 51:271–274.

Tatsumi, Y., A. E. Watada and W. P. Wergin. 1991. "Scanning Electron Microscopy of Carrot Stick Surface to Determine Cause of White Translucent Appearance," *J. Food Sci.,* 56:1357–1359.

Torreggiani, D., E. Forni and A. Rizzolo. 1987. "Osmotic Dehydration of Fruit: Part 2: Influence of the Osmosis Time on the Stability of Processed Cherries," *J. Food Processing Preservation,* 12:27–44.

Torreggiani, D., E. Forni, A. Maestrelli and F. Quadri. 1991. "Influence of Firmness of Raw Fruit on Quality of Osmodehydrofrozen Kiwifruit Disks," *8th World Congress of Food Science and Technology,* Toronto, Canada, September 29–October 4, 1991.

Uritani, I. and T. Asahi. 1980. "Respiration and Related Metabolic Activity in Wounded and Infected Tissues," in *The Biochemistry of Plants, Vol. 2,* P. K. Stumpf and E. E. Conn, eds., Orlando, FL: Academic Press, Chapter 11, pp. 463–485.

Vojandi, F. and J. A. Torres. 1989. "Potassium Sorbate Permeability of Methylcellulose and Hydroxypropyl Methylcellulose Multi-Layer Films," *J. Food Processing and Preservation,* 13:417–430.

Wagner, M. K. and L. J. Moberg. 1989. "Present and Future Use of Traditional Antimicrobials," *Food Technol.,* 43:143–155.

Wang, C. Y. 1982. "Physiological and Biochemical Responses of Plants to Chilling Stress," *Hort. Sci.,* 17:173–186.

Wardale, D. A. and T. Galliard. 1977. "Further Studies on the Subcellular Localization of Lipid-Degrading Enzymes," *Phytochem.,* 16:333–338.

Webb, T. A., and J. O. Mundt. 1978. "Molds on Vegetables at the Time of Harvest," *Appl. Environ. Microbiol.*, 35:655–658.

Wong, D. W. S., A. E. Pavlath and S. J. Tillin. 1992. "Edible Double-Layer Coating for Slightly Processed Fruits and Vegetables," *Symposiumm on Edible Coatings for Food*, American Chemical Society, Washington, DC, August 23–28, 1992.

Yu, Y.-B. and S. F. Yang. 1980. "Biosynthesis of Wound Ethylene," *Plant Physiol.*, 66:281–285.

CHAPTER 4

Edible Coatings and Films for Processed Foods

ROBERT A. BAKER[1]
ELIZABETH A. BALDWIN[1]
MYRNA O. NISPEROS-CARRIEDO[2]

INTRODUCTION

EDIBLE coatings and films are applied to processed foods for many of the same reasons they are used on whole fruits and vegetables – to retard or prevent undesirable movement of water or gases. The two terms, *films* and *coatings,* are often used interchangeably. Generally, however, films are preformed, and coatings are formed directly on food products. Depending on the product, coatings may be used to prevent loss of moisture, moisture absorption, transfer of moisture between components of differing water activity in a heterogeneous food system, formation of package ice in frozen foods, exposure to oxygen, or diffusion of CO_2. Edible coatings also have the potential to reduce flavor and aroma loss in processed foods. Control of water activity and oxygen levels in many processed foods is essential to protect product texture and flavor. While conventional plastic and foil packaging can usually accomplish these objectives, edible coatings offer a viable alternative with certain advantages. Packaging waste is often reduced, edible coatings are perceived as less costly environmentally than plastic, and edible coatings continue to offer protection after the package has been opened.

[1]U.S. Citrus and Subtropical Products Laboratory, P.O. Box 1909, Winter Haven, FL 33880. South Atlantic Area, Agricultural Research Service, U.S. Department of Agriculture. Mention of a trademark or proprietary product is for identification only and does not imply a guarantee or warranty of the product by the U.S. Department of Agriculture. All programs and services of the U.S. Department of Agriculture are offered on a nondiscriminatory basis without regard to race, color, national origin, religion, sex, age, marital status, or handicap.
[2]Food Research Institute, Dept. of Agriculture, Werribee, Victoria, Australia.

In addition to restricting water and gas exchange, edible coatings may reduce movement of cooking oils into food pieces during frying, or leakage of shortening from the fried product. Edible coatings and coatings applied during processing may reduce product shrinkage during later cooking, contribute to improved texture, or improve appearance of the final product. In addition, coatings can function as carriers for antimicrobials and antioxidants, localizing their activity at the surface of the food. While normally designed to be thin and unnoticeable, coatings may be a significant component of a food, or may contribute to its structural integrity. In at least one usage, the traditional oriental meat analog known as yuba, the film itself may constitute the entire product.

MEATS, POULTRY, AND SEAFOOD

While not the earliest use of edible coatings, application of coatings to meats to prevent shrinkage has been practiced since at least the sixteenth century, when meat cuts were coated with fats (Kester and Fennema, 1986). In addition to shrinkage, both primal and retail cuts of meat suffer from loss of quality due to microbial contamination and surface discoloration. Conventional practices for the control of these factors have involved the use of plastic overwraps, chemical applications (Biemuller et al., 1973) or UV irradiation (Reagan et al., 1973). Attempts to ameliorate one quality defect may exacerbate another; e.g., use of plastic coverings reduced carcass shrinkage, but increased cooling time and surface microbial counts (Lazarus et al., 1976).

A number of edible polysaccharide coatings have been used to improve stored meat quality, including alginates, carrageenans, cellulose ethers, pectin, and starch derivatives (Kester and Fennema, 1986). Most of these coatings offer little barrier against moisture transport. However, the water contained within the gel coating serves as a sacrificing agent, and loss of product moisture is delayed until this liquid has been evaporated.

Use of an alginate coating gelled with calcium chloride was somewhat less effective than plastic wrap in reducing weight loss of lamb carcasses, but yielded a better quality product due to reduced surface microbial counts (Lazarus et al., 1976). Cooking loss, flavor, odor and overall acceptability were not affected by the coating. Reduction in shrinkage was more than adequate to cover the cost of the coating. Application of the same coating to beef loin cuts reduced weight loss for 96 hours storage, without affecting flavor, tenderness, appearance, overall acceptability, or cooking loss (Williams et al., 1978). Off-odor and drip were reduced, and maintenance of muscle color was significantly improved for 96 hours. However, after 144 hours storage, loss of moisture in the coated steaks exceeded controls, as the coating dried out.

A coating composed of alginate and corn starch improved uncooked characteristics as well as cooked texture and juiciness of beef steaks, pork chops, and skinned chicken drumsticks (Allen et al., 1963b). However, flavor of the coated products was inferior to the uncoated controls, which was attributed to bitterness imparted by 5 M calcium chloride which served as a gelling agent. Although calcium chloride was found to be the most effective gelling agent (Allen et al., 1963a), other calcium salts could be substituted, or lower levels with less organoleptic impact could be used (Earle, 1968; Lazarus et al., 1976; Williams et al., 1978). If calcium chloride is used, a balance must be struck between the desirable effect from setting of the alginate by the gelling agent and the negative effect of the resultant bitter taste. An edible gum as a thickening agent in the calcium chloride dip enhanced its effectiveness, and shortened dip time (Earle, 1968; Earle and McKee, 1976). Concentration and viscosity of the gelling agent must be adjusted to permit adequate film formation; excessive viscosity increases the potential for bitterness.

Alginate coatings have also been used to retard development of oxidative off-flavors in previously cooked meat patties which were stored before reheating (Wanstedt et al., 1981). The shelf-lives of frozen shrimp, fish and sausage have also been extended with alginate coatings (Earle and Snyder, 1966; Daniels, 1973). Carrageenans have been used to prolong the shelf-life of poultry (Meyer et al., 1959).

Cellulose derivatives, usually in combination with an oil as a plasticizer, have been utilized to coat both fresh and frozen meats (Daniels, 1973). Carboxymethyl cellulose (CMC), added to a commercial breading mix, significantly improved breading adhesion after baking (Suderman et al., 1981). Solutions of methylcellulose (MC) or hydroxypropyl methylcellulose (HPMC) exhibit thermal gelation; i.e., a solution of these derivatives gels above a critical temperature, and reverses when cooled. This unusual property has been used to produce glazed sauces for poultry and seafood which minimize runoff during cooking, as well as reducing moisture loss from the entree (Anonymous, 1993b). Coating meat portions with these solutions prior to breading provides a thermally induced barrier which reduces moisture loss during frying and lessens fat absorption.

Protein coatings composed of collagen and gelatin have been used on meats to reduce gas and moisture transport. While gelatin can be easily applied as a warm aqueous solution which gels upon cooling, the gel lacks strength and requires a drying step to form a durable film (Daniels, 1973). Solubilization of gelatin in polyhydric alcohols such as propylene or ethylene glycol, glycerol, or sorbitol gave coating formulations which were quick-setting and exhibited good barrier and flexibility properties at low temperatures (Whitman and Rosenthal, 1971). Both gelatin and egg albumen were effective at reducing breading loss from coated chicken parts (Suderman et al., 1981).

A number of lipid coating formulations have been used on meats and meat products. These include meat fats, vegetable oils, acetylated monoglycerides, mono-, di- and triglycerides, waxes, and mixtures thereof. Water-in-oil emulsions composed of meat fats or vegetable oils emulsified with any of a variety of lipophilic or hydrophilic emulsifiers reportedly protected frozen chicken pieces and pork chops against dehydration for extended periods (Bauer et al., 1968). Addition of a small amount of MC improved the adhesion of this coating during cooking (Bauer et al., 1969). Other lipid coatings suggested for use on beef, pork, and fish have included mixtures of fatty acids and glycerol, meat fats and triglycerides, and triglycerides from peanut oil with glycerine mono- and distearate (Daniels, 1973).

A potential function of edible coatings, in both meats and other products, is as a carrier of antimicrobial compounds. Dipping meats or other foods in solutions of antimicrobials is of limited benefit, since the preservative diffuses into the bulk of the product, lowering the concentration at the food surface (Torres et al., 1985). Incorporation of sorbic acid into a zein protein film slowed migration of preservative into the product, increasing the duration of effective surface concentration (Torres and Karel, 1985). Edible films with sorbic acid permeability values comparable to zein films were made from MC and HPMC, in combination with fatty acids (Vojdani and Torres, 1990). The mass transfer of these films was determined to be about three orders of magnitude less than that of a typical food system. Permeability of sorbic acid in such films was strongly dependent on coating water activity (a_w), increasing almost two orders of magnitude as a_w increased from 0.65 to 0.80 (Rico-Peña and Torres, 1991). Films were unstable when a_w exceeded 0.8. Film pH affected both sorbic acid permeability and its dissociation—as pH increased, permeability and dissociation decreased. Immobilization of acetic or lactic acids on the surface of lean beef with a calcium alginate coating significantly reduced populations of *Listeria monocytogenes* (Siragusa and Dickson, 1992). Application of acids without the alginate coating was relatively ineffective. The authors hypothesized that the enhanced effect of alginate immobilized acids was due to increased contact time before acid evaporation or neutralization, and prolonged maintenance of a moist surface environment.

An interesting variation of edible coating technology is represented by yuba, a traditional oriental food. While most coatings are formed on the surface of an existing food piece, yuba is produced as a thin film of denatured soy protein on the surface of heated soymilk (Wu and Bates, 1972). This high-protein film is removed, dried, and used as a wrap for ground meats or vegetables, or as an ingredient of soups. Alternatively, the films may be appropriately flavored and rolled or pressed into molds to constitute a meat analog (Wu and Bates, 1975). Evaluation of other seed or animal protein-lipid sources as potential ingredients of edible films made by this process showed that protein quality and oil content of the mixture were critical to effective film formation (Wu and Bates, 1973).

NUTS, CEREALS, AND CEREAL-BASED PRODUCTS

Nuts, because of their high oil content, are particularly susceptible to development of oxidative off-flavors. A primary use of edible coatings on nuts is to prevent or delay onset of rancidity. Other potential functions of coatings are to provide better adhesion of salt, sugars, flavorings, colorants and antioxidants.

Low-methoxyl pectins, plasticized with glycerin and gelled with calcium chloride, provided a coating for almonds which bound salt uniformly to the kernels and gave a nonoily surface (Swenson et al., 1953). Despite the low oxygen permeability of pectinate coatings, coated nuts became rancid faster than uncoated nuts, an effect traced to the salts and their impurities. Coated nuts with an antioxidant incorporated in the coating were much more stable than uncoated nuts when stored at 23–27°C. However, at 38°C onset of rancidity was as rapid in coated nuts with antioxidant as in uncoated nuts. A similar coating for nuts using the corn protein, zein, extended shelf-life of pecans from one to three months at 21°C (Andres, 1984). Coatings composed of zein and acetylated monoglycerides were comparable in appearance to confectioner's glaze when used on sugar-coated burnt peanuts (Cosler, 1957). Addition of an antioxidant to the zein coating delayed rancidity in pecan halves stored at 38°C.

When nuts are granulated for use in food products such as cake mixes, a coating is essential to protect the nut meats from absorption of moisture or reaction with other ingredients (Daniels, 1973). Using total carbonyl content of granulated peanuts as a measure of off-flavor generation, Hoover and Nathan (1981) found that coating with several commercial hydrogenated fats or an acetylated monoglyceride failed to retard development of rancidity. Inclusion of the antioxidant tertiary butylhydroquinone (TBHQ) in the coatings reduced carbonyl formation by 80% during a fifteen-month storage study. Coating nutmeats immediately after roasting with commercial hydrogenated oils or acetylated monoglycerides containing butylated hydroxyanisole (BHA) or butylated hydroxytoluene (BHT) more than doubled the shelf-life of almond pieces (Daniels, 1973).

Conventional roasting of nuts is done in oils or fats, after which salt is applied. Dry roasting is often preferred, because it eliminates the oily appearance, and is somewhat lower in calories. However, dry-roasted nuts lack the flavor of oil-cooked nuts, and do not hold salt well. Coatings to allow satisfactory adherence of salt and flavorings have been developed from mixtures of wheat gluten and dextrin, modified food starches and from gum arabic (Daniels, 1973). A coating of a hydroxypropyl derivative of high amylose starch had no measurable oxygen transmission, and was effective in delaying oxidative rancidity in almonds (Jokay et al., 1967). Some starch and gum coatings for dry-roasted nuts have shown a tendency to flake off during subsequent handling; a coating which reportedly avoids this problem was devised by Wells and Melnick (1969). Nutmeats were coated with a molten mixture of sorbitol and mannitol, and dusted with salt and flavorings before cooling. These coatings prevented off-flavor development in stored peanuts for three months, and adhered even under adverse handling conditions.

A variety of coatings have been developed for milled rice. These differ from the preceding coatings in that they are usually used for cosmetic reasons, such as imparting a glazed appearance to the rice grain. Traditionally, this glazed surface has been achieved with a coating of corn syrup and talc (Daniels, 1973). However, since talc is inedible, rice must be washed before preparation. Alternative coatings using edible ingredients to impart a glaze have been developed, in which calcium salts of certain organic acids are substituted for talc (Ferrel, 1968). Rice coatings have also been utilized as carriers of spices, flavorings and thickeners for quick-cooking rice dishes (Daniels, 1973). Coatings may be used for rice in ready-made dinners to prevent the rice from becoming soggy after prolonged contact with a covering sauce (Anonymous, 1993a).

Breakfast cereals are sometimes coated to impart a sweet flavor and to improve texture (Tribelhorn, 1991). This has traditionally been done with flaked cereals, manufactured mostly from corn and wheat. Since sweet flavors are generally the most acceptable for breakfast cereals, the majority of coating materials have a sugar base. There are problems due to the hygroscopic nature of cereal products and the fact that sugars require water as the solvent. This can result in sticky pieces which are difficult to dry and which adhere to each other. The amount of water in the coating step of cereal processing is, therefore, reduced to a minimum. One technique is to heat sugar with water to a hard candy condition and spray it on cereal products under constant pressure. The enrobed material is then heated to fuse the sugar to the cereal product (Massmann et al., 1954). Another technique is the use of a spun sugar coating where sugar is spun into a blanket. The processed cereal pieces are placed on top of the blanket, and then a second blanket is placed on top of the cereal pieces. The spun sugar is compressed around the cereal pieces and dried (Shoaf et al., 1971). Most sugar coatings use sucrose as the main ingredient. Application of sucrose and sugar syrups (including corn and cereal malt syrup) generally results in cereal pieces that stick together and have an opaque apearance. Incorporation of glucose or invert sugar helps to overcome these problems. Honey is one form of nonsucrose sugar which is used extensively, since it has acceptable flavor characteristics and results in a transparent coating. Additives to coating formulations improve flavor enhancement and adhesion of the coating to the cereal. Acetic acid and sodium acetate provide some flavor enhancement (Fast, 1967). Emulsifiers have been added to create emulsions of oil, water, and sugar (Lyall and Johnston, 1976). This helps reduce the amount of water that will penetrate the cereal piece and, therefore, reduces the amount of drying required. Thickening agents, such as MC, gum alginate, dextrin, pregelatinized starch, and other hydrophilic colloids are sometimes employed to change the consistency of coatings (Christianson, 1972). Other emulsifiers, adhesives, humectants, plasticizers, and thickening agents that are used in cereal coatings include carrageenan, gelatin, glycerine, guar gum, gum arabic, modified corn starch, maltodextrins, sodium alginate, sorbitol, and lecithin (Tribelhorn, 1991). Another method encapsulates colloid material with a solvent, which is used as a carrier for the sweetener and later evaporated, leaving the colloid and sugar material to merge and coat the cereal, resulting in a frosted appearance (Schade et al., 1978). An organic silicone-based additive, which reduces the degree of liquid absorption when milk is added to a flaked cereal, has been proposed, but so far, use is limited (Maloney and Craig, 1971).

FRUITS AND VEGETABLES

Even when dried, fruits and vegetables undergo deteriorative changes during storage which reduce quality and shorten shelf-life. Because of their high sugar content, dried fruits such as raisins, dates, apricots, figs and prunes tend to be sticky and agglomerate during storage. Raisins can exude small droplets of syrup; these serve to clump the fruit together, and in moist environments can serve as a substrate for mold growth (Watters and Brekke, 1961). With extended storage, raisins develop a crystalline surface coating of glucose, fructose and tartaric acid (Miller and Chichester, 1960) which detracts from their appearance. Exposure of dried fruit to low humidity promotes further loss of product moisture, causing the product to become hard and tough.

Mineral oil has traditionally been used to reduce stickiness and clumping of raisins. In addition, oiling reduces further moisture loss and gives the raisins a more attractive appearance (Kochhar and Rossell, 1982). Alternative coatings which have been suggested as replacements for this petroleum-based material include beeswax and blends of beeswax and vegetable oils (Watters and Brekke, 1959, 1961), beeswax and acetylated monoglycerides (Lowe et al., 1962, 1963), pectinate coatings (Swenson et al., 1953), hydroxypropyl derivatives of high amylose starch (Jokay et al., 1967), and even vinyl polymers (Daniels, 1973).

When raisins are coated with wax or mixtures of wax and oils, the droplets of syrup on the surface interfere with coverage; small openings remain where wax fails to adhere to the liquid droplet. Dusting the raisin surface with powdered starch provided a continuous dry surface which allowed for adequate coverage with wax (Watters and Brekke, 1961). Lowe et al. (1963) devised an alternative procedure to eliminate the problem of surface moisture droplets. Raisins were heated in molten wax to drive off the water, then centrifuged to remove excess wax. This deposited a wax coating of approximately 20 μ. Moisture loss from raisins packed in dry cereal was reduced at least fourfold by a beeswax coating (Watters and Brekke, 1961). Coatings of solid wax such as paraffin or beeswax are brittle and are easily damaged by subsequent handling (Bolin, 1976). Mixtures of beeswax and vegetable oils or acetylated monoglycerides are more resilient, but are less effective as moisture barriers (Lowe et al., 1963; Watters and Brekke, 1961).

Attempts have been made to utilize various vegetable oil blends as coatings for dried fruit, but a limiting feature has been the poor flavor stability of such coatings (Goldenberg, 1976). A commercial partially hydrogenated vegetable oil with resistance to rancidity fared somewhat better (Kochhar and Rossell, 1982). Although the flavor of sultanas (small seedless raisins) coated with this product was inferior to uncoated controls after one month's storage, there were no significant differences in flavor after sixteen months. No data were given on moisture content, but uncoated sultanas were reported to clump and appeared dried after eight months' storage.

Low-methoxyl pectin coatings of the type used on nuts have also been applied to dried dates to reduce stickiness and improve appearance (Swenson et al., 1953). Prunes have been successfully coated with powdered formulations of hydroxypropyl starch (Jokay et al., 1967). Similar coatings of amylose starch and plasticizer were used to coat raisins and dates with a transparent glossy coating (Moore and Robinson, 1968). Stored fruit did not become sticky, although the storage period reported was only one week. Egg albumen or solubilized soy protein coatings significantly reduced moisture loss from stored raisins, although the effect was much more modest than that achieved with paraffin (Bolin, 1976).

Many "dried" fruits such as raisins, dates, figs, etc., actually require a certain residual moisture level to maintain proper appearance and texture. As can be seen from the foregoing discussion, one function of edible coatings is to reduce further loss of internal moisture from these products.

However, in freeze-dried fruits or vegetables, internal moisture levels are reduced to near zero. In these products, coatings are used to reduce moisture absorption, retard oxidative deterioration, or serve as carriers for antioxidants and/or preservatives. Although many of the same types of coatings used on other foods may be used on freeze-dried fruits and vegetables, freeze-dried materials present additional requirements. The coating process must not add significant moisture to the dehydrated product, and the coating must not interfere with subsequent rehydration. Additionally, the surface of freeze-dried tissues is usually quite porous, which can significantly alter the efficiency of the edible coating against moisture transfer (Martin-Polo et al., 1992).

Freeze-dried peas, carrots, and apple slices were successfully coated with a laminated coating comprising an initial layer of an amylose ester of fatty acid, followed by a layer of soy or zein protein (Cole, 1965). Such coatings reduced moisture absorption in dried peas to a third of that absorbed by the uncoated product, without affecting rehydration. Coating of dried apple slices with granular sugar was improved by immersing partially dried fruit slices in dried corn syrup solids, exposing them to moist air or steam to liquify the solids, and then vacuum drying. After removal of excess dry-foamed corn syrup coating, a granulated sugar coating could be applied with excellent retention (Williams, 1968).

Fresh, lightly processed fruits and vegetables may also benefit from edible coatings. Dusting cut fruit and vegetable pieces with dry CMC and starch prevented loss of juice and preserved texture in both fresh-stored and processed products (Mason, 1969). The powdered CMC was presumed to swell within fruit or vegetable pores, preventing leakage of fluids. Browning and water loss of cut apple slices has been inhibited by coatings of chitosan and lauric acid (Pennisi, 1992). Another use may be the prevention of darkening of cooked potatoes. Cooking of Irish potatoes forms a chlorogenic acid-ferrous iron complex which oxidizes to the ferric state, producing a harmless but objectionable dark bluish pigment. Precooked french fries are normally treated with sodium acid pyrophosphate to prevent this discoloration, but treated fries may develop a bitter flavor. Coatings of gum acacia, gelatin, and calcium chloride were as effective as sodium acid pyrophosphate in preventing darkening (Mazza and Qi, 1991). A number of other coatings composed of starches, pullulans, gums, and cyclodextrins were ineffective in inhibiting the discoloration.

CONFECTIONERIES

Many candies, including candied fruit, require an edible coating to prevent stickiness, clumping, absorption of moisture, and, in the case of chocolates and other lipid-containing confections, loss of oil. Chocolate itself may be considered an edible coating when it is used to enrobe a candy center, dried fruit piece, or cookie. Model studies of chocolate coatings showed that a 0.6-mm (24-mil) coating of semisweet dark chocolate was a more effective moisture barrier than a 0.025-mm (1-mil) low-density polyethylene coating (Biquet and Labuza, 1988). Pan-coating of candies can be an elaborate process requiring as much as a day for completion. More complete discussions of the technology of enrobing and panning may be found in Chapter 8 and in the books of Alikonis (1979) and Gillies (1979).

As with other foods, many coatings have been suggested for candies and other sweets. Candied diced fruit was successfully coated with low-methoxyl pectins gelled with calcium chloride (Swenson et al., 1953). Water-soluble coatings such as pectinates, alginates, and starch are able to minimize stickiness, but do not protect hygroscopic materials from absorption of water. A further problem is the use of water as the solvent and the need to remove this from the product. Jokay et al. (1967) devised a procedure to minimize exposure of the product to water when applying water-soluble coatings. Candy pieces were slightly dampened, then dusted with a powdered hydroxypropyl derivative of amylose. The candy pieces were then briefly steamed to rehydrate the water soluble amylose to a continuous film, and then dried.

When a smooth, shiny surface is required, candies are often coated with confectioner's glaze, which is largely shellac, or a wax such as carnauba. An alternative to these nonnutritive coatings was developed from zein and a suitable plasticizer. Candies coated with this material remained free-flowing even under conditions causing surface stickiness in candies coated with confectioner's glaze (Cosler, 1957). Zein coatings, which are ethanol based, dry faster than confectioner's glaze, and can be labeled as a corn protein coating (Gennadios and Weller, 1990). However, the use of solvent-based coatings has certain disadvantages, such as solvent recovery.

With high-fat content candies, such as chocolates and those with crushed nut centers, migration of lipids can adversely affect product stability. For example, if lipids from peanut butter centers migrate into a chocolate coating, the coating may soften or develop a chalky appearance (bloom or whitening). Methylcellulose films were suggested as potential barriers to lipid movement between components in chocolate confections, since these films were quite effective in inhibiting permeation of lipid (Nelson and Fennema, 1989). As thickness of films was reduced from 1.6 mil to 0.2 mil, lipid migration increased, largely as a result of increased film defects. Sensory evaluation of films embedded in chocolate determined that even the 0.2-mil film was detectable.

HETEROGENEOUS PRODUCTS

Increasing demand for prefabricated foods containing components of widely differing water activity (a_w) requires new packaging or formulation strategies to prevent transfer of moisture from one ingredient to another. In addition, certain ingredients of a heterogeneous food may undergo deleterious reactions with other ingredients, such as changes in color and/or nutritional quality. Often the only practical method for separating components of a composite food is to provide an edible film barrier between components (Guilbert, 1986).

One of the simplest heterogeneous foods benefiting from edible coatings is dry cereal with added raisins. Raisins contain from 13–18% moisture (Lowe et al., 1963; Watters and Brekke, 1961), while dry cereal contains approximately 2–3% water. Dry cereal and starch-based snack products soften and become organoleptically unacceptable when their a_w exceeds 0.35–0.50 (Katz and Labuza, 1981), while raisins become progressively tougher as their moisture content drops below 13% (Watters and Brekke, 1959). Thus, transfer of moisture from raisins to cereal results in quality deterioration in both components. Coating raisins with beeswax or mixtures of beeswax and lipids extended their shelf-life in cereal by reducing transfer of moisture to the cereal flakes (Lowe et al., 1962; Watters and Brekke, 1959).

Commercially produced sundae ice cream cones have a storage life of three months, after which the cone becomes soggy due to moisture transfer from the ice cream. Edible polysaccharide and polysaccharide–fatty acid films (MC and MC-palmitic acid) were used to provide moisture-impermeable barriers in a simulated ice cream cone test system. Sugar cones of samples with these films showed no detectable moisture increase for ten weeks of storage at $-23°C$ or four weeks at $-4°C$ (Rico-Peña and Torres, 1990).

Dried fruit pieces used in cake mixes must be coated to prevent interaction between the fruit acids and the leavening in the mix. Edible coatings also stabilize the fruit piece density, allowing for a more even spatial distribution of fruit particles in the cake or bread mix. Dried fruit pieces were stabilized by coating with acetylated monoglycerides, hydrogenated coconut oil, or confectioner's coating butter (Shea, 1970). These coating materials were solid at room temperature, providing an effective barrier to moisture. Furthermore, when baked, the coating dissolved. Uncoated apricot pieces in a bread blackened during storage as a result of a reaction between the soda in the mix and the fruit acids, while coated pieces retained their normal color.

Other uses for edible films in heterogeneous food systems include frozen pizzas and filled pies, where a relatively dry crust is juxtaposed with a moist filling. Edible bilayer films composed of a water-soluble component and a lipid have been examined as possible barriers for such foods. In a model system, films of HPMC and palmitic/stearic acid were as effective as certain plastic films (Kamper and Fennema, 1984). A limiting feature of such films was their poor barrier properties at relative humidities above 0.85. However, at $-20°C$ the a_w of tomato paste is only 0.82, and such films prevented migration of water from this substrate to dried crackers for 70 days (Kamper and Fennema, 1985). Coatings developed from MC and beeswax have shown water permeabilities comparable to commercial packaging films (Greener and Fennema, 1989a). These coatings, when applied to a cake product (brownie), significantly reduced moisture absorption (Greener and Fennema, 1989b). This would be of value in a heterogeneous product such as a fudge-coated cake item, where the edible coating could serve to prevent moisture transfer between a high a_w topping and the drier cake.

A somewhat more complex edible film was constructed of MC, HPMC, and saturated C_{16} and C_{18} fatty acids, coated with a thin film of beeswax (Kester and Fennema, 1989a). Unlike the earlier films of HPMC and palmitic/stearic acids, the beeswax coated film exhibited good stability and barrier properties even at a_w of 0.97. In addition, this film, unlike many containing wax, displayed good low-temperature stability. In a model heterogeneous system composed of a high-moisture sauce layered on bread, this film significantly improved sensory properties during frozen storage (Kester and Fennema, 1989b).

Historically, waxes have been used to coat cheese as peelable protection to be removed before consumption. Examples are the red wax-coated Dutch cheeses, Gouda and Edam (Johnson and Peterson, 1974). Cheddar cheese, after pressing into a cheese block, is dipped in hot paraffin. The wax coating, in addition to preventing surface molding, prevents the cheese from excessive dehydration during the long ripening or ageing period. The waxed cheese is then boxed and placed in the curing room for ripening (Potter, 1986). Blue cheese may also be coated with wax or paraffin to reduce drying during the molding period (Johnson and Peterson, 1974), but often these coatings are removed before retail display or before consumption. Paraffin and beeswax are approved for coating of cheese, cheese rinds, or cured cheese in various countries (Baranowski, 1990). Parmesan cheese is cleaned and rubbed with oil during curing (Johnson and Peterson, 1974). Mineral oil and glycerol are approved as film-formers for cheese in the United States and in other countries, while polyvinyl acetate is also approved in Portugal (Baranowski, 1990). Surface-ripened cheeses, such as the mold-ripened varieties Camembert and Brie, or bacterial-ripened such as Limburger, contain an edible rind.

These cheeses contain more moisture in the curd and are allowed to develop a surface growth of the appropriate organism which is sprayed onto the cheese surface. For example, once the surface growth has developed on Camembert cheese, a white feltlike mold growth (*Penicillium camemberti*) covers the entire surface. Ripening is highly proteolytic and breaks down the curd protein, from the surface inward, to the texture of soft butter, creating a rind on the surface. In the case of Limburger cheese, the ripening agent is the surface bacterium, *Brevibacterium linens* (Potter, 1986; Johnson and Peterson, 1974). Coatings can also retain preservatives on food surfaces, where they are needed most to retard spoilage. Torres et al. (1985) used an edible composite film of zein, acetylated monoglycerides and glycerol to maintain a high concentration of sorbic acid at the surface of an intermediate-moisture cheese analog. Diffusion of the preservative into the food bulk was significantly retarded by the coating.

CONCLUSIONS

Edible films and coatings can play a significant role in the preservation of quality in a number of processed foods. The current interest in more "natural" and healthful foods suggests an increased consumption of fresh and dried fruit and nut products. Restrictions on the quantity of preservatives allowed in meat and other products will challenge the manufacturer to find alternative means of extending shelf-life. As more complex processed foods, including ethnic and homestyle foods new to the American market, become more commonplace, innovative techniques must be found to preserve their structural and nutritional integrity. Recent significant advances in edible coating and film technology promise to contribute to this effort.

REFERENCES

Alikonis, J. J. 1979. *Candy Technology.* Westport, CT: AVI Publishing Co.

Allen, L., A. I. Nelson, M. P. Steinberg and J. N. McGill. 1963a. "Edible Corn-Carbohydrate Food Coatings I. Development and Physical Testing of a Starch-Algin Coating," *Food Technol.*, 17:1437.

Allen, L., A. I. Nelson, M. P. Steinberg and J. N. McGill. 1963b. "Edible Corn-Carbohydrate Food Coatings II. Evaluation on Fresh Meat Products," *Food Technol.*, 17:1442.

Andres, C. 1984. "Natural Edible Coating Has Excellent Moisture and Grease Barrier Properties," *Food Process.*, 45(12):48.

Anonymous. 1993a. "Edible Packaging," *Sustain Notes*, 5(1):1–4.

Anonymous. 1993b. "Food Gums Stick It Out in Dry Mix Glazes," *Prepared Foods,* (1):53.

Baranowski, E. S. 1990. "Miscellaneous Food Additives," in *Food Additives,* A. L. Branen, P. M. Davidson and S. Salminen, eds., New York, NY: Marcel Dekker, Inc., pp. 511–578.

Bauer, C. D., G. L. Neuser and H. A. Pinkalla. October 15, 1968. U.S. Patent 3,406,081.

Bauer, C. D., G. L. Neuser and H. A. Pinkalla. December 9, 1969. U.S. Patent 3,483,004.

Biemuller, G. W., J. A. Carpenter and A. E. Reynolds. 1973. "Reduction of Bacteria on Pork Carcasses," *J. Food Sci.,* 38:261.

Biquet, B. and T. P. Labuza. 1988. "Evaluation of the Moisture Permeability Characteristics of Chocolate Films as an Edible Moisture Barrier," *J. Food Sci.,* 53:989.

Bolin, H. R. 1976. "Texture and Crystallization Control in Raisins," *J. Food Sci.,* 41:1316.

Christianson, G. 1972. U.S. Patent 3,686,001.

Cole, M. S. November 18, 1965. U.S. Patent 3,479,191.

Cosler, H. B. May 7, 1957. U.S. Patent 2,791,509.

Daniels, R. 1973. *Edible Coatings and Soluble Packaging.* Park Ridge, NJ: Noyes Data Corp.

Earle, R. D. July 30, 1968. U.S. Patent 3,395,024.

Earle, R. D. and D. H. McKee. 1976. U.S. Patent 3,991,218.

Earle, R. D. and C. E. Snyder. June 7, 1966. U.S. Patent 3,255,021.

Fast, R. B. 1967. U.S. Patent 3,318,706.

Ferrel, R. E. May 21, 1968. U.S. Patent 3,384,493.

Gennadios, A. and C. L. Weller. 1990. "Edible Films and Coatings from Wheat and Corn Proteins," *Food Technol.,* 44(10):63.

Gillies, M. T. 1979. *Candies and Other Confections.* Park Ridge, NJ: Noyes Data Corp.

Goldenberg, N. 1976. "The Oiling of Dried Sultanas," *Chem. Ind.,* 21:956.

Greener, I. K. and O. Fennema. 1989a. "Barrier Properties and Surface Characteristics of Edible, Bilayer Films," *J. Food Sci.,* 54:1393.

Greener, I. K. and O. Fennema. 1989b. "Evaluation of Edible Bilayer Films for Use as Moisture Barriers for Food," *J. Food Sci.,* 54:1400.

Guilbert, S. 1986. "Technology and Application of Edible Protective Films," in *Food Packaging and Preservation. Theory and Practice,* M. Mathlouthi, ed., London, England: Elsevier Applied Science Publishing Co., pp. 371–394.

Hoover, M. W. and P. J. Nathan. 1981. "Influence of Tertiary Butylhydroquinone and Certain Other Surface Coatings on the Formation of Carbonyl Compounds in Granulated Roasted Peanuts," *J. Food Sci.,* 47:246.

Johnson, A. H. and M. S. Peterson, eds. 1974. *Encyclopedia of Food Technology.* Westport, CT: AVI Publishing Co., Inc., pp. 178–183.

Jokay, L., G. E. Nelson and E. L. Powell. 1967. "Development of Edible Amylaceous Coatings for Foods," *Food Technol.,* 21:1064.

Kamper, S. L. and O. Fennema. 1984. "Water Vapor Permeability of Edible Bilayer Films," *J. Food Sci.,* 49:1478.

Kamper, S. L. and O. Fennema. 1985. "Use of an Edible Film to Maintain Water Vapor Gradients in Foods," *J. Food Sci.*, 50:382.

Katz, E. E. and T. P. Labuza. 1981. "Effect of Water Activity on the Sensory Crispness and Mechanical Deformation of Snack Food Products," *J. Food Sci.*, 46:403.

Kester, J. J. and O. R. Fennema. 1986. "Edible Films and Coatings: A Review," *Food Technol.*, 40(12):47.

Kester, J. J. and O. Fennema. 1989a. "An Edible Film of Lipids and Cellulose Ethers: Barrier Properties to Moisture Vapor Transmission and Structural Evaluation," *J. Food Sci.*, 54:1383.

Kester, J. J. and O. Fennema. 1989b. "An Edible Film of Lipids and Cellulose Ethers: Performance in a Model Frozen-Food System," *J. Food Sci.*, 54:1390.

Kochhar, S. P. and J. B. Rossell. 1982. "A Vegetable Oiling Agent for Dried Fruits," *J. Food Technol.*, 17:661.

Lazarus, C. R., R. L. West, J. L. Oblinger and A. Z. Palmer. 1976. "Evaluation of a Calcium Alginate Coating and a Protective Plastic Wrapping for the Control of Lamb Carcass Shrinkage," *J. Food Sci.*, 41:639.

Lowe, E., E. L. Durkee and W. E. Hamilton. July 24, 1962. U.S. Patent 3,046,143.

Lowe, E., E. L. Durkee, W. E. Hamilton, G. G. Watters and A. I. Morgan, Jr. 1963. "Continuous Raisin Coater," *Food Technol.*, 17(11):109.

Lyall, A. A. and R. J. Johnston. 1976. U.S. Patent 3,959,498.

Maloney, J. F., Jr. and T. W. Craig. 1971. U.S. Patent 3,595,670.

Martin-Polo, M., C. Mauguin and A. Voilley. 1992. "Hydrophobic Films and Their Efficiency against Moisture Transfer. I. Influence of the Film Preparation Technique," *J. Agric. Food Chem.*, 40:407.

Mason, D. F. October 14, 1969. U.S. Patent 3,472,662.

Massmann, W. G., E. W. Michael and W. L. Vollink. 1954. U.S. Patent 2,689,796.

Mazza, G. and H. Qi. 1991. "Control of After-Cooking Darkening in Potatoes with Edible Film-Forming Products and Calcium Chloride," *J. Agr. Food Chem.*, 39:2163.

Meyer, R. C., A. R. Winter and H. H. Weiser. 1959. "Edible Protective Coatings for Extending the Shelf Life of Poultry," *Food Technol.*, 13:146.

Miller, M. W. and C. O. Chichester. 1960. "The Constituents of the Crystalline Deposits on Dried Fruits," *Food Res.*, 25:425.

Moore, C. O. and J. W. Robinson. February 13, 1968. U.S. Patent 3,368,909.

Nelson, K. L. and O. R. Fennema. 1989. "Methylcellulose Films to Prevent Lipid Migration in Confectionery Products," *J. Food Sci.*, 54:1383.

Pennisi, E. 1992. "Sealed in Edible Film," *Sci. News*, 141:12.

Potter, N. N. 1986. *Food Science.* Westport, CT: AVI Publishing Co., Inc., pp. 372–389.

Reagan, J. O., G. C. Smith and Z. L. Carpenter. 1973. "Use of Ultraviolet Light for Extending the Retail Case Life of Beef," *J. Food Sci.*, 38:929.

Rico-Peña, D. C. and J. A. Torres. 1990. "Edible Methylcellulose-Based Films as Moisture-Impermeable Barriers in Sundae Ice Cream Cones," *J. Food Sci.*, 55:1468–1469.

Rico-Peña, D. C. and J. A. Torres. 1991. "Sorbic Acid and Potassium Sorbate Permeability of an Edible Methylcellulose-Palmitic Acid Film: Water Activity and pH Effects," *J. Food Sci.*, 56:497.

Schade, H. R., P. A. Baggerly and D. R. Woods. 1978. U.S. Patent 4,079,151.

Shea, R. A. June 23, 1970. U.S. Patent 3,516,836.

Shoaf, M. D., C. W. Groesbeck and D. G. Cowart. 1971. U.S. Patent 3,615,671.

Siragusa, G. R. and J. S. Dickson. 1992. "Inhibition of *Listeria monocytogenes* on Beef Tissue by Application of Organic Acids Immobilized in a Calcium Alginate Gel," *J. Food Sci.*, 57:293.

Suderman, D. R., J. Wiker and F. E. Cunningham. 1981. "Factors Affecting Adhesion of Coating to Poultry Skin: Effects of Various Protein and Gum Sources in the Coating Composition," *J. Food Sci.*, 46:1010.

Swenson, H. A., J. C. Miers, T. H. Schultz and H. S. Owens. 1953. "Pectinate and Pectate Coatings. II. Application to Nuts and Fruit Products," *Food Technol.*, 7:232.

Torres, J. A. and M. Karel. 1985. "Microbial Stabilization of Intermediate Moisture Food Surfaces. III. Effects of Surface Preservative Concentration and Surface pH Control on Microbial Stability of an Intermediate Moisture Cheese Analog," *J. Food Proc. Pres.*, 9:107.

Torres, J. A., M. Motoki and M. Karel. 1985. "Microbial Stabilization of Intermediate Moisture Food Surfaces. I. Control of Surface Preservative Concentration," *J. Food Proc. Preserv.*, 9:75.

Tribelhorn, R. E. 1991. "Breakfast Cereals," in *Handbook of Cereal Science and Technology*, K. J. Lorenz and K. Kulp, eds., New York: Marcel Dekker, Inc., pp. 741–761.

Vojdani, F. and J. A. Torres. 1990. "Potassium Sorbate Permeability of Methylcellulose and Hydroxypropyl Methylcellulose Coatings: Effects of Fatty Acids," *J. Food Sci.*, 55:941.

Wanstedt, K. G., S. C. Seideman, L. S. Donnelly and N. M. Quenzer. 1981. "Sensory Attributes of Precooked, Calcium Alginate–Coated Pork Patties," *J. Food Protec.*, 44:732.

Watters, G. G. and J. E. Brekke. October 20, 1959. U.S. Patent 2,909,435.

Watters, G. G. and J. E. Brekke. 1961. "Stabilized Raisins for Dry Cereal Products," *Food Technol.*, 15(5):236.

Wells, P. R. and D. Melnick. November 11, 1969. U.S. Patent 3,477,858.

Whitman, G. R. and H. Rosenthal. January 19, 1971. U.S. Patent 3,556, 814.

Williams, L. G. October 15, 1968. U.S. Patent 3,406,078.

Williams, S. K., J. L. Oblinger and R. L. West. 1978. "Evaluation of a Calcium Alginate Film for Use on Beef Cuts," *J. Food Sci.*, 43:292.

Wu, L. C. and R. P. Bates. 1972. "Soy Protein–Lipid Films. I. Studies on the Film Formation Phenomenon," *J. Food Sci.*, 37:36.

Wu, L. C. and R. P. Bates. 1973. "Influence of Ingredients upon Edible Protein-Lipid Film Characteristics," *J. Food Sci.*, 38:783.

Wu, L. C. and R. P. Bates. 1975. "Protein-Lipid Films as Meat Substitutes," J. Food Sci., 40:160.

CHAPTER 5

Flavor Encapsulation

GARY A. REINECCIUS[1]

INTRODUCTION

A N important use of edible films in the area of flavor is for the encapsulation of flavorings (Reineccius, 1991). The edible film serves many purposes including permitting the production of a dry free-flowing flavor (most flavors are liquids), protection of the flavoring from interaction with the food or deleterious reactions such as oxidation, confinement during storage, and finally, controlled release. The degree to which the edible film meets these requirements depends upon the process used to form the film around the flavoring and the film composition itself. This chapter will discuss the major processes used in manufacturing encapsulated flavorings, the materials commonly used as the encapsulation matrix, and the factors determining the efficacy of the edible film (hereafter termed the encapsulation matrix).

PROCESSES FOR FLAVOR ENCAPSULATION

While there are numerous processes described for flavor encapsulation in both the scientific and patent literature, only two processes find significant commercial usage: spray drying and extrusion (Reineccius, 1989).

SPRAY DRYING

Spray drying is used to produce more than 90% of the encapsulated flavorings on the market. Spray drying is accomplished by initially making

[1]Dept. of Food Science and Nutrition, University of Minnesota, St. Paul, MN 55108.

an emulsion of the flavoring (20% on a dry basis) in an encapsulation matrix (80% on a dry basis) using a minimal amount of water (preferably 40% water-wet basis). An oil in water emulsion is formed via homogenization to obtain a particle size averaging about 1 μm and then the emulsion is pumped into a spray dryer. The flavor emulsion is atomized into a hot air stream where evaporative cooling brings the inlet air temperature down from 200–325°C to 80–90°C (Figure 5.1). During drying, the flavoring particle never exceeds the exit air temperature.

This process requires the encapsulating matrix to meet several requirements. The matrix must be water-soluble since virtually all flavoring materials are manufactured in and ultimately released by water contact. The infeed emulsion must be low enough in viscosity at high solids concentrations (50–70%) to permit pumping and, more restrictive, atomization. While the final product need not necessarily be a stable emulsion on reconstitution (e.g., cake mix or frosting flavorings), the encapsulation matrix must yield at least a temporary emulsion which should not break from the time of homogenization until it is atomized in the spray dryer. The encapsulation matrix must dry well in a normal spray dryer (e.g., not become sticky at high temperatures) so that a good yield is obtained, and then not be hygroscopic following drying. The matrix must ultimately produce a good quality flavoring (i.e., it must retain the volatile flavor constituents), protect it from degradation or evaporation during storage, and finally provide release in the finished product.

The requirements discussed above largely limit matrix materials to maltodextrins, corn syrup solids, modified starches and gum acacia – the traditional encapsulation matrix.

EXTRUSION PROCESS

The extrusion process consists of initially making a low-moisture carbohydrate melt (ca. 15% moisture), adding flavor (10–20% on a dry weight basis), forming an emulsion, and then extruding the melt through a die (1/64″ holes) under pressure (< 100 psig). The extruded flavoring drops into a cold isopropanol bath where it immediately forms an amorphous glass structure, is broken into small pieces by a mixer, and is then further dried in hot air to yield the finished flavoring material (Figure 5.2).

The encapsulating material used in this process has similar requirements as that used in spray drying. The encapsulating matrix does not have to be atomized but it must be somewhat workable (soluble) at 85% solids, 110°C. Emulsification is also less of an issue since synthetic emulsifiers are added to the carbohydrate melt (ca. 2% on a total weight basis), and there is little opportunity for phase separation during manufacturing. This is due to the short time between emulsion formation and solidification, and

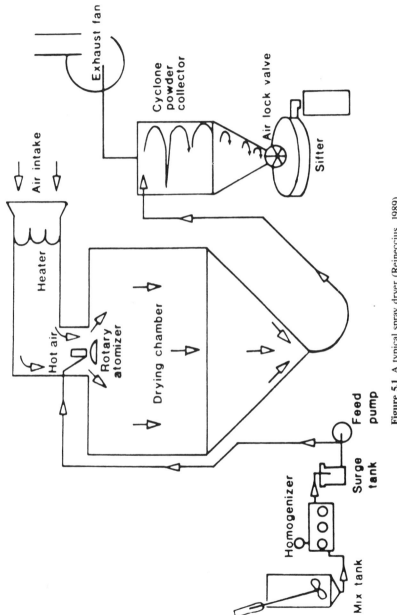

Figure 5.1 A typical spray dryer (Reineccius, 1989).

107

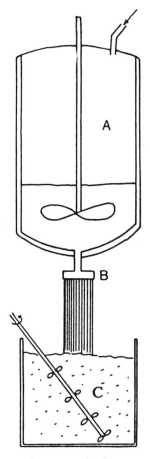

Figure 5.2 Schematic of the extrusion process for flavor encapsulation: A–mixing chamber, B–die and C–cold isopropanol bath. Reprinted with permission from Risch, 1986. Copyright 1986, American Chemical Society.

the extremely high viscosity of the carbohydrate melt. The matrix must form a good amorphous structure and be nonhygroscopic. It must also result in good retention of flavor compounds during manufacturing, offer protection during storage, and release the flavor when placed in water.

The encapsulation materials which meet the criteria for both spray drying and extrusion processing are discussed next.

EDIBLE FILMS FOR FLAVOR ENCAPSULATION

While there are a few examples in the patent literature where noncarbohydrate-based films are used for flavor encapsulation, the vast majority of flavors are encapsulated using water-soluble carbohydrates. These carbohydrates include gum acacia, modified starches, maltodextrin, and corn syrup solids.

GUM ACACIA

Gum acacia is the material traditionally used for the encapsulation of food flavors. It is derived from the gum acacia tree which is grown in the semidesert region of North Central Africa (Thevenet, 1986). While there are several hundred species of the acacia tree, only a few species are used for gum manufacture. Gum is produced by the tree in response to injury inflicted by the people who harvest the gum (Figure 5.3). Each tree will produce about 300 grams of gum, which is allowed to partially dry on the tree before hand harvesting, sorting, grading, and distributing to processors. The processors most often grind the gum, solubilize it in water, filter or centrifuge to remove foreign materials, pasteurize, and then spray dry it. While the gum may be sold in the crude form, the conditions of harvesting and storage result in contamination by microbes and foreign materials so it must be cleaned up prior to use as a food ingredient.

Gum acacia is a polymer made up primarily of D-glucuronic acid, L-rhamnose, D-galactose, and L-arabinose (Figure 5.4). It also contains about 5% protein, which is responsible for its emulsification properties (Thevenet, 1986). Gum acacia is unique among the plant gums since it exhibits low viscosity at relatively high solids (ca. 300 cps at 30–35% solids) and is an excellent emulsifier.

MODIFIED FOOD STARCH

Starches have virtually no intrinsic emulsification properties. If they make a contribution to emulsion stability, it is only through their effects on viscosity. The need for emulsification properties in encapsulation matrices as well as other food applications (e.g., beverage flavor emulsions or cloud emulsions) prompted the development of starches which have been chemically modified for this purpose (Trubiano and Lacourse, 1986). Emulsification properties are attained by the chemical addition of octenyl succinate to partially hydrolyzed starch (Figure 5.5). Octenyl succinate may be used at a 0.02 degree of substitution on the starch polymer and still be approved for food use. However, octenyl succinate is not considered to be a natural product and this may affect claims for food labels (e.g., one cannot put a "100% natural" label on the product).

The octenyl succinate derivatized starches have been chemically engineered to be excellent emulsifiers and exhibit low viscosity at high solids (40–50% solids).

Figure 5.3 Gum exuded from tree in response to injury (courtesy of Colloides Naturels Int.).

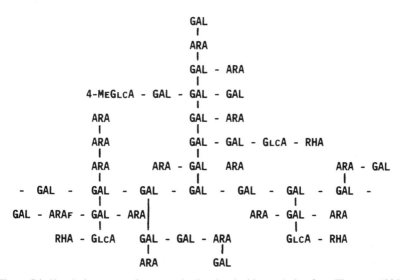

Figure 5.4 Chemical structure of gum acacia. Reprinted with permission from Thevenet, 1986. Copyright 1986, American Chemical Society.

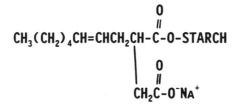

Figure 5.5 Chemical structure of octenyl succinate derivatized starch. Reprinted with permission from Trubiano and Lacourse, 1986. Copyright 1986, American Chemical Society.

MALTODEXTRINS/CORN SYRUP SOLIDS

Both maltodextrins and corn syrup solids are made from the acid, enzyme, or acid/enzyme treatment of starches. If the product has a dextrose equivalent (DE) of less than twenty, it is a maltodextrin. Products with a DE equal to or greater than twenty are classified as corn syrup solids.

Neither maltodextrins nor corn syrup solids have any inherent emulsification properties. While they may be used at sufficiently high solids levels (and therefore high viscosities) during the preparation of "emulsions" for encapsulation so that they will temporarily maintain an emulsion, they offer no emulsion stability to the finished product. The maltodextrins and corn syrup solids are inexpensive compared to either gum acacia or the modified food starches and thus are commonly used as nonfunctional "fillers."

FILM PERFORMANCE

As mentioned earlier, the film used for flavor encapsulation must meet several criteria, such as the following: it must form and stabilize an emulsion, retain flavors during encapsulation, protect the flavor during storage, and then release the flavor to the final food product. The suitability of various edible films will be discussed with these criteria in mind.

EMULSIFICATION PROPERTIES

The importance of the film's ability to stabilize an emulsion depends upon the final application of the encapsulated flavor. Applications such as a dry savory mix (e.g., gravies or sauces), baked goods (e.g., cake mixes or frostings), or confectionery products (e.g., candies) do not require the edible film to provide significant emulsification properties. However, a substantial portion of dry flavorings are used in dry beverage mixes. Dry beverage mixes must give a one week shelf-life (no visible ring of separated flavor) following reconstitution. Thus, the edible film used as an encapsulant must provide good emulsification properties to these beverage products.

As noted earlier, maltodextrins and corn syrup solids impart no emulsion stability other than that which is due to their high viscosity when initially manufactured. When an encapsulated flavor is based on maltodextrin or corn syrup solids, it cannot be used for a dry beverage flavor application unless a secondary emulsifier is incorporated. Secondary emulsifiers are not commonly used with spray dried flavors (although an

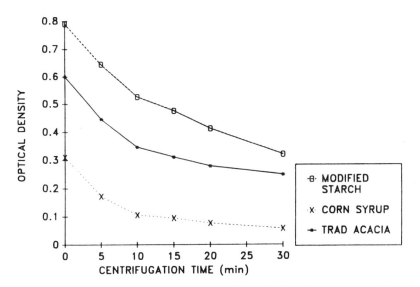

Figure 5.6 Stability of orange oil emulsions made with modified starch, corn syrup solids, and gum acacia (Reineccius, 1991).

emulsifying agent such as a modified starch or gum acacia may be used in combination with the maltodextrin or corn syrup solid flavor carrier) but are essential for the manufacture and stability of extruded flavorings. Both gum acacia and the modified food starches are excellent emulsifiers. The relative ability of corn syrup solids, traditional gum acacia, and a modified food starch to stabilize an orange oil emulsion is shown in Figure 5.6. This figure gives an initial absorption for the emulsion (time = 0) which indicates particle size of the flavor droplets and resultant cloudiness. The modified food starch formed an emulsion that was substantially more cloudy than either the gum acacia or the corn syrup emulsions (0.8 versus 0.6 or 0.3 absorbance units, respectively). Upon centrifugation (500 × g), all the emulsions cleared to some extent due to separation of the emulsion. However, the modified food starch remained more stable than the gum acacia or corn syrup emulsions. The advantage of using modified food starches for flavor encapsulation is their ability to form a stable flavor emulsion.

FLAVOR RETENTION

The ability of an edible film to "capture" or hold onto flavor compounds during the drying process is critical. If flavors are lost during the drying process, the resultant flavor will become imbalanced. The lighter, more volatile constituents will be preferentially lost during the process, with the resulting dry flavor lacking in the very volatile light fresh notes. A secondary concern centered on retention is that a flavor compound not retained in the powder is lost to the drying air (spray drying process). This drying air then contains volatile organic compounds that must be removed by a costly scrubbing process in order to protect the environment. Therefore, materials that offer poor retention result in increased processing costs.

Of the common edible films used in flavor encapsulation, the modified food starch excels in terms of flavor retention during drying (Table 5.1). As noted earlier, modified food starches are excellent emulsifiers and emulsion quality has a strong influence on flavor retention during spray drying (Risch and Reineccius, 1986). Additionally, modified food starches can typically be used at 50–55% solids levels versus 30–35% solids for gum acacia. Thus, although both gum acacia and modified food starches will yield good emulsions (i.e., they should yield good flavor retention), modified food starches yield the best flavor retention because they can be used at higher infeed solids levels, which also improves flavor retention (Reineccius, 1989).

Maltodextrins typically can be used at high infeed solids levels but their poor emulsification properties result in poor flavor retention during drying.

STORAGE STABILITY

The most common problem occurring during the storage of flavors is deterioration due to oxidation. Any flavoring containing cirtus oils or oils based on aldehydes is susceptible to oxidative reactions and development of off-flavors. An important function of an edible film is protection of the flavoring from oxygen.

The films used for flavor encapsulation differ greatly in their ability to protect a flavoring from oxygen. The maltodextrins provide varying protection depending upon their dextrose equivalent (Anandaraman and Reineccius, 1986). In general, the higher the DE the better the protection (Figure 5.7). This presents a problem since higher DE materials tend to be difficult to dry, yield poor flavor retention, and are very hygroscopic. Thus, one can get excellent shelf-life using a high DE maltodextrin (more correctly, a corn syrup solid) but have an otherwise unsuitable product.

While the modified food starches produce an encapsulated flavoring which has excellent flavor retention and emulsion stability, they provide very poor protection against oxidation (Reineccius, 1991).

The acacia products vary greatly in their suitability as encapsulating materials. As was mentioned earlier, there are over 300 different species of

TABLE 5.1. Comparison of Encapsulating Efficiencies of Gum Acacia versus Starch Carriers (Trubiano and Lacourse, 1986).

Encapsulating Agent	Amount of Flavor in Powder (%)		Flavor Lost on Drying (%)	Retained Flavor on Surface (%)	Truly Encapsulated Flavor (%)
	Initial	Retained			
Gum Acacia	30.5	28.7	5.9	16.5	23.9
Conventional Dextrin	30.7	23.5	23.6	25.6	17.4
Low Visc. Starch Octenyl Succinate	30.1	30.0	0.3	1.0	29.4

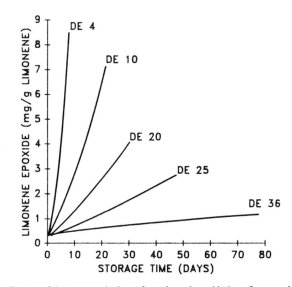

Figure 5.7 Influence of dextrose equivalent of starch on the oxidation of encapsulated single-fold orange peel oil (as demonstrated by oxidation of limonene to limonene epoxide) during storage (Anandaraman and Reineccius, 1986).

the Acacia tree. While only a few of these are used for gum production, there is still substantial variation in the product. Some of the acacias afford excellent protection to the encapsulated flavoring while others offer little protection (Figure 5.8).

The reason for the difference in protection afforded by the various encapsulating materials is not well understood. It has been postulated, however, that shelf-life is related to the permeability of the encapsulation matrix to oxygen (Baisier and Reineccius, 1989). Recently, polymer scientists have been using glass transition theory to explain why materials differ in permeability and also why moisture content influences permeability of a given material (Levine and Slade, 1989). At low temperatures and water activities, food polymers exist in the glassy state, which severely limits molecular diffusion. As either temperature or water activity increases, the material will change to a rubbery state which is known to readily permit molecular diffusion (Figure 5.9). This theory dictates that a flavoring would be stable to oxidation while the encapsulating film remained in the glassy state (no diffusion of oxygen) and that oxidation of flavorings would occur if either temperature or moisture content (water activity) increased such that the film changed to the rubbery state.

This relationship was not found for the oxidation of orange oil which had been spray dried using maltodextrin (10 DE) as the encapsulation matrix (Figure 5.10). Ma et al. (1991) found that the encapsulated orange

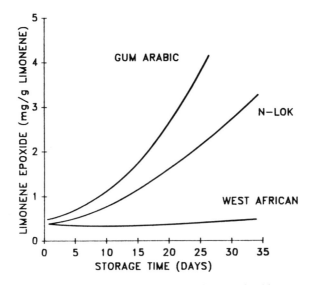

Figure 5.8 Oxidative stability of single-fold orange peel oil encapsulated in a "new generation" gum acacia, traditional gum acacia and N–Lok (product of National Starch, Bridgewater, NJ) (Reineccius, 1991).

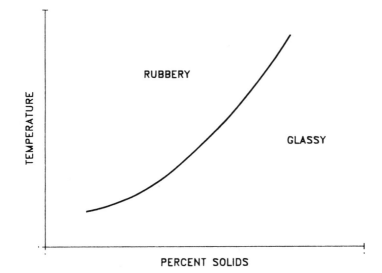

Figure 5.9 Theoretical state diagram for an amorphous polymer. Reprinted with permission from Nelson and Labuza, 1992. Copyright 1992, American Chemical Society.

117

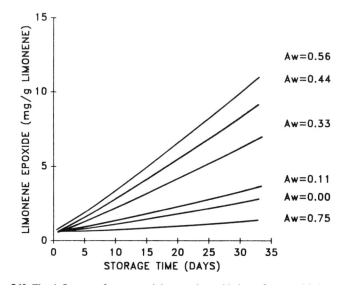

Figure 5.10 The influence of water activity on the oxidation of spray dried orange peel oil–maltodextrin carrier (Ma et al., 1991).

oil was not stable even though the maltodextrin was clearly in the glassy state. However, as water activity was increased, the product was less stable (which does agree with theory) until collapse of the material occurred (a_w = 0.75). Once the material collapsed, shelf-life increased greatly. It appears that collapse of the film structure reduces pore size and leads to substantially reduced oxygen permeability.

Thus the glass transition theory does not appear to adequately predict whether a film will protect an encapsulated flavoring. We are left recognizing that some edible films give excellent protection against oxidation while others give little or none without understanding why.

FLAVOR RELEASE

A shortcoming of the processes used for flavor encapsulation is their inability to use edible films that are water insoluble. Virtually all of the encapsulated flavorings on the market employ aqueous soluble edible films for their encapsulation matrix and, thus, they release flavor upon contact with water. The industry needs encapsulated flavorings that release the flavor very slowly with water contact (e.g., chewing gums) or with heating (e.g., baked and microwave products). Currently, these controlled release properties can only be attained through the use of secondary coating on an already encapsulated flavoring. Secondary coatings (e.g., fats, oils, edible shellacs, modified cellulose) tend to dilute the flavor strength and add additional costs to the product. There is a substantial need for technologies that will encapsulate flavorings in other types of edible films that will provide the desired controlled release.

SUMMARY

The properties of the edible films used for the encapsulation of flavoring materials are critical to the efficacy of the flavoring material. The edible film must often serve as an emulsifier, film former (trap flavoring during the dehydration process), oxygen barrier (to protect against flavor deterioration), and as a controlled release agent. The materials available serve each of these functions to varying degrees. Maltodextrins are relatively inexpensive but lack most of the functional properties desired in an encapsulation material. Corn syrup solids are also inexpensive and give excellent protection against oxidation. However, they also give poor retention during drying and no emulsification. The modified food starches are moderately priced, excellent emulsifiers, and give excellent flavor retention during drying. These materials, unfortunately, provide poor protection to the flavoring during storage. The gum acacias vary greatly in function properties. They generally are slightly more expensive than the modified food starches and somewhat inferior in emulsification properties and flavor retention. They can, however, offer excellent protection against oxidation, and therefore can be an excellent overall choice as an edible film for the encapsulation of flavors. The gum acacias have the additional advantage of being "natural," which is important to some product labels.

REFERENCES

Anandaraman, S. and G. A. Reineccius. 1986. "Stability of Spray Dried Orange Peel Oil," *Food Technol.*, 40(11):88.

Baisier, W. and G. A. Reineccius. 1989. "Spray Drying of Food Flavors V: Factors Influencing Shelf-Life of Encapsulated Orange Peel Oil," *Perfum. Flav.*, 14:48–53.

Levine, H. and L. Slade. 1989. "Interpreting the Behavior of Low-Moisture Foods," in *Fundamental Aspects of the Dehydration of Foodstuffs*, T. M. Hardmann, ed., New York: Elsevier Applied Science, pp. 71–134.

Ma, Y., G. A. Reineccius and T. P. Labuza. 1991. Unpublished data, University of Minnesota, St. Paul, MN.

Nelson, K. and T. P. Labuza. 1992. "Understanding the Relationship between Water and Lipid Oxidation Rates Using Water Activity and Glass Transition Theory," in *Lipid Oxidation in Foods*, A. St. Angelo, ed., Washington, DC: American Chemical Society, pp. 91–101.

Reineccius, G. A. 1989. "Flavor Encapsulation," *Food Rev. Intern.*, 5:147.

Reineccius, G. A. 1991. "Carbohydrates for Flavor Encapsulation," *Food Technol.*, 45:144.

Risch, S. J. 1986. "Encapsulation of Flavors by Extrusion," in *Flavor Encapsulation*, S. J. Risch and G. A. Reineccius, eds., Washington, DC: American Chemical Society, pp. 103–109.

Risch, S. J. and G. A. Reineccius. 1986. *Flavor Encapsulation*. Washington, DC: American Chemical Society.

Thevenet, F. 1986. "Acacia Gums: Stabilizers for Flavor Encapsulation," in *Flavor Encapsulation*, S. J. Risch and G. A. Reineccius, eds., Washington, DC: American Chemical Society, pp. 37–44.

Trubiano, P. C. and N. L. Lacourse. 1986. "Emulsion Stabilizing Starches: Use in Flavor Encapsulation," in *Flavor Encapsulation*, S. J. Risch and G. A. Reineccius, eds., Washington, DC: American Chemical Society, pp. 45–54.

CHAPTER 6

Edible Coatings as Carriers of Food Additives, Fungicides and Natural Antagonists

SUSAN L. CUPPETT[1]

INTRODUCTION

THE use of edible coatings and films by the food industry has become a topic of great interest because of their potential for increasing the shelf-life of many food products. Edible coatings and films are a viable means for incorporating food additives and other substances to enhance product color, flavor, and texture; to control microbial growth; and to improve general coating performance (i.e., strength, flexibility, and adherence). An "Expert Committee of Food Additives" made up of representatives of the Food and Agriculture Organization and the World Health Organization after Rome meetings in 1956, defined food additives as non-nutritive substances added intentionally to food, generally in small quantities to improve its appearance, flavor, texture, or storage properties. The Food and Drug Administration (FDA) defines a food additive as

> any substance, the intended use of which results or may reasonably be expected to result, directly or indirectly, in its becoming a component of or otherwise affecting the characteristics of any food (including any substance intended for use in producing, manufacturing, packing, processing, preparing, treating, transporting, or holding a food; and including any source of radiation intended for any such use), if such substance is not generally recognized, among experts qualified by scientific training and experience to evaluate its safety, as having been adequately shown through scientific procedures (or, in the case of substances used in food prior to January 1,

[1]Department of Food Science and Technology, University of Nebraska–Lincoln, Lincoln, NE.

1958, through either scientific procedures or experience based on common use in food) to be safe under the conditions of its intended use.

A wide variety of materials that have good film-forming properties are often listed as food additives. These include polysaccharides, proteins, lipids, and resins; their chemistry and technology are discussed in other chapters. Additives that are incorporated to improve the functional properties or aesthetic appeal of edible coatings and films are discussed here. Their classifications, regulations, and usage are presented in Table 6.1.

The concept of using an edible coating is not new. Evidence in the literature indicates that the use of coatings on foods has been studied since the 1800s. Harvard and Harmony in 1869 first reported dipping foods in gelatin (Allen et al., 1963). Patents on edible films developed for food use have been available since the 1950s, and the types of films that have been patented are as varied as the products produced by the food industry. Patents for extending the shelf-life of frozen meat (Bauer et al., 1968; Earle, 1968), poultry (Shaw et al., 1980), and seafood (Earle and Snyder, 1966) have utilized coatings based on alginates, fats, gums, or starches. Patents for use in the baking industry for maintaining product crispness and/or preventing dehydration have been based primarily on emulsions of saturated fats (Silva et al., 1981; Werbin et al., 1970) and/or carbohydrate complexed fatty acid esters (Cole, 1969).

The confectionery industry has used coatings for years in their pan-coating process as a means of enhancing the appearance of candies, such as jellybeans and Jordan almonds (Silverwater, 1991). A typical coating for these confectioneries would include gum arabic, modified starch, maltodextrin, xanthan gum, waxes, corn syrup, and sucrose. Gloss can be obtained by polishing the candies with carnauba, beeswax and/or candelilla wax (Anonymous, 1989). Typically, a lacquer is added as the final coating, providing a moisture barrier that prevents the loss of gloss when the product is stored in high humidity (Silverwater, 1991).

ANTIMICROBIALS, FUNGICIDES AND NATURAL ANTAGONISTS IN EDIBLE COATINGS

ANTIMICROBIALS

Antimicrobials are added to edible coatings to retard the growth of yeasts, molds and bacteria during storage and distribution. Common antimicrobials used include benzoic acid, sodium benzoate, sorbic acid, potassium sorbate, and propionic acid.

Food quality can be affected by the ability of a food additive to diffuse from the environment into the food as well as between different regions,

TABLE 6.1. Classification, Regulation, and Allowed Use of Some Food Additives, Fungicides and Natural Antagonists (Baranowski, 1990; CFR, 1991).

Food Additive	Classification	CFR No.	Allowed Use
A. Antimicrobial Agent, Fungicides and Natural Antagonists:			
Benzoic acid	Antimicrobial agent	21 CFR 184.1021	GRAS[1]/GMP[2]
Potassium sorbate	Chemical preservative	21 CFR 182.3640	GRAS/GMP
Propionic acid	Antimicrobial agent	21 CFR 184.1081	GRAS/GMP
Sodium benzoate	Antimicrobial agent	21 CFR 184.1733	GRAS/GMP
Sorbic acid	Chemical preservative	21 CFR 182.3089	GRAS/GMP
Benomyl	Fungicide	40 CFR 180.294	
sec–Butylamine	Fungicide	40 CFR 180.321	
Captan	Fungicide	40 CFR 180.103	
2,6,Dichloro-4-nitroaniline	Fungicide	40 CFR 180.200	
Imazalil	Fungicide	40 CFR 180.413	
Iprodione	Fungicide	40 CFR 180.399	
Metalaxyl	Fungicide	40 CFR 180.408	
Sodium o-phenyl phenate	Fungicide	40 CFR 180.129	
Thiabendazole	Fungicide	40 CFR 180.242	
Thiram	Fungicide	40 CFR 180.32	
Zinc ion and manganese ethylenebisdithiocarbamate	Fungicide	40 CFR 180.176	
Bacillus subtilis	Natural antagonist		
Debaryomyces hansenii	Natural antagonist		
B. Antioxidants:			
Ascorbic acid	Chemical preservative	21 CFR 182.3013	GRAS/GMP
Ascorbyl palmitate	Chemical preservative	21 CFR 182.3149	GRAS/GMP

(continued)

123

TABLE 6.1. (continued).

Food Additive	Classification	CFR No.	Allowed Use
B. Antioxidants: *(continued)*			
BHA	Chemical preservative/antioxidant	21 CFR 182.3169	GRAS/GMP
BHT	Chemical preservative/antioxidant	21 CFR 182.3173	GRAS/GMP
Citric acid	Multipurpose food substance	21 CFR 182.1033	GRAS/GMP
	Sequestrant	21 CFR 182.6033	GRAS/GMP
Propyl gallate	Antioxidant	21 CFR 184.1660	GRAS/GMP
Tocopherols	Chemical preservative	21 CFR 182.3890	GRAS/GMP
TBHQ	Food preservative/antioxidant	21 CFR 172.185	Antioxidant
C. Miscellaneous Food Additives:			
Calcium chloride	Antimicrobial agent/firming agent	21 CFR 184.1193	GRAS/GMP
Glycerol	Multipurpose food substance	21 CFR 182.1320	GRAS/GMP
Polyethylene glycol (200–9500)	Substance for use as component of coatings	21 CFR 175.300	Plasticizer
	Multipurpose additive	21 CFR 172.820	Coating, binder, plasticizer
	Substance for use as component of coatings	21 CFR 175.300	Resinous/polymeric coatings
Propylene glycol	Substance for use as component of coatings	21 CFR 175.300	Plasticizer
Silicone	Substance for use as component of coatings	21 CFR 175.300	Release agent
Sorbitol	Substance for use as component of coatings	21 CFR 175.300	Plasticizer
Squalene	Naturally occurring C_{30} isoprene hydrocarbon	21 CFR 184.1835	GRAS/GMP
2,4-Dichlorophenoxyacetic acid	Growth regulator		
2,4,5-Trichlorophenoxyacetic acid	Growth regulator		
Gibberellic acid	Growth regulator		
Polyamines	Growth regulator		

[1]GRAS = Generally Recognized as Safe.
[2]GMP = Good Manufacturing Practices.

124

i.e., surface vs. bulk of a food system (Giannakopoulos and Guilbert, 1986a; Torres et al., 1985a). This movement of water and/or solutes can be especially important for long-term storage of foods (Guilbert, 1988; Guilbert et al., 1985) or for imparting specific desirable characteristics, such as flavor, to a food system (Kester and Fennema, 1986).

Surface microbial growth is the main cause of spoilage for many refrigerated food products (Vojdani and Torres, 1990). One method for controlling microbial growth has been through reduced water activity (a_w), producing intermediate moisture foods (IMF). Most IMF products are packaged in moisture-proof materials and are not affected by changes in the external relative humidity. However, they are affected if the sample/package is subjected to changing temperatures, which can cause a redistribution of moisture within the package and the development of condensed moisture on the food's surface. This increase in surface moisture/a_w allows surface spoilage (Torres et al., 1985a).

The effectiveness of edible films in helping to control diffusion of sorbic acid and potassium sorbate has been studied extensively by two major research groups. The first research group evaluated the effects of temperature, the agent (glycerol or sodium chloride) used to control a_w, the concentration of the gelling agent, and the method (titration vs. gas chromatography) used to measure the sorbic acid on the diffusivity (D_a) of sorbic acid in three different model systems that had a_w's ranging from 0.64 to 1.0 (Guilbert, 1988; Giannakopoulos and Guilbert, 1986a, 1986b; Guilbert et al., 1985). Their data indicated that as the temperature decreased ($10°C$) there was a marked decrease in the D_a of the sorbic acid. In addition, the authors concluded that the agent used to lower the water activity of the system did not appear to have an effect on the D_a of sorbic acid. The role of the concentration of the gelling agent was not as concise. They theorized that at higher levels of gelling agent, there was increased networking which acted to restrict the movement of the sorbic acid. In addition, higher values for sorbic acid were determined with chromatography analysis because it measured total sorbic acid content rather than only the undissociated form, as occurs with titration.

Guilbert (1988) reported the application of either casein or carnauba wax edible films, with and without sorbic acid, to papaya and apricot cubes having a_w's that ranged from 0.70 to 0.90. At a_w's greater than 0.78 there is a significant risk of yeast and mold spoilage. The coating efficiencies showed that the carnauba wax containing sorbic acid was the best, followed by the carnauba wax alone. The casein coating containing sorbic acid was superior to casein alone.

The second research group studied the permeability of potassium sorbate and sorbic acid in several different types of edible films, as well as the effect of adding fatty acids to films on the sorbic acid/sorbate permeability

(Vojdani and Torres, 1990, 1989; Torres, 1987; Torres and Karel, 1985). Overall, they reported that the permeability of the sorbic acid/sorbate was film-dependent and that the addition of fatty acids reduced permeability.

Torres and Karel (1985) reported the effect of two surface modifications—i.e., reduced surface pH and controlled surface preservative concentration—in an IMF product on the growth of *Staphylococcus aureus* S-6. Improved surface microbial stability was achieved when the IMF product was coated with a zein protein coating with an additional surface application of sorbic acid. Reduction of the surface pH of a low electrolyte IMF with a carrageenan/agarose coating increased the surface microbial stability. The reduced pH gave a 2.5-fold increase in the level of the active form of sorbic acid.

Controlling surface microbial growth has been attempted through the use of surface dips of antimicrobials, such as potassium sorbate (Robach and Sofos, 1982; Cunningham, 1979). The use of potassium sorbate dips has been shown to reduce the total number of viable bacteria in meats at both refrigeration and temperature abuse conditions. However, their effectiveness over time is limited because the preservative diffuses into the food, allowing surface organisms to grow and cause spoilage (Torres et al., 1985b).

Edible films have been extensively studied for their moisture barrier properties, and most research indicates that incorporation of lipids reduces the moisture transport of the films (Hagenmaier and Shaw, 1991, 1990; Kester and Fennema, 1989b, 1989c, 1989d). The effect of altered barrier properties on the ability of a film to control the gaseous environment of food has only been partially investigated. Available data indicate that it is possible to retard the ability of oxygen to gain access to the food (Kester and Fennema, 1989a, 1989c, 1989d) possibly creating anaerobic conditions on the food surface. If an anaerobic condition is created, the growth of *Clostridium botulinum* would be favored. Therefore, the incorporation of antimicrobials capable of controlling this pathogen into the edible film becomes important. Although the cited research findings indicate edible films can carry antimicrobials, e.g., sorbic acid, capable of controlling *Clostridium botulinum*, more research is needed to fully define the impact of antimicrobial edible films on food safety.

FUNGICIDES

Postharvest diseases of fruits and vegetables are a major expense in food production. Harvested food has a higher value than the same crop in the field because it carries the cumulative cost of production, harvesting, storage, distribution, and sales (Wilson and Pusey, 1985). Therefore, point of sale loss of such high-value commodities as fruits and vegetables has a tremendous impact on the total food production budget. Postharvest diseases of fruits and vegetables can be suppressed by low-temperature storage, a low-oxygen atmosphere, treatment with growth regulators that delay tissue senescence (Eckert and Ogawa, 1985), and various fungicidal treatments, including use of fungicides with and without waxes (Wilson and Pusey, 1985). A fungicidal treatment is not a substitute for a satisfactory storage environment, since these treatments rarely influence the rate of physiological deterioration of the perishable product (Eckert and Ogawa, 1985). Selection of an antimicrobial compound for a specific postharvest application depends upon: (1) sensitivity of the pathogen to the chemical agent, (2) ability of the chemical agent to penetrate through host surface barriers to the infection site, and (3) crop tolerance to the chemical agent from the standpoint of phytotoxicity and any adverse effect upon product quality (Eckert and Ogawa, 1985).

Although the chemical industry is mostly involved in developing fungicides for use in the field, about twenty organic compounds have been extensively evaluated as postharvest treatments during the last thirty years (Eckert, 1983). Their uses are regulated by the Environmental Protection Agency (Table 6.1). Fungicides may be applied to the harvested products either as a gas or liquid, depending on the chemical and physical properties of the chemical and the treatment compatibility with the usual crop handling procedures (Eckert and Ogawa, 1985). Today, most postharvest fungicides are applied in water-based formulations that can be sprayed or flooded onto the crop as it is conveyed through the packinghouse. After treatment, a light rinsing with fresh water removes unabsorbed fungicide but leaves a significant, effective inhibitory level of fungicide at wound sites (which, unprotected, are entrances for pathogen invasion). Organic fungicides are generally applied in hydrocarbon/solvent-based waxes. Waxes have the advantage of reducing moisture loss, which is extremely important to thin-skinned fruits, such as peaches and nectarines (Wells, 1971).

Citrus Fruit

Application of fungicides in wax-based emulsions or water suspensions has been studied primarily on citrus fruit. Citrus fruit are subject to a variety of postharvest diseases. The predominant diseases of citrus include the blue and green molds caused by *Penicillium* species, stem end rot caused by *Diplodia natalensis* and *Phomopsis citri* (Eckert and Brown, 1988; Chalutz et al., 1988; Eckert and Ogawa, 1985), and brown rot caused by *Phytophthora citrophthora* (Cohen, 1981). The fungicides that have been studied for their efficacy as surface inhibitors of the green and blue molds in citrus fruit include benomyl, imazalil, and thiabendazole (TBZ). Factors that affect the efficacy of these mold-inhibiting fungicides when applied to citrus fruit have been reported in the literature. A primary factor is the mode of application, i.e., water suspension vs. wax emulsion systems. Although higher levels of fungicide residues are found when applied in a wax emulsion, fungicides applied in a wax emulsion have reduced ability to inhibit mold growth (Brown, 1974; Eckert and Kolbezen, 1977; Brown et al., 1983; Brown, 1984). Eckert and Kolbezen (1977), in their research with TBZ, theorized that the reason for the reduced activity of a fungicide when applied in a wax-based system could be due to the fungicide being encapsulated in the wax, making it unavailable.

Brown et al. (1983), working with imazalil, found that the wax suspension containing the fungicide had greater viscosity and, therefore, the fungicide had less efficacy due to reduced mobility into the tissue of the fruit. In addition, the high pH (9.5) of the fungicide containing wax suspension reduced the lipophilicity of the fungicide and reduced the mobility of the compound into the peel of the fruit. Brown (1984) later confirmed this theory by showing that as the pH of the wax suspension increased from 6.8 to 9.5, imazalil penetration of the orange peel decreased. Although there was greater penetration of fungicides into citrus fruit when they were applied in water (Eckert and Ogawa, 1985; Brown, 1984, 1974; Brown et al., 1983), a doubling of fungicide in a wax system overcomes this loss of effectiveness (Brown, 1984). In addition, the method used for the inoculation of the fruit affected the ability of the fungicide system to penetrate the wound. When small puncture holes (0.3 mm in diameter) were used as the means of inoculating fruit, wax-based fungicide systems were unable to penetrate the wound and inhibit mold growth. However, when applied in a water-based system, the fungicide was effective regardless of puncture size (0.3 mm to 1.0 mm in diameter). If the peel was injured by sandpaper, there was equal penetration of the fungicide whether in water- or wax-based solutions. It was again theorized that the wax system's greater viscosity reduced the fungicides' ability to penetrate the small wound openings (Brown, 1984).

Radina and Eckert (1988) studied the mobility/permeability of imazalil into Valencia orange epicuticular wax. Mobility was related to wax formulation and storage temperature. No evidence was found for binding between the fungicide and the coating resin and no partitioning of the fungicide was found in the wax. They also found that the effectiveness of imazalil in controlling *Penicillium* decay of fruit varied with wax and fungicide formulations. Imazalil effectiveness increased as the emulsion solids were increased in the wax formulation.

Fungicides commonly used to protect citrus fruit against stem end rot and blue and green molds have no effect against brown rot. Cohen (1981) found that metalaxyl was effective in protecting citrus fruit against brown rot caused by *Phytophthora citrophthora*. This fungicide, however, provided limited control against the fungus inside the infected peel. When incorporated in a wax coating, metalaxyl prevented spread of the organism's mycelium to adjacent sound fruit, eliminating contact infection, but promoted the development of *Penicillium digitatum* in the infected fruit. A combination of metalaxyl and imazalil controlled both *P. citrophthora* and *P. digitatum* (Cohen, 1981).

Sorbic acid has been used in commercial citrus packinghouses to control *Penicillium* decays, usually when isolates of the pathogen are resistant to the benzimidazoles, sodium *ortho*-phenylphenate (SOPP), and *sec*-butylamine (Eckert, 1975). The addition of 2% sorbic acid to a water-based wax formulation containing thiabendazole or benomyl brought about a significant reduction in *Penicillium* decay in commercial shipments of lemons and grapefruit from packinghouses that were contaminated with benzimidazole-resistant *Penicillium* spores. However, sorbic acid must be considered a weak postharvest fungicide when compared to thiabendazole and benomyl, although it can be used when others fail due to fungicide-resistant *Penicillium* isolates (Eckert, 1975).

Stone Fruits

Storage and marketing of peaches and nectarines are limited by the incidence of postharvest diseases, including *Monilinia fructicola* and *Rhizopus stolonifer* (Wells, 1971). Postharvest decay of peaches and nectarines has been reduced with the application of fungicidal dips, fungicide-impregnated wrappers, and the combination of hot water fungicide dips. A more widely used method is a spray with a wax emulsion containing a fungicide. Waxes have been used on peaches and nectarines as a means to prevent weight/moisture loss and to enhance the fruit's appearance. Wells (1971) treated fruits with 2,6-dichloro-4-nitroaniline (DCNA) in water or in wax emulsions either as a dip or a spray and found smaller mean lesion diameters when fruits inoculated with *Monilinia fructicola* and *Rhizopus stolonifer* were treated with the wax emulsions. The application of benomyl in a wax emulsion on inoculated fruit was also more effective in reducing lesion diameters than benomyl applied in water. Within the wax emulsion treatments, it was found that the use of heated (52°C) as opposed to unheated (24°C) emulsion for the application of the fungicides improved the effectiveness of the fungicide in controlling the outgrowth of the test organisms.

Wells and Reaver (1976) studied the level of postharvest decay and weight loss of peaches as a function of hydrocooling and coating with a nonemulsified, water-insoluble blend of mineral oil, petrolatum, and paraffin wax containing 5000 ppm Botran and 2500 ppm benomyl. Results showed that fruit treated with wax containing the Botran and benomyl had a lower rate (one-sixth) of decay than uncoated fruit. Hydrocooling in combination with waxing reduced moisture loss.

Papayas

Papayas are subject to a variety of postharvest diseases. A heat treatment developed by Akamine and Arismi (1953) provided adequate protection of fruit sold locally; however, it was insufficient for fruit being transported to distant markets.

Couey and Farias (1979) applied TBZ in a carnauba wax formulation to papaya fruit for effective control of postharvest diseases, e.g., anthracnose (*Colletotrichum glocosporiodes,* Penz.), stem-end rot (*Ascochyta caricaepapayae* Tarr and *Botryodiplodia theobromae* Pat.), peduncle rot (*Fusarium* sp. and *Cladosporium* sp.), and rhizopus rot (*Rhizopus stolonifer*, Lind) for up to fourteen days at 10°C followed by ripening at room temperature. Slightly greater control of postharvest diseases of papaya was achieved when TBZ in wax was applied immediately after a short hot water dip (54°C, 1.5 minutes). SOPP used alone or in combination with TBZ did not reduce decay of papaya fruit (Couey and Farias, 1979).

Strawberries

Ghaouth et al. (1991) compared the effect of a chitosan coating containing the fungicide, iprodione (Rovral®), to an aqueous dipping into the same fungicide, on the perishability of strawberries inoculated with *Botrytis cineria* and stored at 4°C and 13°C. The application of fungicide within the coating significantly reduced decay incidence when compared to the untreated control. The coating also prevented outgrowth of mold for a longer period of time than did the use of the aqueous dip.

Tomatoes

Captan (*N*-[(trichloromethyl)thio]-4-cyclohexene 1,2-dicarboximide), dithane M45 (coordination of zinc and manganese ethylenebisdithiocarbamate), sodium *o*-phenyl phenate (OPP), and thiram [tetramethylthiuramdisulfide: bis(dimethylthiocarbamoyl disulfide)] showed potential as fungicide treatments to control *Colletotrichum, Phytophthora, Rhizopus, Botrytis, Penicillium* and other pathogens on tomatoes when applied in a wax solution. Benomyl prevented growth of the most damaging pathogens alone or in combination with other fungicides (Domenico et al., 1972).

NATURAL ANTAGONISTS

Although the application of fungicides onto fruits, whether in a wax- or water-based coating, provides protection against postharvest loss of crops, there is a growing concern among consumers about the level of chemicals in the food supply. An alternative technique for controlling the postharvest loss of produce that has been receiving attention is the use of antagonistic microorganisms, i.e., biological control (Pusey et al., 1986). A successful example of the use of biological control of a major postharvest disease is the control of brown rot (*Monilinia fructicola,* Wint. Honey) of peaches by *Bacillus subtilis* (Janisiewicz, 1987). Other instances of biological control of postharvest diseases include control of blue mold (*Penicillium expansum*) by a bacterium and a yeast. Chalutz et al. (1988) found that a microbial antagonist, *Debaryomyces hansenii,* a yeast, was effective against *Penicillium digitatum, Penicillium italicum,* and *Geotrichum candidum* on grapefruit and lemons.

Wilson and Pusey (1985) and Pusey et al. (1986) used a biological control agent, *Bacillus subtilis,* in combination with commercial fruit waxes to control brown rot in peaches. They found that although the wax in some instances appeared to inhibit biological agent activity against brown rot, success was achieved in controlling brown rot in the microbe-treated fruits. Their work indicates a good potential for the use of biological control agents in combination with fruit waxing procedures that are traditionally used in commercial packinghouses.

The use of coatings to protect fruits is not a new concept. However, the use of edible coating materials to carry antagonistic microorganisms is an area that needs more research. Ideally, coatings should control the level of the agent on the product surface so that less agent would be required to maintain product acceptability during storage.

ANTIOXIDANTS IN EDIBLE COATINGS

Antioxidants are added to edible coatings to increase stability and maintain the nutritional value and color of food products by protecting against oxidative rancidity, degradation, and discoloration. The two types of food antioxidants are acids (as well as their salts and esters) and phenolic compounds (Sherwin, 1990). Acids such as citric, ascorbic, and their esters, act as metal chelating agents or synergists when used alone or in combination with phenolic antioxidants. Phenolic compounds, such as butylated hydroxyanisole (BHA), butylated hydroxytoluene (BHT), tertiary butylated hydroxyquinone (TBHQ), propyl gallate, and tocopherols inhibit oxidation related to the fats and oils of foods. BHA, BHT, and citric acid were added to pectinate, pectate, and zein coatings to prevent rancidity and maintain desirable texture of nuts (Andres, 1984; Swenson et al., 1953).

Guilbert (1988) investigated the retention of α-tocopherol in gelatin films. α-Tocopherol (5%) was incorporated into a film containing gelatin (76%) and glycerol (19%). The blend was then left alone or was cross-linked with tannic acid solution (30% w/w) to increase the retention properties of the protein. The films were placed in contact with either margarine or an aqueous model food composed of saccharose (44%), agar (1.5%), and water (54.5%). The systems were held at 25°C and 95% relative humidity for fifty days. Results indicated that there was a rapid diffusion of the α-tocopherol from the non–cross-linked gelatin in contact with margarine; however, α-tocopherol retention was maintained in the non–cross-linked gelatin in contact with the aqueous model food. This is easily explained by the high fat solubility of α-tocopherol, especially when there was no gelatin cross-linking. Cross-linking of the gelatin significantly improved the retention of α-tocopherol in film in contact with margarine. These data indicate that it is possible to control the surface concentration of antioxidants in foods, and since lipid oxidation is a surface phenomenon, antioxidant containing films could be very effective in preventing the development of rancidity in a processed food system (Guilbert, 1988).

The infusion or aqueous dip application of antioxidants reduced the incidence of anthracnose (*Colletotrichum gloeosporioides*) or stem end rot (*Diplodia natalensis*) in avocados. It may be possible to increase avocado storability even further through controlled release of antioxidants from edible coatings (Prusky, 1988). There is even potential for correcting metabolic disorders. Edible coatings reduced enzymatic browning in whole and sliced mushrooms. Further improvement in the anti-browning property of the coating was accomplished with the incorporation of an antioxidant and a chelator (1% ascorbic acid, 0.2% calcium disodium ethylenediamine tetraacetic acid) (Nisperos-Carriedo et al., 1991).

MISCELLANEOUS FOOD ADDITIVES

There are food additives that provide functions other than those mentioned above. Plasticizers such as glycerol, polyethylene glycol, propylene glycol and sorbitol are compounds of low volatility which are added to impart flexibility and elongation to polymeric substances. They are added to films or coatings made from zein, wheat and corn proteins, amylose, amylose fatty ester, soya-protein-lipid, starch, gelatin, gum arabic, cellulose derivatives, shellac, and pectin for various applications (Gennadios and Weller, 1990; Andres, 1984; Chuah et al., 1983; Cole, 1969; Jokay et al., 1967; Schultz et al., 1949). Plasticizers function by weakening intermolecular forces between adjacent polymer chains, resulting in decreased tensile strength and increased film flexibility (Banker, 1966).

The addition of surface-active agents (surfactants) and emulsifiers to edible coatings reduces superficial a_w and rate of moisture loss in food products (Roth and Loncin, 1984). This group of food additives is discussed in Chapter 10.

Release agents and lubricants are substances added to prevent coated food products from sticking. These include fats and oils, emulsifiers, petrolatum, polyethylene glycol, and silicone. Polyethylene glycol is used as a release agent in food tablets while silicone is a component of resinous/polymeric coatings.

Calcium chloride, a firming agent, is incorporated in coatings to improve the texture and color of food products. Solutions of calcium chloride induce gelation of alginate and low-methoxyl pectin coatings (Earle, 1968; Miers et al., 1953; Swenson et al., 1953). Adding calcium chloride to gum acacia or gum acacia/gelatin coatings controlled after-cooking darkening of potatoes (Mazza and Qi, 1991).

There is evidence that edible films/coatings could be improved by the addition of plant hormones, such as 2,4-dichlorophenoxyacetic acid (2,4-D), 2,4,5-trichlorophenoxyacetic acid (2,4,5-T), gibberellic acid, and polyamines to fresh produce. These compounds have been shown to delay ripening and senescence in citrus and tomato fruits (Evans and Malmberg, 1989; Sharma et al., 1989; Kader et al., 1966; Lodh et al., 1963).

Squalene, a natural component of the epicuticular wax of grapefruit, helped prevent chilling injury when applied to grapefruit surfaces (Nordby and McDonald, 1991, 1990). The use of a coating to maintain squalene on the surface of chilling-sensitive produce could minimize losses encountered during low-temperature storage.

In summary, this research area, using edible films and coatings to carry food additives, fungicides, and natural antagonists is just beginning. There are no boundaries—only possibilities.

REFERENCES

Allen, L., A. I. Nelson, M. P. Steinberg and J. N. McGill. 1963. "Edible Corn-Carbohydrate Food Coatings. II. Evaluation of Fresh Meat Products," *Food Technol.*, 17(11):104–108.

Akamine, E. K. and T. Arisumi. 1953. "Control of Post-Harvest Storage Decay of Fruits of Papaya (*Carica papaya* L.) with Special Reference to the Effect of Hot Water," *Proc. Amer. Soc. Hort. Sci.*, 61:270–274.

Anonymous. 1989. "How to Optimize Your Glazing Process," in *Sweet Times, the Confectioner's Communique, Vol. 1, Number 2.* Tarrytown, NY: Centerchem, Inc., pp. 1–3.

Banker, G. S. 1966. "Film Coating Theory and Practice," *J. Pharm. Sci.*, 55:81.

Baranowski, E. S. 1990. "Miscellaneous Food Additives," in *Food Additives*, A. L. Branen, P. M. Davidson and S. Salminen, eds., New York, NY: Marcel Dekker, Inc., pp. 511–579.

Bauer, C. D., G. L. Neuser and H. A. Pinkalla. October 15, 1968. U.S. patent 3,406,081.

Brown, G. E. 1974. "Benomyl Residues in Valencia Oranges from Postharvest Applications Containing Emulsified Oil," *Phytopathol.*, 64:539–542.

Brown, G. E. 1984. "Efficacy of Citrus Postharvest Fungicides Applied in Water or Resin Solution Water Wax," *Plant Disease*, 68:415–418.

Brown, G. E., S. Nagy and M. Maraulia. 1983. "Residues from Postharvest Nonrecovery Spray Applications of Imazalil to Oranges and Effects on Green Mold Caused by *Penicillium digitatum*," *Plant Disease*, 67:954–959.

Chalutz, E., L. Cohen, B. Weiss and C. L. Wilson. 1988. "Biocontrol of Postharvest Diseases of Citrus Fruit by Microbial Antagonists," *Citriculture Proceedings of the Sixth International Citrus Congress*, R. Cohen and K. Mendel, eds., Philadelphia, PA: Balaban Publishers, pp. 1467–1470.

Chuah, E. C., A. Z. Idrus, C. L. Lim and C. C. Seow. 1983. "Development of an Improved Soya Protein-Lipid Film," *J. Food Technol.*, 18:619–627.

Cohen, E. 1981. "Metalaxyl for Postharvest Control of Brown Rot of Citrus Fruit," *Plant Disease*, 65:672–675.

Cole, M. S. November 18, 1969. U.S. Patent 3,479,191.

Couey, H. M. and G. Farias. 1979. "Control of Postharvest Decay of Papaya," *HortSci.*, 14(6):719–721.

Cunningham, F. E. 1979. "Shelf Life and Quality Characteristics of Poultry Meat Parts Dipped in Potassium Sorbate," *J. Food Sci.*, 44:863–866.

Domenico, J. A., A. R. Rahman and D. E. Westcott. 1972. "Effects of Fungicides in Combination with Hot Water and Wax on the Shelf Life of Tomato Fruit," *J. Food Sci.*, 37:957–960.

Earle, R. D. July 30, 1968. U.S. Patent 3,395,024.

Earle, R. D. and C. E. Snyder. June 2, 1966. U.S. patent 3,255,021.

Eckert, J. W. 1975. "Postharvest Pathology. Part 1. General Principles," in *Postharvest Physiology, Handling and Utilization of Tropical and Subtropical Fruits and Vegetables*, E. R. Pantastico, ed., Westport, CT: AVI Publishing Co., pp. 393–414.

Eckert, J. W. 1983. "Control of Post-Harvest Diseases with Antimicrobial Agents," in *Postharvest Phyiology and Crop Preservation*, M. Lieberman, ed., New York, NY: Plenum Press, pp. 265–285.

Eckert, J. W. and E. G. Brown. 1986. "Postharvest Citrus Diseases and Their Control," in *Fresh Citrus Fruits*, W. F. Wardowski, S. Nagy and W. Grierson, eds., Westport, CT: AVI Publishing Co., pp. 315–353.

Eckert, J. W. and M. J. Kolbezen. 1977. "Influence of Formulation and Application Method on the Effectiveness of Benzimidazole Fungicides for Controlling Postharvest Diseases of Citrus Fruits," *Neth. J. Plant Path.*, 83(supp):343–352.

Eckert, J. W. and J. M. Ogawa. 1985. "The Chemical Control of Postharvest Diseases: Subtropical and Tropical Fruits," *Ann. Rev. Phytopathol.*, 23:421–454.

Evans, P. T. and R. L. Malmberg. 1989. "Do Polyamines Have Roles in Plant Development," *Annu. Rev. Plant Physiol. Plant Mol. Biol.*, 40:235–269.

FDA. 1991. "Title 21, Food and Drugs," Washington, DC: Federal Register, National Archives and Records Administration.

Ghaouth, A. E., J. Arul, R. Ponnampalam and M. Boulet. 1991. "Chitosan Coating Effect on Storability and Quality of Fresh Strawberries," *J. Food Sci.*, 56(6):1618–1631.

Giannakopoulos, A. and S. Guilbert. 1986a. "Determination of Sorbic Acid Diffusivity in Model Food Systems," *J. Food Technol.*, 21:339–353.

Giannakopoulos, A. and S. Guilbert. 1986b. "Sorbic Acid Diffusivity in Relation to the Composition of High and Intermediate Moisture Model Gels and Foods," *J. Food Technol.*, 21:477–485.

Grennadios, A. and C. L. Weller. 1990. "Edible Films and Coatings from Wheat and Corn Proteins," *Food Technol.*, 43:63–69.

Guilbert, S. 1988. "Use of Superficial Edible Layer to Protect Intermediate Moisture Foods: Application to the Protection of Tropical Fruit Dehydrated by Osmosis,"in *Food Preservation by Moisture Control*, C. C. Seow, ed., London, UK: Elsevier Applied Science Publishers, Ltd., pp. 119–219.

Guilbert, S., A. Giannakopoulos and J. C. Cheftel. 1985. "Diffusivity of Sorbic Acid in Food Gels at High and Intermediate Water Activities," in *Properties of Water in Foods*, D. Simatos and J. L. Multon, eds., Dordrecht, Netherlands: Martinus Nijhoff Pub., pp. 343–356.

Hagenmaier, R. D. and P. E. Shaw. 1990. "Moisture Permeability of Edible Films Made with Fatty Acids and (Hydroxypropyl)-Methylcellulose," *Agric. Food Chem.*, 38:1799–1803.

Hagenmaier, R. D. and P. E. Shaw. 1991. "The Permeance of Shellac Coatings," *J. Agric. Food Chem.*, 39:825–829.

Janisiewicz, W. J. 1987. "Postharvest Biological Control of Blue Mold on Apples," *Phytopathol.*, 77:481–485.

Jokay, L., G. E. Nelson and E. L. Powell. 1967. "Development of Edible Amylaceous Coatings for Foods," *Food Technol.*, 21:12–14.

Kader, A. A., L. L. Morris and E. C. Maxie. 1966. "Effect of Growth-Regulating Substances on the Ripening and Shelf-Life of Tomatoes," *HortSci.*, 1:90–91.

Kester, J. J. and O. R. Fennema. 1986. "Edible Films and Coatings. A Review," *Food Technol.*, 40(12):47–59.

Kester, J. J. and O. R. Fennema. 1989a. "Resistance of Lipid Films to Water Vapor Transmission," *J. Amer. Oil Chem. Soc.*, 66:1129–1138.

Kester, J. J. and O. R. Fennema. 1989b. "Resistance of Lipid Films to Water Vapor Transmission," *J. Amer. Oil Chem. Soc.*, 66:1139–1146.

Kester, J. J. and O. R. Fennema. 1989c. "The Influence of Polymorphic Form on Oxygen and Water Vapor Transmission through Lipid Films," *J. Amer. Oil Chem. Soc.*, 66:1147–1153.

Kester, J. J. and O. R. Fennema. 1989d. "Tempering Influence on Oxygen and Water Vapor Transmission through Stearyl Alcohol Film," *J. Amer. Oil Chem. Soc.*, 66:1154–1157.

Lodh, B. B., S. De, S. K. Mukherjee and A. W. Bose. 1963. "Storage of Mandarin Oranges. II. Effects of Hormones and Wax Coatings," *J. Food Sci.*, 28:519–524.

Mazza, G. and H. Qi. 1991. "Control of After-Darkening in Potatoes with Edible Film-Forming Products and Calcium Chloride," *J. Agric. Food Chem.*, 39:2163–2166.

Nisperos-Carriedo, M. O., E. A. Baldwin and P. E. Shaw. 1991. Development of an

Edible Coating for Extending Postharvest Life of Selected Fruits and Vegetables," *Proc. Fla. State Hort. Soc.*, 104:122–125.

Nordby, H. E. and R. E. MacDonald. 1990. "Squalene, a Possible Protectant from Chilling Injury in Grapefruit," *Lipids*, 25:807–810.

Nordby, H. E. and R. E. MacDonald. 1991. Relationship of Epicuticular Wax Composition of Grapefruit to Chilling Injury," *J. Agric. Food Chem.*, 39:381–384.

Prusky, D. 1988. "The Use of Antioxidants to Delay the Onset of Anthracnose and Stem End Decay in Avocado Fruit after Harvest," *Plant Disease*, 72:381–384.

Pusey, P. L., C. L. Wilson, M. W. Hotchkiss and J. D. Franklin. 1986. "Compatibility of *Bacillus subtilis* for Postharvest Control of Peach Brown Rot with Commercial Fruit Waxes, Dichloran, and Cold Storage," *Plant Disease*, 70:587–590.

Radina, P. M. and J. W. Eckert. 1988. "Evaluation of Imazalil Efficacy in Relation to Fungicide Formulation and Wax Formulation," *Citriculture Proceedings of the Sixth International Citrus Congress*, R. Cohen and K. Mendel, eds., Philadelphia, PA: Balaban Publishers, pp. 1427–1434.

Robach, M. C. and J. N. Sofos. 1982. "Use of Sorbates in Meat Products, Fresh Poultry and Poultry Products. A Review," *J. Food Protect.*, 41:284–293.

Roth, T. and M. Loncin. 1984. "Superficial Activity of Water," in *Engineering and Food, Vol. 1*, B. M. Mckenna, ed., London, England: Elsevier Applied Science Publishers, p. 433.

Sharma, R. K., R. Singh, J. Kumar and S. S. Sharma. 1989. "Shelf Life of Baramasi Lemon as Affected by Some Chemicals," *Res. and Dev. Reporter*, 6:78–82.

Shaw, C. P., J. L. Secrist and J. M. Tuomy. April 1, 1980. U.S. patent 4,196,219.

Sherwin, E. R. 1990. "Antioxidants," in *Food Additives*, A. L. Branen, P. M. Davidson and S. Salminen, eds., New York, NY: Marcell Dekker, Inc., pp. 139–193.

Silva, R., D. J. Ash and C. E. Scheible. October 6, 1981. U.S. patent 4,293,572.

Silverwater, B. 1991. Personal communication.

Smith, S., J. Geeson and J. Stow. 1987. "Production of Modified Atmospheres in Deciduous Fruits by the Use of Films and Coatings," *Hort Sci.*, 22(5):772–776.

Torres, J. A. 1987. "Microbial Stabilization of Intermediate Moisture Food Surfaces," in *Water Activity: Theory and Applications to Food*, L. B. Rockland and L. R. Beuchat, eds., New York, NY: Marcel Dekker, Inc., pp. 329–368.

Torres, J. A. and M. Karel. 1985. "Microbial Stabilization of Intermediate Moisture Food Surfaces. III. Effects of Surface Preservative Concentration and Surface pH Control on Microbial Stability of an Intermediate Moisture Cheese Analog," *J. Food Proc. Preserv.*, 9:107–119.

Torres, J. A., M. Motoki and M. Karel. 1985a. "Microbial Stabilization of Intermediate Moisture Food Surfaces. I. Control of Surface Preservative Concentration," *J. Food Proc. Preserv.*, 9:75–92.

Torres, J. A., J. O. Bouzas and M. Karel. 1985b. "Microbial Stabilization of Intermediate Moisture Food Surfaces. II. Control of Surface pH," *J. Food Proc. Preserv.*, 9:93–106.

Torres, J. A., J. O. Bouzas and M. Karel. 1989. "Sorbic Acid Stability during Processing of an Intermediate Moisture Cheese Analog," *J. Food Proc. Preserv.*, 13:409–415.

Vojdani, F. and J. A. Torres. 1989. "Potassium Sorbate Permeability of Polysaccharide Films, Chitosan, Methylcellulose and Hydroxypropyl Methylcellulose," *J. Food Proc.*, 12:33–48.

Vojdani, F. and J. A. Torres. 1990. "Potassium Sorbate Permeability of Methylcellulose and Hydroxypropyl Methylcellulose Coatings," *J. Food Sci.*, 53:841–846.

Wells, J. M. 1971. "Heated Wax-Emulsions with Benomyl and 2,6-Dichloro-4-Nitroaniline for Control of Postharvest Decay of Peaches and Nectarines," *Phytopath.*, 62:129–133.

Wells, J. M. and L. G. Reaver. 1976. "Hydrocooling Peaches after Waxing: Effects on Fungicide Residues, Decay Development and Moisture Loss," *HortSci.*, 11:107–108.

Werbin, S., I. H. Rubenstein, S. Weinstein and D. Weinstein. September 1, 1970. U.S. patent 3,526,515.

Wilson, C. L. and P. L. Pusey. 1985. "Potential for Biological Control of Postharvest Plant Diseases," *Plant Disease*, 69:375–378.

CHAPTER 7

Permeability Properties of Edible Films

TARA HABIG McHUGH[1]
JOHN M. KROCHTA[2]

INTRODUCTION

THE importance of accurate methodologies for determining edible film permeability properties cannot be overemphasized; permeability values can be used in product shelf-life predictions, as well as in tailoring film permeability properties for specific food applications. Therefore, this chapter compares various methodologies employed for the determination of water vapor, gas, and lipid permeabilities. Detailed descriptions of various mathematical equations used in permeability analysis are presented, followed by an examination of the mechanisms of moisture and permanent gas transfer through edible films. Solubility and diffusivity processes are also reviewed, as are factors affecting permeability. Comparisons of the gas and water vapor permeability values of various edible films are also presented.

By employing an integrated perspective of mass transfer theory, polymer science, and food science, it is hoped that this chapter will shed some light on a number of anomalous behaviors commonly observed upon examination of edible film permeability properties. The majority of these anomalous behaviors arise due to the hydrophilic nature of most edible films, as well as the presence of inhomogeneities in their film structures.

This chapter is by no means comprehensive. It is an overview of complications arising from attempts to determine permeability properties of

[1]U.S. Dept. of Agriculture, Agricultural Research Service, Western Regional Research Center, 800 Buchanan St., Albany, CA 94710.
[2]Departments of Food Science and Technology, and Biological and Agricultural Engineering, University of California, Davis, CA 95616.

edible films and suggestions for researchers as they deal with these complications in an attempt to better characterize edible film permeability properties.

PERMEABILITY OF EDIBLE FILMS TO WATER VAPOR

WATER VAPOR PERMEABILITY (WVP)

Water vapor permeability (WVP) is a measure of the ease with which a material can be penetrated by water vapor. ASTM E96-80 further defines permeability as the rate of water vapor transmission through a unit area of flat material of unit thickness induced by a unit vapor pressure difference between two specific surfaces, under specified temperature and humidity conditions (ASTM, 1980). Permeability should not be confused with transport through pores. True permeability consists of a process of solution and diffusion where the vapor dissolves on one side of the film and then diffuses through to the other side. In the presence of macroscopic cracks or holes, transport through pores occurs and the water vapor flows directly through pinholes (Sacharow, 1971).

Water vapor permeability is generally expressed by composite units as shown in Equation (7.1).

$$\frac{(\text{amount of water vapor})(\text{thickness of film})}{(\text{film area})(\text{time})(\text{differential partial pressure of water vapor})} \quad (7.1)$$

Numerous combinations of different units are found in the literature. This poses difficulties when comparing permeability data from different sources. Common conversion values are listed in Table 7.1.

MEASUREMENT OF BARRIER PROPERTIES OF EDIBLE FILMS TO WATER VAPOR

Gravimetric Techniques

Gravimetric techniques are commonly employed for determining WVP of edible films. These tests have been standardized by ASTM E96 (1980). This chapter will focus on the ASTM E96 method, as well as improvements to it. Two versions of this technique, the Desiccant Method and the Water Method, are available for the measurement of WVP. In the Desiccant Method, the test specimen is sealed over the mouth of a test dish containing desiccant, and the assembly is placed in a controlled temperature and relative humidity (RH) chamber. In the Water Method, the test dish

TABLE 7.1. Conversion Factors for Commonly Used Water Vapor Permeability Units. All Conversions Assume Standard Temperature and Pressure Conditions.

Units Given (A)	Units Desired (Multiply A by the Values Listed Below)				
	$g \cdot mm/m^2 \cdot d \cdot kPa$	$g \cdot mil/m^2 \cdot d \cdot mmHg$	$g \cdot mil/100\ in.^2 \cdot s \cdot cmHg$	$g/m \cdot s \cdot Pa$	$cm^3 \cdot cm/cm^2 \cdot s \cdot cmHg$
$g \cdot mm/m^2 \cdot d \cdot kPa$	1	5.25	3.91×10^{-5}	1.16×10^{-11}	1.92×10^{-7}
$g \cdot mil/m^2 \cdot d \cdot mmHg$	1.90×10^{-1}	1	7.46×10^{-6}	2.20×10^{-12}	3.66×10^{-8}
$g \cdot mil/100\ in.^2 \cdot s \cdot cmHg$	2.25×10^4	1.34×10^5	1	2.95×10^{-7}	4.90×10^{-3}
$g/m \cdot s \cdot Pa$	8.62×10^{10}	4.54×10^{11}	3.39×10^6	1	1.66×10^4
$cm^3 \cdot cm/cm^2 \cdot s \cdot cmHg$	5.20×10^6	2.73×10^7	2.04×10^2	6.02×10^{-5}	1

contains distilled water or saturated salt solution. For both cases, periodic weighings determine the rate of water transmission through the specimen. These weighings must be taken after a steady state rate of moisture transmission is obtained, usually several hours after the test cells have equilibrated in the chambers. Once steady state moisture transfer is obtained, weights can be taken at three hour intervals until the steady state rate is accurately determined. Eight measurements are usually sufficient (ASTM, 1980).

Figure 7.1 depicts a model test dish. It must be made of a noncorroding, water-impermeable material. Furthermore, the mouth of the dish must be at least 4.65 in.^{2}, and the desiccant or water area should not be less than the mouth area (ASTM, 1980). Finally, it is important to emphasize the importance of maintaining the RH conditions desired inside the test cell.

The stagnant air gap between the test specimen and the solution or desiccant in the cup can result in significant differences between the solution or desiccant vapor pressure, p_{A0}, and the true experimental vapor partial pressure condition, p_{A1}, that the underside of the film experiences inside the cup. ASTM assumes that resistance to mass transfer inside the cup is negligible. However, this is not true in the case of some edible films which have low resistance to mass transfer. McHugh et al. (1993) developed the WVP Correction Method, which enables elimination of such errors. Use of this correction will be discussed in further detail later in this chapter.

It is also important that the vapor pressure, p_{A0}, at the surface of the test solution or desiccant inside the cup remains constant throughout the test. Therefore, the test solution or desiccant inside the test cell must not lose or gain enough moisture to change p_{A0} during the experiment. Care must be taken that the test dish is deep enough to contain adequate solution or desiccant in order to allow for the maintenance of a constant p_{A0}. Test cups

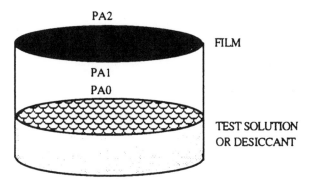

Figure 7.1 Test cup.

can either be produced in a mechanical shop or purchased from Fisher Scientific, Inc. (Fair Lawn, NJ).

Another important aspect of test dish specifications concerns sealing the film specimen to the test dish. It is crucial that proper sealing take place. If not, significant errors may result, as was shown in a 1985 round-robin test series using the ASTM E96-80 procedure (Toas, 1989). It is recommended by ASTM that films be sealed to test dishes using molten wax. However, McHugh and Krochta found that sealing with silicon vacuum grease is equally effective, provided a sufficient amount of vacuum grease is employed so as to prevent leakage of moisture through seals (unpublished research). Furthermore, silicon vacuum grease is much easier to work with than molten wax when sealing edible film materials.

The controlled temperature and RH test chamber constitutes the second critical part of the test apparatus for the determination of WVP. Researchers often fail to comply with established ASTM standards, resulting in significant errors in permeability determinations. Adequate air circulation is crucial during testing in order to maintain uniform conditions at all test locations. Air circulation is also necessary to increase the mass transfer coefficient above the cup and insure that the ASTM assumption of negligible resistance above the film is valid. ASTM recommends that air velocities over the specimen be at least 500 ft/min. McHugh et al. (1993) verified this requirement while investigating the WVP properties of edible films. Air circulation rates in the range of 200–300 ft/min resulted in lowered experimental permeability values using the Water Method, due to inadequate maintenance of vapor partial pressure, p_{A2}, at the upper side of the film. Data collected using conventional glass desiccators without air circulation as test chambers resulted in up to 35% errors in permeability determinations, due to inadequate control of relative humidity within the desiccators and low mass transfer coefficients above the cups. These results emphasize the importance of maintaining 500 ft/min air velocities and monitoring RH conditions inside the test chamber during the experiment. It is recommended that an accurate hygrometer be placed in the chamber as a monitoring device and that the relative humidity be maintained within ±2%. It is also important that the temperature be maintained to within ±2°C throughout the experiment. Adherence to the above guidelines is crucial for accurate WVP determinations of edible films (McHugh et al., 1993; Toas, 1989; ASTM, 1980).

Several test cabinets have been described in the literature. Packaging and Allied Trades Research Association (PATRA) cabinets have been employed for WVP determinations (Davis, 1965). General Foods, Inc. developed a GFMVT cabinet that meets the requirements discussed above (Unknown, 1942). The use of glass desiccators containing fans was described by Schultz et al. (1949); however, they failed to monitor RH conditions during their experiments. Modified Labmaster Model D4 desiccator cabinets housed in constant temperature rooms have also been used (Davis, 1965). These are fitted with an electric fan to promote air circulation. Shallow polyethylene trays are placed on each shelf to hold saturated salt solutions or desiccant. McHugh et al. (1993) developed a system similar to that described above using Fisher Scientific, Inc. (Fairlawn, NJ) desiccator cabinets. Fans and variable speed motors were installed to control air circulation. Once again, the importance of adequate air circulation cannot be overemphasized. The use of laboratory desiccators with no provision for air circulation is liable to give erroneous results (Toas, 1989; Davis, 1965).

It is also important to call attention to the accurate control of RH using saturated salt solutions. The use of desiccant (e.g., calcium sulfate) or distilled water to control RH at 0% or 100%, respectively, is relatively simple. Also, water is especially resistant to temperature fluctuations. Preparation of saturated salt solutions is more complicated. ASTM defines standard procedures for maintaining constant RH by means of aqueous solutions and it is recommended that researchers follow these guidelines when preparing salt solutions for the determination of WVPs (ASTM, 1985). Tables of RH values exhibited by various saturated salt solutions at different temperatures can be found in a variety of sources (Kitic et al., 1986; Stoloff, 1978; Rockland, 1960; Wink and Sears, 1950).

The final point that needs to be mentioned concerns specifications for the balance used for determining test cell weight changes. The balance should be sensitive to changes less than 1% of the weight change during the period when moisture transfer attains a steady state.

The advantages of gravimetric techniques are (1) allowance for control of experimental parameters, (2) use of the WVP Correction Method (to be discussed later in the chapter), and (3) relatively low cost. The greatest disadvantage of the gravimetric technique is the time required to obtain results. Tests take anywhere from days to several weeks, depending on the permeability of the test specimen. Most edible films are relatively poor moisture barriers; therefore, test times range from one to four days. Provided all guidelines discussed above are followed, the gravimetric technique can result in highly accurate permeability data.

In the section "Mathematical Models for Permeability Analysis," methods for analysis and calculation of results will be discussed, as well as several complications arising when one employs gravimetric techniques to determine the WVPs of edible films.

Infrared Sensor Technique

Instruments capable of rapidly yielding WVP results are produced by Modern Controls, Inc. (MOCON) (Minneapolis, MN). The two main

MOCON systems are the Infrared Diffusometer (IRD) series and the Permatran-W series. The IRD method was approved by ASTM in 1973 for the testing of flexible barrier materials in film or sheet form (ASTM, 1973). This test method consists of a dry chamber (0% RH) separated by the material to be tested from a chamber of known temperature and RH. The differential between two bands in the infrared spectral region, one in which water molecules absorb and the other where they do not, is monitored to determine the time for a given increase in water vapor concentration of the dry chamber. Next, the water vapor movement per unit area of material is calculated. It is important to note that ASTM reminds users of this method that the inverse relationship between water vapor transmission rate (WVTR) and thickness as well as the direct relationship between WVTR and partial pressure differential of the water vapor are not always linear. Therefore, water vapor permeability values must be used with caution.

Recently, ASTM approved the Permatran-W method for determining the WVP through flexible barrier materials (ASTM, 1990). This method also consists of a permeability cell containing a dry chamber (0% RH) separated by the material to be tested from a chamber of known temperature and RH. The dry chamber is swept with a stream of dry air. Water vapor diffusing through the film enters the dry air stream and is carried to a pressure-modulated infrared sensor, which measures the fraction of infrared energy absorbed by the water vapor. In turn, an electrical signal is produced with an amplitude proportional to the water vapor concentration. This signal amplitude is compared to that of a standard film of known transmission rate from which the rate of moisture transmitted is calculated (ASTM, 1990). The Permatran-W series of instruments improved upon the IRD system and is therefore more reliable for the testing of edible film materials. However, the system does have limitations which arise because many edible films are poor water barriers. Limits on the test range of WVTRs are shown in Table 7.2. The standard test sample size is a 3-in. diameter disc.

By masking the sample area, greater WVTRs can be determined without overloading the sensor. For films of even greater WVTRs, a high transmitter switch can be installed onto any Permatran-W series instrument.

TABLE 7.2. Specifications for WVTR Test Range of
MOCON Permatran-W Series of Instruments.

Method	WVTR (g/m²/day)
Standard	0.01–60
Masked	60–600
Hi-Transmitter Switch	600–5000

This switch functions by decreasing the intensity of the light source, thereby allowing for the testing of WVTR up to 5000 g/m²/day. Many edible films exhibit transmission rates between 1000–2000 g/m²/day, making this Hi-Transmitter Switch a useful option for data acquisition. However, MOCON has warned Permatran-W operators of the possibility, at very high permeation rates, that water contacting the sensor could condense or block the tubing. To avoid these pitfalls, MOCON recommends that operators maintain slightly elevated room temperatures and high flow rates. Condensation would result in inaccurate WVP values.

Several edible film researchers have used Permatran-W series equipment to determine WVPs. Gennadios et al. (1990) tested poor water barrier wheat protein films using the Permatran-W6000 instrument. Low RH test conditions of 11.1% were necessary to prevent overloading of instrument sensors, limiting applicability of the data. Kester and Fennema (1989c), found that the Permatran-W1A could be used to determine WVTRs of relatively good water barrier lipid films using masking to reduce the film area tested.

In conclusion, the Permatran-W series of instruments offers an alternative method to the traditional gravimetric techniques for determining water vapor permeabilities of edible films. The major advantage is the rapidity with which data can be collected. However, high cost and possibility of on-line condensation problems represent major disadvantages of the system. Masking of films and lowering of experimental RH gradients may solve condensation problems. These modifications also potentially solve the limitations of the sensors to high WVTR conditions, as does the installation of a high transmitter switch. Accurate control and monitoring of RH conditions in MOCON diffusion cells is questionable. Unfortunately, MOCON has installed no hygrometer sensors to insure that the dry air steam gas maintains 0% RH. Furthermore, RH conditions on the wet sponge used in the wet chamber of the permeability cell are not experimentally monitored and therefore may be subject to error when testing low barrier materials such as many edible films. The low water vapor resistance of many edible films may result in inaccurate WVP measurements, due to p_{A1} differing from p_{A0} as was apparent in the gravimetric methods. Another disadvantage to using the Permatran-W system is that it is limited to experimental conditions where $p_{A2} = 0$. Furthermore, the sponge used inside the test cup may also exhibit effects on p_{A0} when testing hydrophilic edible films, if solution relative humidity conditions are not properly maintained. Inadequate temperature control in MOCON systems represents one further disadvantage of these systems. The Permatran-W systems are built for the testing of polymeric barrier materials and must be carefully examined when testing hydrophilic edible films.

Availability of accurate standards is essential for proper examination of instrument accuracy. The most commonly used standard for examination of WVPs is the National Bureau of Standards (NBS) 1470 standard polyester reference material.

Coulometric Cells

The Minneapolis Honeywell Tester developed by the St. Regis Company can rapidly determine water vapor barrier properties of packaging materials. The Honeywell line is manufactured by Thwing-Albert Co., Philadelphia, PA. This instrument operates on the principle of a changing humidity in a controlled air space above the barrier material. As in the MOCON systems, the barrier material is placed between two chambers. The bottom chamber contains a layer of water which maintains a 100% RH. The top chamber contains a RH sensor consisting of two platinum wire electrodes wound around a glass tube coated with a thin layer of phosphorus pentoxide. Water vapor permeating through the specimen into the upper chamber is absorbed by the phosphorus pentoxide. The absorbed moisture is then electrolyzed and decomposed into hydrogen and oxygen. Changes in electrical resistance are measured to determine RH changes. The rate of water vapor movement from the bottom chamber through the test material to the upper chamber is monitored by the rate of increase in RH in the upper chamber over time. Cost and sensor WVTR limitations must be considered before judging the acceptability of this method. It also should be noted that hydrogen and oxygen in the upper cell may recombine to water vapor and increase the pressure in the cell during preconditioning and testing (Sacharow, 1971; Lelie, 1969).

Spectrophotometric Techniques

Holland and Santangelo (1984) developed a spectrophotometric technique for determining WVPs of films. In this method, a detector film is clamped between two pieces of test film in a simple cell. The detector film is a bright blue film made of regenerated cellulose soaked in a 2.6–2.8 M aqueous solution of cobalt chloride, air dried, and stored in a desiccator. The blue color fades in proportion to the amount of water vapor present, and the fading can be monitored with a spectrophotometer at 690 nm. This method works well for nonpolar polymetric water barrier materials. However, one limitation to its use is the fact that the detector film has to be more permeable than the test film, or measurement of permeation rates may be impossible. Another disadvantage is the requirement that the test film be translucent, allowing light to reach the detector film. The major advantage claimed for this method is its supposed sensitivity. However, the above considerations must be taken into account before using this method to determine WVPs of edible films (Holland and Santangelo, 1984).

Gas Chromatographic Systems

Gas chromatography (GC) can be employed to determine WVPs of edible films. Currently, GC is used primarily to determine gas permeabilities of films. Therefore, this topic will be discussed in more detail later in this chapter. However, it is important to mention at this point that similar systems are available for determining WVPs of films, and some of these systems have the added ability to examine gas and WVPs simultaneously.

Summary

Most of the methods available for determining WVPs of edible films have been described in this section, along with their respective advantages and disadvantages. Accurate control and monitoring of RH, proper sealing procedures and vigorous air circulation are crucial if accurate permeability data are to be obtained. This section is by no means comprehensive; therefore, researchers are urged to carefully examine all methods as to their applicability for testing edible films, because edible films often pose complications not observed when testing nonpolar polymeric packaging materials.

PERMEABILITY OF EDIBLE FILMS TO PERMANENT GASES

GAS PERMEABILITY

Permanent gases are substances that exist in the gaseous form under standard conditions of temperature and pressure (0°C, 1 atm). Most edible film researchers are interested in the oxygen and carbon dioxide permeabilities because these properties have a bearing on rates of oxidation and respiration in the enclosed foods.

Permeability of films to permanent gases is defined in the same way as that for WVP in the previous section. However, the units for gas permeability differ from those for WVP. A conversion table for commonly used units is provided in Table 7.3.

MEASUREMENT OF BARRIER PROPERTIES OF EDIBLE FILMS TO PERMANENT GASES

Manometric Methods

Prior to 1975, the main methods for determining oxygen permeabilities of films were described in the ASTM D1434. These were manometric and

TABLE 7.3. Conversion Factors for Commonly Used Oxygen Permeability Units. All Conversions Assume Standard Temperature and Pressure Conditions.

Units Given (A)	Units Desired (Multiply A by the Values Listed Below)				
	$g/m \cdot s \cdot Pa$	$mol/m \cdot s \cdot Pa$	$cm^3 \cdot \mu m/m^2 \cdot d \cdot kPa$	$cm^3 \cdot mil/m^2 \cdot d \cdot atm$	$cm^3 \cdot mil/100 \ in.^2 \cdot d \cdot atm$
$g/m \cdot s \cdot Pa$	1	3.12×10^{-2}	6.05×10^{16}	2.41×10^{17}	1.56×10^{16}
$mol/m \cdot s \cdot Pa$	3.2×10^{1}	1	1.94×10^{18}	7.72×10^{18}	4.98×10^{17}
$cm^3 \cdot \mu m/m^2 \cdot d \cdot kPa$	1.65×10^{-17}	5.15×10^{-19}	1	3.99	2.57×10^{-1}
$cm^3 \cdot mil/m^2 \cdot d \cdot atm$	4.14×10^{-18}	1.30×10^{-19}	2.51×10^{-1}	1	15.5
$cm^3 \ mil/100 \ in.^2 \cdot d \cdot atm$	6.42×10^{-17}	2.01×10^{-18}	3.89	6.45×10^{-2}	1

volumetric methods, both of which are applicable only when testing flat materials at 0% RH conditions. The manometric method is commonly referred to as the Dow Cell technique (ASTM, 1988). Various Dow gas transmission cell consoles are available for purchase from Custom Scientific Instruments, Inc. (Cedar Knolls, NJ).

The sample is mounted in a gas transmission cell between two chambers. One side is pressurized with the test gas and the other side is evacuated. The transmission rate through the film is determined by measuring changes in the capillary pressure and making a suitable interpretation based on calibration data. The latter is obtained by measuring the volume of gas through a specimen at a specified temperature (and at a static and essentially constant pressure) transmitted through a standard specimen into the initially evacuated calibrated manometer. A detailed description of this method can be found in the ASTM D1434 (ASTM, 1988).

This method has several drawbacks. Calibration can be very difficult. Furthermore, the method is prone to temperature drift, and the unnatural hydrostatic pressure conditions placed on the sample cause complications when applying data to real situations. Pressure conditions may cause problems when testing many edible films, due to their fragility. Results from the Dow Cell manometric method are also very operator-dependent and they often lack sensitivity. Manometric methods are seldom capable of testing oxygen barriers with transmission rates of less than 1.0 cc/m²/day (Demorest, 1989).

Volumetric Method

The second absolute pressure method available is the volumetric differential method. This method is also described in ASTM D1434 (ASTM, 1988). The volumetric method is commonly referred to as the Linde Cell. The test cell is similar to the Dow Cell, except that the low pressure chamber is maintained near atmospheric pressure and the transmission of gas through the test specimen is indicated by a change in volume. The Linde Cells are manufactured by Custom Scientific Instruments, Inc. (Cedar Knolls, NJ). Results from the Linde Cell lack precision and are strongly operator-dependent (Demorest, 1989). Therefore, it is not commonly used.

TAPPI also offers a standard volumetric method for determining gas transmission through sheets and films. The sensitivity of this method is approximately 0.019 cc/100 in.²/day. However, this method suffers from the same problems as the Linde Cell and is not commonly employed (Todd, 1944).

Isostatic Concentration Methods

Isostatic methods maintain constant atmospheric pressures in both chambers of the permeation cell. Because of this, isostatic methods offer an advantage over manometric and volumetric techniques. They do not impose an absolute pressure gradient on the specimen; therefore, samples are tested under more real conditions. Gas permeability is determined by creating a permeant partial pressure or concentration gradient between the two chambers. Isostatic concentration methods also provide the advantage of being able to determine film gas permeabilities at any desired RH, not just 0% RH.

Landrock and Proctor (1952) developed an isostatic concentration-increase method for determining oxygen and carbon dioxide transmission rates simultaneously. This method is based on Dalton's law of partial pressures, which stipulates that gases in mixtures behave independently of one another such that the total pressure is the sum of the partial pressures. They constructed permeability cells from chrome-plated brass. Each cell was divided into two chambers separated by the test film. Nitrogen carrier gas passed through the lower chamber, while an oxygen and/or carbon dioxide test gas passed through the upper chamber. Screw clamps and rubber gaskets were employed to seal a film between the two chambers. To prevent distortion of the film during a trial, heavy wire screens were placed above and below the test films. Gas flowed through glass humidity towers to control the RH of the chamber. The modified gaseous mixture accumulated in the lower chamber of the cell was analyzed using an Orsat gas-analysis burett system. The mixture was analyzed for oxygen using "Oxsorbent," and carbon dioxide using potassium hydroxide.

Following the Landrock and Proctor method, two other categories of isostatic concentration methods were developed for determining gas permeabilities of films. These methods involve gas chromatographic and coulometric techniques.

Gas Chromatographic

Several gas chromatographic methods have been established for determining gas permeabilities of films. Gas concentrations can be measured either by directly injecting samples or passing flows of samples through a specific detector in a gas chromatograph.

Karel et al. (1963) improved upon the concentration-increase method of Landrock and Proctor by application of a gas chromatograph to measure the permeating gas. This method is also based on measurement of the amount of gas diffusing through the film over time. The test cell used in this study consisted of two chambers separated by the film to be tested. Both chambers had an inlet and an outlet for gas flushing. The lower test compartment was also equipped with a sampling port. The film was first sealed into the test cell. Nitrogen gas dried over calcium chloride was passed through both sides of the test cell until no detectable test gas was apparent. Next, a stream of test gas (oxygen or carbon dioxide) was passed through the upper compartment of the cell, while the lower compartment was sealed. At suitable intervals, gas samples were withdrawn from the lower compartment and analyzed by gas chromatography. To avoid total pressure changes, 1 cc of nitrogen was injected into the lower compartment before each sample was withdrawn. In this study, an Aerograph A-90P (Wilkens Instrument and Research, Walnut Creek, CA) gas chromatograph with a thermal conductivity detector and two columns was employed. The first column was packed with silica gel to separate carbon dioxide from oxygen and nitrogen. A second molecular sieve column separated the oxygen and the nitrogen. No provisions for relative humidity control were provided in this study. The method was proven to be accurate to within ±5%.

Several advantages are apparent for this system. Application of gas chromatography for analysis of gas permeating through the film expanded the versatility of the system and its accuracy. Gas chromatographic systems allow for repeated sampling of cell test gases, increased sensitivity, more rapid collection of data and the possibility of determining the time needed to establish steady-state conditions of gas transfer (Karel et al., 1963).

Gilbert and Pegaz (1969) developed a similar gas chromatographic system for analyzing gas permeating through films. They improved upon the method of Karel et al. (1963) by humidifying the gas streams. Gilbert and Pegaz also modified the test cell. They designed cells to accommodate two film samples clamped between stainless steel devices to form a pair of outer chambers and one inner chamber. This doubled the exposed film surface, thereby improving the sensitivity of the method. Humidified nitrogen was passed through the center test cell until it was freed of test gas. Then the center cell inlet and outlet were sealed, leaving it filled with nitrogen. Meanwhile, a humidified test gas such as oxygen was passed through the outer chambers. The operator periodically withdrew 0.5 ml samples from the inner chamber and analyzed them by gas chromatography. To avoid error, precise temperature control was essential. Technique variance due to manual gas chromatographic injections often resulted in errors. Furthermore, cell leaks were a problem.

Davis and Huntington (1977) described in detail a sample test cell for measuring the permeabilities of films. The cell operated on the Landrock and Proctor (1952) concentration-increase principle and was used in combination with a gas chromatograph. The cell and its fittings were made of stainless steel. Support screens were placed on each side of the film, which was situated between two test cell compartments. Each compartment had an inlet and an outlet for gas flow. A clamping device was used to seal the film in place and to prevent leakage.

An obvious potential area of improvement for these methods involves automation of the gas sampling procedure, to avoid the variation present when manual injection is used. This improvement was subsequently implemented by Baner et al. (1986). In this method, constant low concentrations of permeant flowed through the upper chamber of the cell, while nitrogen of a known flow rate continually passed through the lower chamber. A computer-controlled sampling valve transferred the lower chamber gas stream into the gas chromatograph for analysis.

The use of gas chromatographic methods for determining the barrier properties of edible films has several advantages and disadvantages. The major advantage is low cost. These systems are much less costly than most coulometric systems. Furthermore, the sensitivity of gas chromatographic methods is better than that of most manometric and volumetric methods. Complete gas chromatographic systems for determining film permeabilities are not available commercially. Therefore, considerable set-up time may be needed prior to accurately determining gas permeabilities.

Several edible film researchers have sought to determine film gas permeabilities using gas chromatographic systems. Lieberman et al. (1972) were the first to test the gas permeability of collagen films using the manual injection gas chromatographic system of Gilbert and Pegaz (1969). McHugh and Krochta (unpublished research) have developed a similar system, using a Shimadzu GC-8A system (Columbia, MD), to determine the carbon dioxide permeability of whey protein–based edible films. This system has the additional advantage of being automated, thereby avoiding manual injection variations and allowing for testing overnight. The basic program in the Shimadzu integrator regulates flows from different test cells into the gas chromatograph for analysis.

Most of the equipment necessary for constructing gas permeability test systems as outlined above is available commercially. R. J. Harvey Co. (Hillsdale, NJ) manufactures test cells that allow for permeation measurements. The cell has two outer and one inner chambers and is heatable with a separate power supply if necessary. Gas streams can be humidified if desired using gas wash bottles. This unit costs between $1000 and $2000, depending on the desired test conditions. Most university researchers fabricate their own test cells in an effort to avoid this expense. The two main gas chromatographs currently available for analysis of oxygen and carbon dioxide are available through Hewlett-Packard Corp. (Sacramento, CA) and Shimadzu, Inc. (Concord, CA). These systems cost between $10,000 and $20,000. Both of these gas chromatographs operate with thermal conductivity detection systems, and both contain a series of columns to separate oxygen from carbon dioxide.

Calculations necessary to determine gas permeabilities of films will be presented later in this chapter.

Coulometric

The most commonly used method to determine the oxygen barrier properties of films is coulometric in nature. MOCON Corporation (Minneapolis, MN) is the major producer of these commerical units (Oxtran). Several Oxtran units with varying capabilities are available for purchase. All of the Oxtran sensors operate by the principle of Faraday's law. Four electrons are released electrochemically for every oxygen molecule passing through the sensor. These electrons produce a current which is passed through a resistor, creating a voltage that can be quantified. The LC-700F is a compact unit capable of measuring the oxygen concentration in both flexible and rigid packages. The Oxtran 100, 200, 300, and 1000 units measure oxygen transmission rates through flat barrier films. Oxtran 200H and 300H units enable film testing at different relative humidities. Recently, a new modular Oxtran instrument, the 2/20 series, became available commercially. The sensor for this unit consists of a cadmium electrode to generate the electrons necessary to determine oxygen concentrations. The Oxtran 2/20 also contains an improved relative humidity control system based on controlling the total pressure above a reservoir of water in the gas flow path. Temperature control is available in all Oxtran systems.

In 1981, ASTM approved the MOCON coulometric method for testing oxygen transmission rates through plastic films and sheeting. The oxygen transmission rates obtained with this method correlated well with those using manometric methods. Several precautions were given by ASTM. Extended use of the Oxtran 100 was found to result in sensor dehydration. The new Oxtran 2/20 sensors are humidified to prevent sensor dehydration. Furthermore, careful control of room temperature is necessary to avoid variation in results. Long times may be required to outgas oxygen from the system.

The sensitivity of the Oxtran coulometric sensors represents their major advantage over other methods for determining oxygen transmission rates. MOCON claims the Oxtran 2/20 system is able to detect oxygen levels down to 0.001 cc/m²·day. However, leaks in the system often result in an inability to control baseline levels sufficiently to obtain reliable data at such low transmission rates. Another advantage of the Oxtran units is their precise temperature and RH control capabilities. This is particularly advantageous when examining the barrier properties of edible films to be used on food systems under different temperature and RH conditions. The major disadvantage of the Oxtran units is their cost. The Oxtran 2/20 ML system with one test module for two films costs about $60,000.

Several edible film research groups have taken advantage of the MOCON instruments for measuring gas transmission rates. Rico-Peña and Torres (1990) measured the oxygen permeability of methylcellulose–palmitic acid films by first flushing the lower cell compartment with nitrogen and the upper compartment with oxygen. After two hours, gas flow through the cell was stopped and the cell's inlet and outlet valves were turned off. At various time intervals, samples of gases (5 ml) were taken from the lower test compartment, after which 5 ml of nitrogen were added back into the lower compartment. The oxygen concentration increase was monitored using a MOCON LC-700F (Minneapolis, MN) head space analyzer. Aydt et al. (1991) used an Oxtran 100 to test the oxygen barrier properties of various protein films. Masks were necessary to provide mechanical support for the films. Testing can only be performed at 0% relative humidity using this unit. Kester and Fennema (1989b) used an Oxtran 100 to test the resistance of lipid films to oxygen transmission, again using 0% RH. McHugh and Krochta (1994a) used an Oxtran 2/20 to measure the effect of RH on whey protein film oxygen permeability values. These data are shown later in this chapter.

Other Methods

Infrared Detection

Carbon dioxide permeability properties are often measured using infrared detectors. Carbon dioxide absorbs infrared radiation at a specific wavelength. MOCON produces equipment to measure the carbon dioxide transmission rates of films. The MOCON Permatran-CIV employs an infrared sensor for detecting carbon dioxide at low concentrations. This method is much faster and more accurate than the Dow or Linde Cell techniques, but is also limited to 0% RH test conditions. The cost of the Permatran-CIV is rather high. Specific infrared analyzers are also available for determining the carbon dioxide permeabilities of films (Clegg, 1978).

Bioluminescence

Miltz and Ulitzur developed a bioluminescence method for determining the oxygen transmission rates through plastic films (1980). The method is based on luminous bacteria emitting light *in vivo*. The bacteria are confined in a pouch made of the film to be tested. The light intensity emitted is measured and quantified. The light emitted is proportional to the oxygen transmitted through the film.

Summary

Most of the methods available for determining transmission rates of permanent gases through edible films are described in this section, along with several of their respective advantages and disadvantages. Proper test cell and system sealing procedures are necessary to prevent leakage. Furthermore, accurate RH control and sensor calibration are crucial for obtaining accurate permeability measurements. This section is by no means comprehensive, and researchers are urged to carefully examine all methods as to their applicability for testing edible films. Edible films often pose complications not observed when testing nonpolar polymeric packaging materials, such as RH effects on the gas barrier properties of the film.

PERMEABILITY OF EDIBLE FILMS TO LIPIDS

LIPID BARRIER PROPERTIES

The barrier properties of edible films to lipids are of interest for a variety of applications. Because many edible films are hydrophilic they would be expected to be good lipid barriers (at low relative humidities). Commercial equipment for quantifying these barrier properties is not available; therefore, a variety of custom methods have been devised.

MEASUREMENT OF BARRIER PROPERTIES OF EDIBLE FILMS TO LIPIDS

Rate Method

ASTM (1988) approved a standard test method for measurement of the rate of grease penetration of flexible barrier materials. Using this method, flexible barrier materials, creased or uncreased, are exposed to a cotton patch covered with grease on one side. Weights are placed on top of the cotton patch and the time required to show a visual change on a ground-glass back-up plate is measured. This visual change is caused by wetting.

This method has several advantages and disadvantages. The test is rapid and sensitive to small quantities of oil (6 μg). Furthermore, the apparatus is simple and inexpensive. However, the reproducibility of data is rather low. A 31% standard deviation was observed (ASTM, 1988). Another disadvantage of this method is that the results are highly operator-dependent. Finally, determination of permeability values is not possible.

Permeability Method

Nelson and Fennema (1991) were the first edible film researchers to examine the lipid barrier properties of methylcellulose edible films. Their apparatus consisted of a film permeability cell containing two chambers. Inner and outer chambers contained different agitated oils. A water jacket enclosed the chambers for temperature control. Films were mounted and sealed between two stainless steel screens separating the inner and outer chambers. Peanut oil was added to the inner chamber and tricaprylin was added to the outer chamber. The cell was maintained at $30 \pm 1°C$, and 5 psig of nitrogen was applied to the inner chamber. Oil samples from the outer chamber were analyzed at the beginning and end of the test using the AOAC method (AOAC, 1984). Testing took approximately twenty-four hours. The rate of lipid migration was measured using a spectrophotometer at 234 nm, and the results were expressed in μg linoleic acid/24 hours at 30°C.

This method represents the first attempt by edible film researchers to determine lipid barrier properties of films. Film transmission rates increased with decreasing thickness, due to the presence of imperfections and pores in thin films. Therefore, film imperfections must be minimized in this and other permeability tests. This method has an advantage over the ASTM (1988) method in that it allows for the application of a defined pressure on the lipid permeating through the film.

It is expected that many new methods will be developed in upcoming years to determine the lipid barrier properties of edible films. The promising lipid barrier properties of edible films make this area deserving of research.

MATHEMATICAL MODELS FOR PERMEABILITY ANALYSIS

In the past, various mathematical methods have been developed to calculate and subsequently compare the barrier properties of film materials. Most of these calculations were developed for commercial nonpolar polymeric film analyses. As mentioned previously in this chapter, hydrophilic films present new challenges for these types of classical calculations. Discussed in this section are classical methods for calculating the barrier properties of film materials, and several nontraditional methods for calculating film permeabilities for hydrophilic films.

WATER VAPOR PERMEABILITY ANALYSIS

Fick's First Law

In 1855, Adolf Eugen Fick, a German physiologist, formulated a fundamental law of diffusion, known as Fick's First Law. Fick's First Law is commonly used to describe the permeation of water vapor and other permeants through a film. For unidirectional diffusion of component A through film B, Fick's First Law is given by Equation (7.2):

$$J_{Az} = -D_{AB}\frac{\partial C_A}{\partial_z} \qquad (7.2)$$

where

J_{Az} flux or amount of a permeant (n) diffusing per unit area (A) over time (t) in the z direction, $(1/A)\,(\partial n/\partial t)$

D_{AB} the binary molecular mass diffusivity of permeant A through film B

$\partial C_A/\partial_z$ the concentration gradient in the z direction

The unintegrated form of Fick's First Law applies to most film systems; however, for Fick's First Law to be used, it must be integrated. In the case of: (1) negligible convective transport, (2) steady-state mass transport, (3) constant diffusion coefficient, and (4) homogeneous films, Equation (7.2) can be integrated to give Equation (7.3):

$$N_{Az} = D_{AB}\frac{(C_{A1} - C_{A2})}{z} \qquad (7.3)$$

N_{Az} describes the steady-state transmission rate of permeant A through unit area of film B, C_{A1} and C_{A2} are the concentrations of species A at the film surfaces, and z is film thickness.

Permeation of vapors and gases across films is usually compared using permeability (P_{AB}) as the parameter. P_{AB} is an objective value used for comparing the barrier properties of films having different thicknesses. Permeability equals the product of permeant solubility and diffusivity in a film. If Henry's law is obeyed, the concentration of A can be expressed as the product of the solubility coefficient (S) and the partial pressure (p_A). Use of Henry's law assumes equilibrium at the air-film interface. Combining Fick's First Law and Henry's law gives Equations (7.4), (7.5), and

(7.6). Equation (7.6) is commonly employed to calculate film permeabilities.

$$N_{Az} = D_{AB} \frac{(S_{AB}p_{A1} - S_{AB}p_{A2})}{z} \qquad (7.4)$$

$$N_{Az} = D_{AB}S_{AB} \frac{(p_{A1} - p_{A2})}{z} \qquad (7.5)$$

$$N_{Az} = P_{AB} \frac{(p_{A1} - p_{A2})}{z} \qquad (7.6)$$

ASTM Method

The ASTM E96 cup method, described previously in this chapter, uses Fick's First Law and Henry's law to calculate film water vapor permeabilities (1980). One first determines the water vapor transmission rate (WVTR) using Equation (7.7):

$$\text{WVTR} = \frac{n}{tA} \qquad (7.7)$$

WVTR equals the slope of the plot of amount of water lost over time (n/t) divided by the film area (A). The steady-state weight loss over time should yield a regression coefficient of 0.99 or greater. The WVTR rate of a barrier is determined under known conditions of thickness, temperature, and water vapor partial pressure gradient.

Permeance is the second parameter specified using the ASTM E96 method (1980). Permeance equals the WVTR divided by the partial vapor pressure gradient across the film, as shown in Equation (7.8):

$$\text{permeance} = \frac{\text{WVTR}}{(p_{A1} - p_{A2})} \qquad (7.8)$$

ASTM E96 assumes that $p_{A1} = p_{A0}$ (Figure 7.1). This is true for good barriers to mass transfer. However, this assumption is often false for edible films. Therefore, alternative methods such as the Schwartzberg method and the WVP Correction Method outlined below should be employed in these cases. Permeance provides an objective value for comparing barrier properties of different films, provided the films tested are of equal thicknesses.

Finally, Equation (7.9) can be used to determine the permeability of films. Permeability values are independent of film thickness.

$$\text{permeability} = \text{permeance} \times \text{thickness} \qquad (7.9)$$

It is important to stress that the permeability equations listed above involve numerous assumptions. The first assumption is that the films are homogeneous. Hydrophilic films exposed to water vapor partial pressure gradients are nonhomogeneous by nature, due to the presence of varying water concentrations within different film sections. Permeability equations also assume that film solubility and diffusivity coefficients are constant. However, the water sorption isotherms of hydrophilic edible films are highly nonlinear, and their diffusivities are known to depend strongly on the water content in the film. Therefore, neither solubility nor diffusivity is constant for most hydrophilic edible films. It is also assumed that equilibrium is established at the film surface and that the mass transfer resistance at the film surfaces is negligible. This latter assumption is also often false.

For hydrophilic films, where the assumptions listed above are not met, permeability can be used for comparative purposes only under the selected p_{A1} and p_{A2} test conditions. In these cases, the expression "effective" permeability constant, P_{eff}, should be used, and this relates only to the test conditions. The related solubility and diffusivity coefficients are "effective" as well. Consequently, it is extremely important when working with hydrophilic films to report the test partial pressures, film thicknesses, and experimental temperatures, as well as effective permeability constants, in order to facilitate more accurate comparisons of permeability data.

Methods to correct for some of the above false assumptions have been developed and will be described in the following sections.

Schwartzberg Model

The ASTM E96-80 method relies upon the assumption that negligible resistance to mass transfer exists in the gas phases on both sides of the test film. This is true of mass transfer resistance above the cup, provided convection is induced by a fan moving air at 500 ft/min, as recommended by ASTM (1980) and confirmed by McHugh et al. (1993) for hydrophilic films. However, the small distance between the mounted film and the water, saturated salt solution, or desiccant in the cup, results in low levels of natural convection within the cup. As a result, $p_{A0} \neq p_{A1}$ (see Figure 7.1). Thus, large errors in permeability calculations can occur when the film resistance is small, as with hydrophilic films. Schwartzberg (1986) developed a method to deal with these errors. He treated mass transfer through the stagnant air layer like the classical problem of water diffusion through stagnant air, except that he simplified the expression by assuming equimolar counter diffusion in the cup. He ignored convective flux induced by diffusion. Thus, the following relationship was employed:

$$N_w = K_w(p_{A0} - p_{A1}) = K_w(p_{A0} - p^*a_w) \qquad (7.10)$$

N_w is the steady-state transmission rate of water per unit area of film. K_w is the mass transfer coefficient which equals D/RTz, D is the diffusivity of water through air, R is the universal gas constant, T is the experimental temperature, z is the depth of the stagnant air layer, p^* is the vapor pressure of pure water, p_{A0} is the vapor pressure at the solution surface inside the test cup, p_{A1} is the vapor pressure at the inner film surface and a_w is the water activity in equilibrium with the respective film moisture content at surface one. Equation (7.10) can be rearranged to Equation (7.11), which can be used to determine the true partial pressure at the underside surface of the film, p_{A1}.

$$p_{A1} = p_{A0} - N_w RTz/D \qquad (7.11)$$

Subsequently, the calculated p_{A1} conditions can be employed in Equation (7.8) to calculate film permeance and permeability.

Schwartzberg also developed a model to determine diffusivity and solubility values for hydrophilic films (1986). These values can be multiplied together to yield film permeability. Measurements over a range of humidities, as well as a complete knowledge of the dependence of equilibrium relative humidity and diffusivity on the film's water content, are employed to accurately predict transport behavior. This method has yet to be employed by edible film researchers.

In developing this model, Schwartzberg made the following assumptions: (1) equilibrium is always maintained due to rapid sorption and desorption, (2) heat of sorption equals heat of desorption, and (3) hysteresis with respect to sorption equilibrium can be neglected. It is important to realize that hysteresis is actually quite common for hydrophilic films. Employing the above assumptions allows a general expression for the sorption isotherm, Equation (7.12), to be developed, where X is moisture content (a function of water activity).

$$X = f(a_w) = f(p_{A1}/p^*) \qquad (7.12)$$

In the analysis of diffusion through a film, steady-state transport and applicability of the free-volume model for diffusivities in polymeric films are assumed, and a modified form of Fick's First Law is used.

$$N_w = -D_w \frac{\partial X}{\partial s} \tag{7.13}$$

D_w is the diffusivity of water in the film and s is the mass of dry polymeric film per unit area between surface one and the level where the moisture content is X. This equation is employed to avoid the computational problems associated with hydrophilic film swelling upon absorption of large amounts of water.

Next, an empirical relationship is used to correlate D_w in terms of X:

$$D_w = D_o \exp(AX) \tag{7.14}$$

Typically, D_o and A are around 2.3×10^{-5} (kg solids^2/m^{4}s) and 2.9 m^2, respectively. This relationship does not always apply; therefore, alternate relationships often need to be developed.

If Equation (7.14) applies and Equation (7.13) is integrated, Equation (7.15) results, where S is the total mass of dry film per unit area of film.

$$N_w = (D_o/AS)[\exp(AX_1) - \exp(AX_2)] \tag{7.15}$$

N_w is needed to determine the water vapor permeability of the film. Using this model, Equations (7.10), (7.12), and (7.15) must be simultaneously solved, and trial and error computation is often necessary. These calculations are outlined by Schwartzberg (1986).

Using this method, the film mass per area (S), the sorption isotherm, and the realtionship between D_w and X must be known. Both S and the sorption isotherm are relatively easy to determine, and the relationship between D_w and X can be determined using a trial and error procedure. K_w can be determined using the relationship $K_w = D/RT_z$. From K_w and a_w, the water vapor flux (N_w) can be calculated. Then through a series of tests carried out under different p_{A2}, a set of N_w and X data is established. By combining these data, Equation (7.13) and a nonlinear regression routine, the best fit values for D_o and A are determined. A goodness of fit for D_w versus X is also developed. D_o and A can then be plugged into Equation 7.14 to determine D_w (Schwartzberg, 1986). D_w can be multiplied by solubility to calculate the film permeability more accurately. It is suggested that edible film researchers examine the possibility of combining sorption isotherm data with diffusivity values to determine permeability characteristics of films, using the method of Schwartzberg (1986) or related methods.

Water Vapor Permeability (WVP) Correction Method

McHugh et al. (1993) suggested several improvements to the ASTM E96 method (ASTM, 1980) for determining the WVPs of hydrophilic edible films. Accurate determination of RH conditions and maintenance of 500 ft/min air speeds were essential inside the test cabinet in order to satisfy the assumption of negligible mass transfer in the gas phase outside the test cups. Rapid air speeds also allow for maintenance of desired RH conditions in the test cabinets. Below 500 ft/min, values of WVP decrease exponentially with air velocity.

McHugh et al. (1993) suggested the use of the Water Vapor Permeability (WVP) Correction Method to account for the water vapor partial pressure gradient in the stagnant air layer within the test cups. Up to 35% errors were found when the WVP Correction Method was not employed. The importance of the correction method lessened as the barrier properties of the film increased, becoming unnecessary for synthetic polymer films due to their high resistance to mass transfer.

The correction method involves a classical calculation for diffusion of water vapor through air (Bird et al., 1960) to calculate mass transfer through the stagnant air layer in the test cup. Water vapor transmission rates are first calculated using the ASTM E96 (1980) method shown in Equation (7.7). Next, the water vapor partial pressure at the inner surface of the film, p_{A1}, is calculated using the equation for diffusion of water vapor through a stagnant air layer, as shown in Equation (7.16) (Krochta, 1992).

$$\text{WVTR} = \frac{P \times D \times \ln\left[(P - p_{A1})/(P - p_{A0})\right]}{R \times T \times z} \qquad (7.16)$$

Where P is the total pressure, D is the diffusivity of water vapor through air at the experimental temperature, T, and R is the universal gas constant.

Water vapor permeance and permeability can then be calculated using the true p_{A1} and Equations (7.8) and (7.9).

The WVP Correction Method improved upon the method developed by Schwartzberg (1986) by taking into account the convective flux induced by diffusion within the cup. Although the convective flux induced by diffusion within the cup is small, it should not be ignored. For a film composed of 62.5% whey protein isolate and 37.5% glycerin, a 1.3% difference between the WVPs corrected using the Schwartzberg method versus the WVP Correction Method was observed.

The method of McHugh et al. (1993) also provided an explanation for commonly observed thickness effects on the water vapor permeability properties of hydrophilic edible films. These effects were attributed to the increasing RH to which the underside of the film was exposed when thickness was increased. The increased RH at the inner surface of the film resulted in larger WVPs due to the nonlinear nature of the moisture sorption isotherms for hydrophilic films.

Most of the above discussion refers to the ASTM E96 (1980) gravimetric technique for determining WVPs. It is important to realize that these same models can be employed using data retrieved from other methods of analysis. For example, the Permatran-W yields data on water vapor transmission rates of films. The WVPs of these films can subsequently be calculated using the methods outlined above.

ANALYSIS OF PERMANENT GAS PERMEABILITY OF FILMS

Absolute Pressure Systems

Manometric

Calculations necessary for manometric determination of film permeabilities are outlined in this section. First, the mercury height in the cell capillary leg must be measured after at least six time intervals. Then, the function $g(h)$ defined in Equation (7.17) must be plotted versus the time. Next, film permeance (perm) can be calculated using Equation (7.18). Permeability equals permeance multiplied by the film thickness (ASTM, 1988).

$$g(h) = -\frac{1}{ART}\left[[V_f + a(p_u + h_B - h_L)]\right.$$

$$\left.\times \ln\left\{1 - \frac{h_0 - h}{p_u - (h_L - h_0)}\right\} + 2a(h_0 - h)\right] \quad (7.17)$$

$$\text{perm} = g\,\frac{h}{t - t_0} \quad (7.18)$$

a area of capillary AB, mm^2
A area of transmission, cm^2
h_0 height of mercury in the capillary leg at the start of the run, after attainment of steady state, mm
h height of mercury in cell capillary leg at given time, mm
h_B maximum height of mercury in the cell manometer from datum plane to top of mercury miniscus, mm
h_L height of mercury in cell reservoir leg from datum plane to top of mercury meniscus, mm

p_u upstream pressure of gas to be transmitted
R universal gas constant
t_0 time at start of the run after attainment of steady state
t time
T absolute temperature
V_{BC} volume of the cell manometer leg, μl
V_{CD} void volume of depression, μl
V_f $(V_{BC} + V_{CD})$, μl

Volumetric

A volumetric method can also be used to calculate the gas permeability of edible films. First, the slope of the capillary slug position versus elapsed time must be obtained. The volume-flow rate, V_r, of transmitted gas can then be calculated using Equation (7.19) where a_c is the cross-sectional area of the capillary (ASTM, 1988).

$$V_r = \text{slope} \times a_c \tag{7.19}$$

Next, the gas transmission rate (GTR) can be calculated using Equation (7.20) where p_0 is the ambient pressure.

$$GTR = 10^{-6} \times p_0 \frac{V_r}{ART} \tag{7.20}$$

Gas permeance can then be calculated using Equation (7.21) where p is the upstream pressure. Permeability can be obtained by multiplying permeance by film thickness.

$$\text{permeance} = \frac{GTR}{p - p_0} \tag{7.21}$$

Isostatic System Analyses

Isostatic system analyses most commonly employ final steady-state concentrations of penetrated test gas to calculate film permeability using Equation (7.22) (based on Fick's First Law).

$$\text{permeability} = \frac{C \times f \times z}{A \times p^1} \tag{7.22}$$

C steady-state concentration of penetrant conveyed to the detector, mass/volume

f rate of carrier gas flowing through the cell, volume/time
A film surface area, area
z film thickness, length
p^1 test gas partial pressure, pressure

METHODS FOR PREDICTING AND INTERPRETING RELATIVE HUMIDITY EFFECTS ON FILM PERMEABILITIES

Prediction Model of Hauser and McLaren

Hauser and McLaren (1949) developed a model to predict the WVPs of hydrophilic films under any RH gradient. This method is extremely useful for modeling mass transfer through coatings on food systems exposed to different water activity gradients.

Using corrected values of WVP (McHugh et al., 1993), plots of WVP versus RH (%) can be developed for films tested using 0% RH on one side of the film and x% RH on the other. These curves must be developed using as large a range of RH gradients as possible. A sample curve is shown in Figure 7.2 for films consisting of 62% whey protein isolate (WPI) and 38% glycerin (McHugh and Krochta, 1994a).

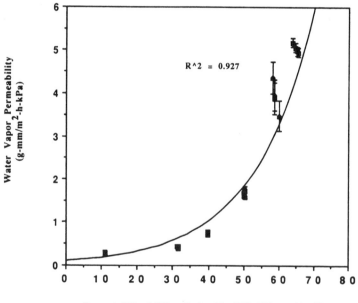

Figure 7.2 Effect of relative humidity gradient on the water vapor permeability of 62% WPI/38% glycerin films.

From Figure 7.2 it is evident that RH has an exponential effect on the WVP of whey protein films. Since WVP varies with RH, the diffusing water vapor encounters varying partial pressure environments during passage through the film. In such cases, the measured value of permeability is an average which can be expressed as an integral:

$$\int_{p_{A1}}^{p_{A2}} P\,dp = P_{avg}(p_{A2} - p_{A1}) \qquad (7.23)$$

Thus, $P_{avg}(p_{A2} - p_{A1})$ or $P_{avg}(RH_2 - RH_1)$ represents an area under the plot of WVP versus RH. If one obtains P_{avg} values for a range of RH_2 values where RH_1 equals 0%, the $P_{avg}(RH_2 - 0)$ values can be plotted and then used to determine the P_{avg} between any two RH_2, by subtracting the area between the smaller $RH_{2'}$ and 0 from the area between the larger RH_2 and 0. Equation (7.24) illustrates such a calculation.

$$P_{2'/2}(RH_{2'} - RH_2) = P_{2'}(RH_{2'} - 0) - P_2(RH_2 - 0) \qquad (7.24)$$

The Hauser and McLaren (1949) method was tested using the data shown in Figure 7.2 under conditions of 30% RH at the outside surface of the film and 51% RH at the inside surface of the film. The average measured experimental WVP value was 3.33 ± 0.20 g·mm/m²·h·kPa. Using the method of Hauser and McLaren, the expected WVP value would be 3.50 g·mm/m²·h·kPa, as is shown calculated in Equation (7.25).

$$P_{2'/2} = [1.69(51 - 0) - 0.42(30 - 0)]/(51 - 30) = 3.50 \qquad (7.25)$$

This represents only a 4.8% difference from the experimental values, which is within the expected deviation of the data. Therefore, the Hauser and McLaren model for prediction of the WVPs of edible films under any RH gradient appears to be an accurate one.

It is also important to note the exponential relationship between WVP and RH. This relationship indicates the importance of accurately determining the true, corrected RH conditions to which the film is exposed during testing. If correction is not made, significant errors will be present for hydrophilic films.

Graphs such as Figure 7.2 can serve as an excellent approach for comparison of film permeability properties under a range of RH gradients. Currently, it is difficult to accurately compare film permeability properties tested under varying RH gradients. Therefore, this method deserves consideration by all edible film researchers. Hagenmaier and Shaw (1991) took this approach for examining the WVPs of shellac coatings under variable RH gradients.

Cluster Function Interpretive Model of Zimm-Lundberg

The Zimm-Lundberg cluster function (1956) enables calculation of the distribution of water molecules in edible films. In synthetic polymer films, rigid site–type sorption theory works well; however, it breaks down when applied to edible films, such as collagen films. It was found that a water clustering theory could be applied to all types of films. This function operates on the basis of statistical mechanics and functions to interpret isotherms in molecular terms. For specific details concerning analyses using the Zimm-Lundberg function, the reader is referred to their article (1956).

This theory enables measurement of the tendency of absorbing molecules to cluster. The theory compares the anticipated number of molecules in the vicinity of a similar molecule, due to random behavior, to the actual number of molecules found clustering. Positive cluster functions indicate that initial solvent molecules loosen the polymer structure, increasing the free volume and the probability that subsequent molecules will enter near the first molecule's entrance site rather than elsewhere. A cluster function of two indicates a cluster of three water molecules, two in excess of the original molecule. When the cluster function is negative, the solvent molecules are said to be attached to specific sites dispersed throughout the polymer matrix.

Zimm and Lundberg (1956) determined cluster functions for a variety of hydrophilic protein polymers over a large range of RHs. At low RHs, cluster functions were negative, indicating that initial water molecules absorbed on specific sites. However, as RH increased, the cluster function also rapidly increased, indicating that clustering occurred at higher water vapor concentrations. Clustering leads to increases in the available free volume and in the permeability of the polymeric matrix to test gases.

Lieberman et al. (1972) and Lieberman and Gilbert (1973) utilized the Zimm-Lundberg cluster theory to analyze the sorption behavior of collagen-based films. Collagen films exhibit exponentially increasing gas permeability properties as RH increases. Results on collagen films confirmed those of Zimm and Lundberg (1956). At low RHs, a negative cluster function indicated that unclustered water molecules were unavailable for gas diffusion. Then as RH increased, the number of clustered water molecules increased, causing a subsequent increase in the free volume of the polymer and in the polymer permeability.

Lieberman et al. (1972) and Lieberman and Gilbert (1973) further applied the Zimm-Lundberg cluster theory to interpret the effects of cross-linking on collagen film permeabilities. Cross-linkage of collagen films with tanning agents resulted in an unexpected increase in oxygen permeability values with increasing levels of cross-linkage, particularly at high

RHs. Clustering was observed above 75% RH in highly cross-linked films, whereas clustering did not occur until above 90% RH in the control. Consequently, at the same RH, one would expect their cross-linked films to exhibit higher permeability values than control films, due to increased clustering and free volume in their cross-linked films. The tanning reaction, by combining formaldehyde with basic amino groups, had reduced the number of available water bonding sites, thereby initiating clustering at lower RHs. Oxygen permeability values increased as cross-linking increased.

Plasticizer effects on oxygen permeability properties at various RHs were similarly interpreted using cluster theory (Lieberman et al., 1972; Lieberman and Gilbert, 1973). Polyol plasticizers compete with water for active sites on polymers, thus promoting water clustering at low moisture levels. The result of increasing glycerin concentrations decreased the negative cluster function and the RH required to exhibit positive clustering, thereby confirming the clustering enhancement of polyols. Increased clustering with increasing levels of plasticizers, resulted in increases in free volume within the polymer and subsequent increases in permeability.

The Zimm-Lundberg cluster theory allows for accurate interpretation of the effects of relative humidity on the permeability properties of hydrophilic edible films. Thus, it may be useful for edible film researchers to adopt this approach for explaining observed permeability behaviors.

MECHANISMS OF WATER VAPOR AND PERMANENT GAS MASS TRANSFER

The diffusion of a water vapor or a gas through a polymer is a consequence of random motions of individual molecules of the species. This is known as the "random walk" theory (Frisch and Stern, 1981). Net transport occurs due to the presence of more molecules on one side of the polymer than on the other.

Mass transfer through polymeric films has been classified in various categories. Alfrey et al. (1966) developed a classification system for diffusion of vapor through films based on the relative rates of penetrant mobility and polymer relaxation. The different diffusion mechanisms are strongly dependent on the glass transition temperatures of the polymers. Three classes are distinguished: Case I (Fickian), Case II, and Anomalous diffusion. These categories are described in further detail below.

CASE I (FICKIAN) DIFFUSION

Penetrant diffusion rates are smaller than rates of polymer segment relaxation in Case I, Fickian diffusion systems. These systems exist in a "rubbery" state, above their glass transition temperatures, and therefore respond rapidly to changes in their physical condition (Frisch and Stern, 1981). Sorption equilibrium is rapidly established, and time-dependent boundary conditions are independent of swelling kinetics (Crank, 1975).

The presence of Case I diffusion can be determined by the shape of the sorption-time curve. Sorption is a linear function of the square root of time. For longer times, sorption curves become concave to the abscissa. Reduced sorption curves for films of different thicknesses should remain superimposable (Marom, 1985). The amount of water sorbed over time can be described by Kt^n, where n equals 1/2 in Case I systems (Crank, 1975).

CASE II DIFFUSION

Diffusion is very rapid compared to polymer relaxation rates in Case II systems. These systems exist in a "glassy" state (below their glass transition temperatures), and therefore polymer mobility is not rapid enough to bring the penetrant environment to complete equilibrium (Frisch and Stern, 1981). Furthermore, sorption processes are strongly dependent on swelling kinetics.

Case II diffusion can be used to characterize diffusion through glassy polymers. Advancing boundaries are present when the penetrant diffuses through the polymer. The boundary exists between the outer swollen shell surrounding the inner glassy core of the polymer. The core diminishes in size until the polymer is at equilibrium (Marom, 1985). No experimental data is available on the nature of the final equilibrium state of sorption in these systems (Frisch and Stern, 1981). The amount of water sorbed over time can be characterized by Kt^n, where n equals 1 in Case II systems (Crank, 1975).

ANOMALOUS DIFFUSION

Anomalous diffusion occurs when penetrant diffusion and polymer relaxation rates are comparable. These systems lie between Case I and Case II, with n values between 1/2 and 1 (Crank, 1975). This type of diffusion occurs in polymers that are hard or glassy, and disappears above the glass transition temperature (Marom, 1985). Sorption is affected in anomalous systems by the presence of "pre-existing" microcavities in the polymer matrix. Resultant dual-mode sorption isotherms contain additive contributions from Henry's law and Langmuir absorption (Frisch and Stern, 1981; Fenelon, 1974).

Penetrant sorption and diffusion in rubbery polymers is generally Fickian and can be described by either molecular or free-volume models. Molecular models assume that thermal energy–induced polymer segmental motions lead to the formation of microcavities or holes. Penetrant molecules jump from hole to hole, resulting in a net transfer of penetrant from the high to the low concentration side of the film. Free-volume models do not examine the microstructure of the films. Instead they rely on statistical estimates of the free (empty) volume of the system and relate these to diffusion coefficients. Various molecular and free-volume models are discussed in detail by Frisch and Stern (1981).

Little research has been performed to characterize the glass transition temperatures of hydrophilic edible film systems. It is felt that many of the anomalous effects observed in edible films could be due to glass transition–induced structural changes within films exposed to high RH gradients. Protein films, for example, at 0% RH would be expected to have extremely low glass transition temperatures. However, as RH increases, the glass transition temperatures are expected to increase greatly. Conceivably, when films are exposed to large RH gradients, a glass transition could occur at some point in the film during testing. This transition could then result in microcracks in films, leading to anomalous effects. More research on edible films is needed in this area.

SOLUBILITY AND DIFFUSIVITY ANALYSES

Analyses of the solubility and diffusivity of penetrants in edible barriers can provide essential information for determining the mechanisms of mass transfer through films. These values are also extremely useful in developing models for predicting the permeability properties of edible films under a variety of conditions. Earlier, this chapter dealt with the Schwartzberg (1986) model for determining solubility and diffusivity coefficients in order to more accurately determine film permeability values. In this section, other methods for analyzing the solubility and diffusivity properties of edible films will be examined.

DETERMINING THE SOLUBILITY OF PENETRANTS IN EDIBLE FILMS

The solubility of water vapor in edible films is very complex, due to the plasticizing and/or clustering tendencies of sorbed water. Consequently, both solubility and diffusivity values for hydrophilic edible films are highly dependent on the RH conditions during testing. (Myers et al., 1962).

The first step in determining the solubility of water in edible films is to examine their sorption isotherms. One of the simplest methods for obtaining a sorption isotherm is described by Labuza (1984). This method has been employed by several edible film researchers to examine the water sorption behavior of edible films (Rico-Peña and Torres, 1990; Biquet and

Labuza, 1988). In this method, film samples are placed in sealed containers at controlled RH and temperature. Enclosed saturated salt solutions must be prepared according to ASTM E104 (1985). Constant temperature conditions are esstenial within 0.01 °C. Inadequate temperature control is one of the most common errors in water activity measurements. Equilibrium is assumed to be established when the weight gain or loss is less than 0.001 g/g dry solids for three consecutive weighings at greater than five-day intervals (Biquet and Labuza, 1988). After equilibrium is reached, samples are immediately tested for their respective moisture contents using either the Karl Fisher (Biquet and Labuza, 1988) or gravimetric methods (Rico-Peña and Torres, 1990). Sorption isotherms for methylcellulose-palmitic acid (Rico-Peña and Torres, 1990) and chocolate (Biquet and Labuza, 1988) edible films were highly nonlinear, depending heavily on the RH conditions of the test. Biquet and Labuza (1988) applied the Guggenheim-Anderson-DeBoer (GAB) equation to the sigmoidal shapes of chocolate isotherms.

Estimates of sorption capacity can also be made from the chemical structure of functional groups using the Group Contribution Method (Barrie, 1968). This method assumes that each functional group contributes additively to the total water sorption of the structure. The sorption capacities of various functional groups have been calculated at various water activities and the sorption capacity of the structure can be determined by the sum of these values divided by the molecular weight of the compound. These values can then be plotted as a function of the partial vapor pressure to obtain an isotherm.

The solubility of gases in polymers is fairly difficult to determine since it cannot be measured gravimetrically (Bixler and Sweeting, 1971). In 1958, Michaels and Parker developed a volumetric method for determining the solubility of gases in polymer films, and this technique could be applied to edible films. Films are wrapped in a coil with nonsorbing spacers to maximize gas access to all film surfaces. The film is first degassed and then allowed to sorb the test gas at constant pressure. Once the system reaches equilibrium, the void volume is quickly evacuated to less than 0.1 mmHg pressure. The system is then isolated from the pump and desorption equilibrium is determined. Isothermal conditions must be maintained throughout the system. The solubility constant is then calculated using Equation (7.26) where p_1 is the initial sorption equilibrium pressure, p_2 is the desorption equilibrium pressure, V_p is the polymer volume, and V_v' is the void volume of the system.

$$Sp_1V_p = p_2(V_v' + SV_p) \qquad (7.26)$$

Manometric methods have been traditionally employed to measure equilibrium vapor pressures of food materials (Legault et al., 1948). Sorption isotherms have also been measured by the quartz helix and tritiated water techniques (Myers et al., 1961, 1962). Researchers interested in exploring the sorption properties of edible films are recommended to review the above listed papers for a more detailed description of the methods.

Film isotherms can be analyzed in various ways to determine the solubility of penetrants in films. Case I, Fickian films have the simplest isotherms to evaluate since they are linear. Henry's law can be used to determine S, the solubility coefficient of these films. The slope of the isotherm multiplied by the density of the film and the molecular weight of water yields the solubility coefficient, $g/cm^3 \cdot mmHg$. In general, as mentioned earlier, edible films are hydrophilic and therefore exhibit nonlinear sorption isotherms. Therefore, numerous alternative models exist for analysis of hydrophilic film solubilities. These methods are reviewed by Bixler and Sweeting (1971). Langmuir, Braunauer, Emmett and Teller (BET) and Guggenheim-Anderson-DeBoer (GAB) expressions can be used in combination with Henry's law to predict the solubility coefficients of hydrophilic films (Stannett et al., 1970; Peppas and Khanna, 1980).

DETERMINING THE DIFFUSIVITY OF PENETRANTS IN EDIBLE FILMS

The diffusion coefficients of barrier films can be determined from the time lag period in a plot of water vapor transmission rate. Theoretically, diffusivity can be calculated using Equation (7.27) where x is the film thickness and θ is the time lag. The concentrations of penetrant on both sides of the film prior to testing must be zero for this equation to be valid. This relationship assumes that there is no interaction between the permeant and the film (Crank, 1975). When examining edible film systems that interact with the permeant, this method can only be applied to determine the effective diffusivity (D_{eff}) for a given concentration gradient.

$$D_{eff} = \frac{x^2}{6\theta} \qquad (7.27)$$

Diffusivity can also be determined using Equations (7.28) and (7.29) for a finite slab (Crank, 1975). Rates of water vapor loss can be determined gravimetrically. Then the effective diffusion coefficient can be calculated from the sorption and desorption curves at long times. Koelsch (1992) used this method to determine the diffusivity of water vapor in methylcellulose films.

$$\ln \Gamma = \ln \frac{(m_i - m_{eq})}{(m_t - m_{eq})} = (\phi_{int})t + \ln \frac{\pi^2}{8} \qquad (7.28)$$

$$\phi_{int} = \frac{D_{eff}\pi^2}{4L_o^2} \qquad (7.29)$$

m_i initial moisture content of the barrier (g H_2O/g solids)

m_{eq} equilibrium moisture content of the barrier (g H_2O/g solids)

t time (sec)

D_{eff} effective diffusion coefficient (m^2/sec)

L_o half thickness of the edible barrier (m)

Γ unaccomplished moisture change

FACTORS AFFECTING PERMEABILITY

CHEMICAL NATURE OF THE POLYMER

Chemical nature plays a major role in the barrier properties of edible films. For example, highly polar polymers, such as many proteins and polysaccharides, exhibit large degrees of hydrogen bonding, resulting in extremely low gas permeability values at low RHs. Unfortunately, chemical nature that leads to good gas barrier properties often results in poor moisture barrier properties. Highly polar polymers containing hydroxyl groups are often poor moisture barriers. On the other hand, nonpolar hydrocarbon-based materials such as lipids exhibit the reverse effects, acting as excellent moisture barriers and less effective gas barriers. Table 7.4 illustrates for a repeating $(CH_2\text{-}CHX)_n$ polymeric structure the relative effects of various functional groups on oxygen permeability (Ashley, 1985). The presence of double bonds is also known to increase the permeability properties of films (Pascat, 1985).

Most good oxygen barriers contain at least one of the first five functional groups listed in Table 7.4. These structures create strong polymer chain interactions, thereby restricting chain motion. With the exception of Cl-, these functional groups also readily hydrogen bond with water and result

TABLE 7.4. Effect of Functional Groups on Oxygen Permeability.

Nature of X in $(CH_2\text{-}CHX)_n$	Oxygen Permeability[a]
-OH	0.01
-CN	0.04
-Cl	8.0
-F	15.0
-COOCH₃	17.0
-CH₃	150.0
-C₆H₅	420.0
-H	480.0

[a]Relative values of oxygen permeability with units of cm³·mil/d·100 in.²·atm.

in water absorption at high RHs. The high solubility coefficients of water in these polymers are responsible for their high rates of water vapor permeation. Furthermore, absorption of water disrupts intermolecular chain interactions, leading to increased diffusivity.

Low molecular weight additives are commonly added to polymer film. These additives, depending on their chemical structure, can either improve or reduce the barrier properties of edible films. Most edible film plasticizers are polyols. Because they disrupt polymer chain hydrogen bonding, they tend to increase WVP of films. The oxygen permeability properties of plasticized films, can, however, remain very low at low RHs (McHugh and Krochta, 1994a). The potential of filler materials has not been explored by edible film researchers and may have potential for reduction of film moisture and/or gas permeabilities.

STRUCTURAL NATURE OF THE POLYMER

The structure of the polymer chains also plays a significant role in the permeability of the films. The effect of chain packing is important. Simple linear polymer chains, such as high-density polyethylene, pack tightly, thereby lowering film permeability values. Polymer chains with bulky side groups pack poorly, resulting in increased permeabilities (Ashley, 1985).

Polymer crystallinity also dramatically affects permeability. Crystalline structures are extremely tightly packed and tend to be impermeable. Materials, such as lipids, can exist in numerous crystalline states. Each of these states is expected to cause films to display different barrier properties. In general, the higher the degree of crystallinity, the lower the permeability.

The glass transition temperature of the film also affects permeability. Above the glass transition temperature, polymers exist in a "rubbery" or "plastic" amorphous state where chain mobility lessens barrier properties. At temperatures below the glass transition temperature, polymers exist as "glasses" in a brittle form. Therefore, the permeability of films below their glass transition temperatures tends to be extremely low.

Orientation of polymers is a practice commonly employed in the synthetic film industry which has not yet been examined by edible film researchers. Orientation of synthetic packaging materials can result in up to 50% reductions in permeation. Therefore, it is expected that orientation of edible films could result in similar decreases in film permeabilities.

Cross-linking of polymer chains, enzymatically or chemically, could also produce a lowering in permeability values due to a resultant decrease in diffusivity of the permeant. The effect is particularly prominent for vapors of large molecular size.

NATURE OF THE PERMEANT

The nature of the permeant also affects its transfer through the film. Small molecules generally diffuse faster than larger ones. Straight chain molecules diffuse much faster than bulky molecules of the same molecular weight. Polar molecules diffuse faster than nonpolar molecules, especially through polar films. This is due to the increased solubility of polar molecules in polar films. The solubility of the permeant depends in large part on its compatibility with the polymer. Differences in the solubility of gases like N_2, O_2, CO_2, and H_2O relate to the facility of condensation of these gases at a given temperature. More easily condensed permeants are generally more soluble. In many cases the permeabilities and solubilities of different gases in polymers relate to their critical temperatures and boiling points respectively (Pascat, 1985).

EXPERIMENTAL CONDITIONS

Experimental test conditions affect film permeabilities. Both temperature and RH are known to exhibit exponential effects on the permeability of edible films.

Permeability, diffusivity, and solubility coefficients vary exponentially with temperature according to the Arrhenius law, as in Equations (7.30), (7.31), and (7.32) (Pascat, 1985). In these equations, E is the activation energy of permeability (P), diffusion (D) or sorption (S), R is the universal gas constant, and T is absolute temperature. At temperatures near the glass transition temperature, pronounced variations in P, D, and S can occur. The temperature dependence of the permeability of edible films has been studied by numerous researchers.

$$P = P_o \exp(-E_p/RT) \qquad (7.30)$$

$$D = D_o \exp(-E_D/RT) \qquad (7.31)$$

$$S = S_o \exp(-E_s/RT) \qquad (7.32)$$

Permeability is independent of the partial pressure of the diffusing gas in systems where there is no interaction between the penetrant and the polymer. However, permeation through edible films often involves interactions of this sort. This effect is illustrated in Figure 7.2 in this chapter. The partial pressure gradient of water vapor is shown to have an exponential effect on the WVP of whey protein based edible films. This occurs because of increased diffusivity caused by a plasticizing effect and increased solubility due to the shape of the films sorption isotherm (Pascat, 1985).

Once again, accurate control of experimental conditions is essential for accurate determination of film permeabilities.

WATER VAPOR AND GAS PERMEABILITY VALUES

WATER VAPOR PERMEABILITY

One of the most important properties of an edible film is its water vapor permeability. Table 7.5 lists the WVPs of several edible films. When comparing these values, it is important to recognize two items. First, different conditions were employed for the testing of these films. Therefore, experimental differences must be taken into account when examining these data. For example, RH effects must be considered. Table 7.5 shows that the effect of RH on WVP of edible films is substantial. RH was shown to have an exponential effect on WVP of whey protein films in Figure 7.2. Therefore, small differences in RH during testing can result in drastic changes in permeability. Second, assumed test RHs are not always true experimental conditions, resulting in errors when RH test conditions are not corrected for by using the WVP Correction Method.

Protein-based edible films have relatively high WVPs. The water vapor permeability of a gluten:glycerin (3.1:1) film is several times larger than that of a zein:glycerin (4.9:1) film at the same test conditions. This is probably attributable to both the larger relative amount of plasticizer used in the gluten:glycerin film and the greater hydrophobicity of zein. The effect of RH on WVP is shown in Table 7.5 for gluten and whey protein films. Both exhibit increased WVPs at elevated RH. Sodium caseinate films, adjusted to their isoelectric points, were shown to have lower WVP than films formed at neutral pH. Furthermore, increasing concentrations of glycerin in whey protein films resulted in increased WVPs. The addition of lipid to protein films was found to decrease the WVP of both caseinate and whey protein films. Films formed from protein emulsion systems exhibit good water barrier properties.

Polysaccharide films are generally rather poor water barriers, due to their hydrophilic nature. However, most pure polysaccharide films do not require the addition of plasticizers, and therefore tend to exhibit lower WVPs than most protein films. The addition of lipids to polysaccharide-based films resulted in films that are exceptionally good barriers to water vapor.

TABLE 7.5. Comparison of Water Vapor Permeabilities of Edible and Nonedible Films.

	Film	Thickness mm	Conditions[a]	Permeability g·mm/m²·d·kPa
Protein-Based Edible Films				
Park & Chinnan (1990)	Zein:Glycerin (4.9:1)	0.12–0.33	21°C, 85/0% RH	7.69–11.49
Park & Chinnan (1990)	Gluten:Glycerin (3.1:1)	0.38–0.42	21°C, 85/0% RH	52.1–54.4
Gennadios et al. (1990)	Gluten:Glycerin (2.5:1)	0.101	23°C, 0/11% RH	4.84
Aydt et al. (1991)	Gluten:Glycerin (2.5:1)	0.140	26°C, 50/100% RH	108.4
Avena-Bustillos & Krochta (1993)	Sodium Caseinate (SC)	0.083	25°C, 0/81% RH[b]	36.7
Avena-Bustillos & Krochta (1993)	Buffer Treated (pH 4.6) SC	0.072	25°C, 0/86% RH[b]	22.3
Avena-Bustillos & Krochta (1993)	SC:Acetylated Monoglyceride (2:8)	0.040	25°C, 0/84% RH[b]	15.8
McHugh et al. (1994)	Whey Protein:Glycerin (1.6:1)	0.106	25°C, 0/11% RH[b]	6.64
McHugh et al. (1994)	Whey Protein:Glycerin (1.6:1)	0.121	25°C, 0/65% RH[b]	119.8
McHugh et al. (1994)	Whey Protein:Sorbitol (1.6:1)	0.129	25°C, 0/79% RH[b]	62.0
McHugh et al. (1994)	WP:Beeswax:Sorbitol (3.5:1.8:1)	0.162	25°C, 0/94% RH[b]	20.4
Polysaccharide-Based Edible Films				
Hagenmaier & Shaw (1990)	Hydroxypropylmethylcellulose	0.019	27°C, 0/85% RH	9.12
Hagenmaier & Shaw (1990)	HPMC[c]/Stearic Acid (0.8/1)	0.019	27°C, 0/85% RH	0.026
Kamper & Fennema (1984b)	HPMC:PEG (9:1)	0.036	25°C, 85/0% RH	6.48
Kamper & Fennema (1984a)	SA:PA:HPMC:PEG[d]	0.041	25°C, 85/0% RH	0.048
Kamper & Fennema (1984b)	SA:PA:HPMC:PEG[d]	0.041	25°C, 97/0% RH	0.336
Kamper & Fennema (1984b)	SA:PA:HPMC:PEG[d]	0.041	25°C, 97/65% RH	1.92
Kester & Fennema (1989a)	BW/SA:PA:MC:HPMC:PEG	0.056	25°C, 97/0% RH	0.058
Greener & Fennema (1989)	M.BW/MC:PEG (11.3:4)[e]	0.05	25°C, 0/100% RH	0.096
Greener & Fennema (1989)	M.BW/MC:PEG (11.3:4)	0.05	25°C, 97/65% RH	1.48
Hagenmaier & Shaw (1990)	SA:HPMC (1.1:1)	0.019	27°C, 0/97% RH	0.106

TABLE 7.5. (continued).

Film		Thickness mm	Conditions[a]	Permeability g·mm/m²·d·kPa
Lipid Edible Films				
Lovegren & Feuge (1954)	Acetylated Monoglyceride		25°C, 100/0% RH	2.00–5.36
Lovegren & Feuge (1954)	Paraffin Wax		25°C, 100/0% RH	0.0190
Biquet & Labuza (1988)	Chocolate		20°C, 81/0% RH	1.06
Hagenmaier & Shaw (1991)	Shellac		30°C, 100/0% RH	0.40–0.571
Greener (1992)	Beeswax	0.04–0.05	25°C, 0/100% RH	0.0502
Greener (1992)	Microcrystalline	0.09–0.11	25°C, 0/100% RH	0.0292
Greener (1992)	Carnauba Wax	0.09–0.11	25°C, 0/100% RH	0.0285
Greener (1992)	Candelilla Wax	0.09–0.11	25°C, 0/100% RH	0.0152
Nonedible Packaging Films				
Smith (1986)	LDPE		38°C, 90/0% RH	0.079
Smith (1986)	HDPE		38°C, 90/0% RH	0.02
Karel (1975)	Saran		38°C, 95/0% RH	0.0086–0.034
Foster (1986)	EVOH (68% VOH)		38°C, 90/0% RH	0.252
Taylor (1986)	Cellophane		38°C, 90/0% RH	7.27

[a]RHs are those on top and bottom sides of film (top/bottom).
[b]Used corrected RH for bottom side of film.
[c]Hydroxypropylmethycellulose.
[d]Stearic acid:palmitic acid:HPMC:polyethylene glycol emulsion film.
[e]Beeswax applied in molten state.

179

Pure lipid films have extremely low WVPs. This is due to their hydrophobic, crystalline nature. Candelilla wax was shown to have the lowest permeability to water vapor of any of the lipids tested.

In Table 7.5, WVP values for various edible films are compared with those of nonedible packaging films. The WVP values of protein and polysaccharide films, shown in Table 7.5, indicate that these films do not possess good water barrier properties, compared to films like high-density polyethylene. Their overall hydrophilic nature renders them poor water barriers. However, by incorporating lipids into these films, WVPs can be reduced. Pure lipid and many composite films exhibit water barrier properties comparable to those of nonedible packaging films.

OXYGEN PERMEABILITY PROPERTIES

Oxygen barrier properties of several edible films are compared in Table 7.6. As with WVP, one must use caution when comparing data, due to experimental variations in temperature and RH.

Protein films are excellent barriers to gases, particularly at low RHs. However, as RH is increased, gas permeability increases as well. Whey protein films have extremely low oxygen permeabilities, even at relatively high RHs. Table 7.6 indicates that protein films have oxygen permeabilities lower than that of polyethylene at 23°C and 0% RH. Unfortunately, no data exists on the influence of RH on the oxygen permeabilities of zein, gluten or soy protein films. Polysaccharide films are also good barriers to oxygen at low RHs. However, no data is available concerning the effect of RH on these values. Finally, lipid films exhibit relatively poor oxygen barrier properties.

These results demonstrate the value of multicomponent edible films in order to optimize film barrier properties to water vapor and gas.

CONCLUSIONS

The importance of both accurate and precise methodologies cannot be overemphasized in edible film research. In this chapter an overview has been presented of currently available methodologies for determining barrier properties of edible films, as well as the relative advantages and disadvantages of these methods. Various means for calculating these barrier properties have also been compared. The hydrophilic nature of most edible films challenges the limits of methods developed to determine the permeabilities of synthetic polymer films. Therefore, alternative approaches for examining edible films have been presented.

TABLE 7.6. Comparison of Oxygen Permeabilities of Edible and Nonedible Films.

	Film	Thickness mm	Conditions[a]	Permeability cm³·μm/m²·d·kPa
Protein-Based Edible Films				
Lieberman & Gilbert (1973)	Collagen		23°C, 0% RH	1.2
Lieberman & Gilbert (1973)	Collagen		23°C, 63% RH	23.3
Lieberman & Gilbert (1973)	Collagen		23°C, 93% RH	890
Park & Chinnan (1990)	Zein:Glycerin (4.9:1)	0.10–0.31	30°C, 0% RH	13.0–44.9
Park & Chinnan (1990)	Gluten:Glycerin (3.1:1)	0.23–0.42	30°C, 0% RH	9.6–24.2
Gennadios et al. (1990)	Gluten:Glycerin (2.5:1)	0.101	23°C, 0% RH	3.82
Gennadios et al. (1990)	AM:Gluten:Glycerin (0.1:2.5:1)	0.066	23°C, 0% RH	2.67
Brandenburg (1992)	Soy Protein Isolate:Glycerin (1.7:1)	0.064–0.089	25°C, 0% RH	4.75
McHugh & Krochta (1994a)	Whey Protein:Sorbitol (1.5:1)	0.118	23°C, 30% RH	1.03
McHugh & Krochta (1994a)	Whey Protein:Sorbitol (1.5:1)	0.118	23°C, 75% RH	144.92
McHugh & Krochta (1994a)	Whey Protein:Sorbitol (1:1)	0.018	23°C, 40% RH	2.35
Polysaccharide-Based Edible Films				
Park & Chinnan (1990)	MC:PEG (9:1)[b]	0.04–0.07	30°C, 0% RH	149–226
Park & Chinnan (1990)	HPC:PEG (9:1)[c]	0.05	30°C, 0% RH	308
Park & Chinnan (1990)	HPC:PEG (9:3)[c]	0.10	30°C, 0% RH	910
Park & Chinnan (1990)	AM:HPC:PEG (9:9:3)[d]	0.15	30°C, 0% RH	298
Greener & Fennema (1989)	M.BW/MC:PEG (11.3:4)[e]	0.05	25°C, 0% RH	960
Greener & Fennema (1989)	S.BW/MC:PEG (11.3:4)[f]	0.05	25°C, 0% RH	319
Rico-Peña & Torres (1990)	PA:MC:PEG (3:9:1)	0.054	24°C, 0/57% RH	365–401
Rico-Peña & Torres (1990)	PA:MC:PEG (3:9:1)	0.054	24°C, 79% RH	785
Allen et al. (1963)	Starch		24°C	13,130
Mark et al. (1966)	Amylomaize		25°C	1480

(continued)

181

TABLE 7.6. (continued).

Film	Thickness mm	Conditions[a]	Permeability $cm^3 \cdot \mu m/m^2 \cdot d \cdot kPa$
Lipid Edible Films			
Greener (1992) Microcrystalline Wax	0.04–0.05	25°C, 0% RH	1540
Greener (1992) Beeswax	0.04–0.05	25°C, 0% RH	931.7
Greener (1992) Candelilla Wax	0.04–0.05	25°C, 0% RH	175.4
Greener (1992) Carnauba Wax	0.04–0.05	25°C, 0% RH	157.2
Lovegren & Feuge (1956) Acetostearin		26°C	1360
Hagenmaier & Shaw (1991) Shellac		30°C	54–157
Nonedible Packaging Films			
Salame (1986) Low Density Polyethylene		23°C, 50% RH	1870
Salame (1986) High Density Polyethylene		23°C, 50% RH	427
Salame (1986) Cellophane		23°C, 0% RH	0.7
Taylor (1986) Cellophane		23°C, 50% RH	16
Taylor (1986) Cellophane		23°C, 95% RH	252
Salame (1986) EVOH (70% VOH)		23°C, 0% RH	0.1
Salame (1986) EVOH (70% VOH)		23°C, 95% RH	12
National Bureau of Stds. Polyester SRM 1470	0.0254	23°C, 0% RH	17.3

[a]RHs are those on top and bottom sides of film (top/bottom).
[b]Methylcellulose:polyethylene glycol film.
[c]Hydroxypropylcellulose:polyethylene glycol.
[d]Acetylated monoglyceride:hydroxypropylcellulose:polyethylene glycol.
[e]Beeswax applied in molten state.
[f]Beeswax applied in solvent.

REFERENCES

Alfrey, F., E. F. Gurnee and W. G. Lloyd. 1966. "Classification System for Diffusion of Vapor through Films," *Journal of Polymer Science,* 12:249.

Allen, L., A. I. Nelson and M. P. Steinberg. 1963. "Edible Corn Carbohydrate Food Coatings. I. Development and Physical Testing of a Starch-Algin Coating," *Food Technology,* 17:1437–1442.

AOAC. 1984. *Official Methods of Analysis, 14th Ed.* Arlington, VA: Assoc. of Official Analytical Chemists.

Ashley, R. J. 1985. "Permeability and Plastic Packaging," *Polymer Permeability,* J. Comyn, ed., New York: Elsevier Applied Science Publishers, pp. 269–307.

ASTM. 1973. "Standard Test Method for Water Vapor Transmission Rate of Flexible Barrier Materials Using an Infrared Detection Technique," *ASTM Book of Standards,* F372-73:854–858.

ASTM. 1980. "Standard Test Methods for Water Vapor Transmission of Materials," *ASTM Book of Standards,* E96-80.

ASTM. 1981. "Standard Test Method for Oxygen Gas Transmission Rate through Plastic Film and Sheeting Using a Coulometric Sensor," *ASTM Book of Standards,* D3985-81:607–612.

ASTM, 1982. "Standard Test Method for Rate of Grease Penetration of Flexible Barrier Materials (Rapid Method)," *ASTM Book of Standards,* F119-82:825–827.

ASTM. 1985. "Standard Practice for Maintaining Constant Relative Humidity by Means of Aqueous Solutions," *ASTM Book of Standards,* E104-85:695–696.

ASTM. 1988. "Standard Test Method for Determining Gas Permeability Characteristics of Plastic Film and Sheeting," *ASTM Book of Standards,* D1434-82:255–266.

ASTM. 1990. "Standard Test Method for Water Vapor Transmission Rate through Plastic Film and Sheeting Using a Modulated Infrared Sensor," *ASTM Book of Standards,* F1249-90:1–5.

Avena-Bustillos, R. J. and J. M. Krochta. 1993. "Water Vapor Permeability of Caseinate-Based Edible Films as Affected by pH, Calcium Crosslinking, and Lipid Content," *Journal of Food Science,* 58(4):904–907.

Aydt, T. P., C. L. Weller and R. F. Testin. 1991. "Mechanical and Barrier Properties of Edible Corn and Wheat Protein Films," *American Society of Agricultural Engineers,* 34(1):207–211.

Baner, A. L., R. J. Hernandez, K. Jayaraman and J. R. Giacin. 1986. "Isostatic and Quasi-Isostatic Methods for Determining the Permeability of Organic Vapor through Barrier Membranes," *Current Technologies in Flexible Packaging, ASTM STP 921,* M. L. Troedel, ed., Philadelphia: American Society for Testing and Materials, pp. 49–62.

Barrie, J. A. 1968. "Water in Polymers," *Diffusion in Polymers,* J. Crank and G. S. Park, eds., London: Academic Press, pp. 259–314.

Biquet, B. and T. P. Labuza. 1988. "Evaluation of the Moisture Permeability Characteristics of Chocolate Films as an Edible Moisture Barrier," *Journal of Food Science,* 53(4):989–997.

Bird, R. B., W. E. Stewart and E. N. Lightfoot, eds. 1960. *Transport Phenomena.* New York: John Wiley and Sons.

Bixler, H. J. and O. J. Sweeting. 1971. "Barrier Properties of Polymer Films," *The Sci-*

ence and Technology of Polymer Films, O. J. Sweeting, ed., New York: John Wiley & Sons, Inc. pp. 1–130.

Brandenburg, A. H., C. L. Weller and R. F. Testin. 1992. "Edible Films and Coatings from Soy Protein," presented at the *1992 Annual Meeting of the Institute of Food Technologists,* June 20–24, New Orleans, LA.

Clegg, M. D. 1978. "A Sensitive Technique for the Rapid Measurement of Carbon Dioxide Concentrations," *Plant Physiol.,* 62:924–926.

Crank, J. 1975. *The Mathematics of Diffusion.* Oxford, England: Clarendon Press.

Davis, E. G. 1965. "Water Vapor Permeability of Food Packaging Materials," *CSIRO Food Preservation Quarterly,* 25(2):21–26.

Davis, E. G. and J. N. Huntington. 1977. "New Cell for Measuring the Permeability of Film Materials," *CSIRO Food Research Quarterly,* 37:55–59.

Demorest, R. L. 1989. "Measuring Oxygen Permeability through Today's Packaging Barriers," *Journal of Packaging Technology* (October).

Fenelon, P. J. 1974. "Experimantal Support of 'Dual Sorption' in Glassy Polymers," *Permeability of Plastic Films and Coatings to Gases, Vapor, and Liquids,* H. B. Hopfenberg, ed., New York: Plenum Press, pp. 285–299.

Foster, R. 1986. "Ethylene-Vinyl Alcohol Copolymers (EVOH)," *The Wiley Encyclopedia of Packaging Technology,* M. Bakker, ed., New York: John Wiley & Sons, pp. 270–275.

Frisch, H. L. and S. A. Stern. 1981. "Diffusion of Small Molelcules in Polymers," *CRC Critical Reviews in Solid State and Materials Sciences,* II(2):123–187.

Gennadios. A., C. L. Weller and R. F. Testin. 1990. "Modification of Properties of Edible Wheat Gluten Films," *American Society of Agricultural Engineers Meeting Presentation,* Paper No. 90-6504:1–18.

Gilbert, S. G. and D. Pegaz. 1969. "Find New Way to Measure Gas Permeability," *Package Engineering* (January):66–69.

Greener, I. K. 1992. "Physical Properties of Edible Films and their Components," Ph.D. thesis, University of Wisconsin, Madison.

Greener, I. K. and O. Fennema. 1989. "Barrier Properties and Surface Characteristics of Edible, Bilayer Films," *J. of Food Science,* 54:1393–1399.

Hagenmaier, R. D. and P. E. Shaw. 1990. "Moisture Permeability of Edible Films Made with Fatty Acid and Hydroxypropylmethylcellulose," *J. Agricultural Food Chemistry,* 38:1799–1803.

Hagenmaier, R. D. and P. E. Shaw. 1991. "Permeability of Shellac Coatings to Gases and Water Vapor," *J. of Agricultural and Food Chemistry,* 39(5):825–829.

Hauser, P. M. and A. D. McLaren. 1948. "Permeation through and Sorption of Water Vapor by High Polymers," *Industrial and Engineering Chemistry,* 40(1):112–117.

Holland, R. V. and R. A. Santangelo. 1984. "Packaging Films: New Techniques in Permeability Measurements," *CSIRO Food Research Quarterly,* 44:20–22.

Kamper, S. L. and O. Fennema. 1984a. "Water Vapor Permeability of Edible Bilayer Films," *J. Food Science,* 49:1478–1481, 1485.

Kamper, S. L. and O. Fennema. 1984b. "Water Vapor Permeability of an Edible, Fatty Acid, Bilayer Film," *J. Food Science,* 49:1482–1485.

Karel, M. 1975. "Protective Packaging of Foods," *Principles of Food Science, Part II. Physical Principles of Food Preservation,* O. Fennema, ed., New York: Marcel Dekker, pp. 399–464.

Karel, M., P. Issenberg, L. Ronsivalli and V. Jurin. 1963. "Application of Gas Chromatography to the Measurement of Gas Permeability of Packaging Materials," *Food Technology* (March):91–94.

Kester, J. J. and O. Fennema. 1989a. "An Edible Film of Lipids and Cellulose Ethers: Barrier Properties to Moisture Vapor Transmission and Structural Evaluation," *J. of Food Science*, 54:1383–1389.

Kester, J. J. and O. Fennema. 1989b. "Resistance of Lipid Films to Oxygen Transmission," *J. of Americal Oil Chemists Society*, 66(8):1129–1138.

Kester, J. J. and O. Fennema. 1989c. "Resistance of Lipid Films to Water Vapor Transmission," *J. of American Oil Chemists Society*, 66(8):1139–1146.

Kitic, D., D. C. P. Jardim, G. J. Favetto, S. L. Resnik and J. Chirife. 1986. "Theoretical Prediction of the Water Activity of Standard Saturated Salt Solutions at Various Temperatures," *J. of Food Sci.*, 51(4):1037–1040.

Koelsch, C. M. 1992. "The Functional, Physical and Morphological Properties of Edible Fatty Acid and Methyl Cellulose Based Barriers," Ph.D. thesis, University of Minnesota.

Krochta, J. M. 1992. "Control of Mass Transfer in Foods with Edible Coatings and Films," *Advances in Food Engineering*, R. P. Singh and M. A. Wirakartakusumah, eds., Boca Raton, FL: CRC Press, Inc., pp. 517–538.

Labuza, T. P. 1984. *Moisture Sorption: Practical Aspects of Isotherm Measurement and Use*. St. Paul, MN: American Association of Cereal Chemists.

Landrock, A. H. and B. E. Proctor. 1952. "Gas Permeability of Films," *Modern Packaging*, 25:131–201.

Legault, R. R., B. Makower and W. F. Talburt. 1948. "Apparatus for Measurement of Vapor Pressure," *Analytical Chemistry*, 20(5):428–430.

Lelie, H. J. 1969. "Coulometric Cell Determination of WVT," *Modern Packaging*, 42(1):91–96.

Lieberman, E. R. and S. G. Gilbert. 1973. "Gas Permeation of Collagen Films as Affected by Cross-Linkage, Moisture, and Plasticizer Content," *Journal of Polymer Science*, 41:33–43.

Lieberman, E. R., S. G. Gilbert and V. Srinivasa. 1972. "The Use of Gas Permeability as a Molecular Probe for the Study of Cross-Linked Collagen Structures," *Transactions NY Academy of Sciences II*, 34:694–707.

Lovegren, N. V. and R. O. Feuge. 1954. "Permeability of Acetostearin Products to Water Vapor," *J. of Agric. Food Chemistry*, 2:558–563.

Lovegren, N. V. and R. O. Feuge. 1956. "Permeability of Acetostearin Products to Carbon Dioxide, Oxygen, and Nitrogen," *J. of Agric. Food Chemistry*, 4:634–638.

Mark, A. M., W. B. Roth, C. L. Mehitretter and C. E. Rist. 1966. "Oxygen Permeability of Amylomaize Starch Films," *Food Technology*, 20(1):75–77.

Marom, G. 1985. "The Role of Water in Composite Materials," *Polymer Permeability*, J. Comyn, ed., New York: Elsevier Science Publishing Co., Inc., pp. 341–374.

McHugh, T. H. and J. M. Krochta. 1994a. "Sorbitol- vs. Glycerol-Plasticized Whey Protein Edible Films: Integrated Oxygen Permeability and Tensile Property Evaluation," *Journal of Agricultural and Food Chemistry*, 42(4).

McHugh. T. H. and J. M. Krochta. 1994b. "Water Vapor Permeability Properties of Edible Whey Protein-Lipid Emulsion Films," *Journal of American Oil Chemists Society*, 71(3).

McHugh, T. H., J. F. Aujard and J. M. Krochta. 1994. "Plasticized Whey Protein Edible Films: Water Vapor Permeability Properties," *Journal of Food Science,* 59(1).

McHugh, T. H., R. Avena-Bustillos and J. M. Krochta. 1993. "Hydrophilic Edible Films: Modified Procedure for Water Vapor Permeability and Explanation of Thickness Effects," *Journal of Food Science,* 58(4):899–903.

Michaels, A. S. and R. B. Parker. 1958. "The Determination of Solubility Constants for Gases in Polymers," *J. of Physical Chemistry,* 62:1604.

Miltz, J. and S. Ulitzur. 1980. "A Bioluminescence Method for the Determination of Oxygen Transmission Rates through Plastic Films," *Journal of Food Technology,* 15:389–396.

Myers, A. W., J. A. Meyer, C. E. Rogers, V. Stannett and M. Szwarc. 1961. "Studies in the Gas and Vapor Permeability of Plastic Films and Coated Papers," *TAPPI,* 44(1):58–64.

Myers, A. W., J. A. Meyer, C. E. Rogers, V. Stannett and M. Szwarc. 1962. "The Permeation of Water Vapor," *Permeability of Plastic Films and Coated Paper to Gases and Vapors,* V. Stannett, M. Szwarc, R. Bhargave, J. A. Meyer, A. W. Myers and C. E. Rogers, eds., Easton, PA: Mack Printing Company, pp. 62–77.

Nelson, K. L. and O. R. Fennema. 1991. "Methylcellulose Films to Prevent Lipid Migration in Confectionery Products," *Journal of Food Science,* 56(1):504–509.

Park, H. J. and M. S. Chinnan. 1990. "Properties of Edible Coatings for Fruits and Vegetables," *American Society of Agricultural Engineers Paper No. 90-6510,* St. Joseph, MI.

Pascat, B. 1985. "Study of Some Factors Affecting Permeability," *Polymer Permeability,* J. Comyn, ed., New York: Elsevier Applied Science Publishing, pp. 7–24.

Peppas, N. A. and R. Khanna. 1980. "Mathematical Analysis of Transport Properties of Polymer Films for Food Packaging," *Polymer Engineering,* 20(17):1147–1156.

Rico-Peña, D. C. and J. A. Torres. 1990. "Oxygen Transmission Rate of an Edible Methylcellulose-Palmitic Acid Film," *Journal of Food Processing Engineering,* 13:125–133.

Rockland, L. B. 1960. "Saturated Salt Solutions for Static Control of Relative Humidity between 5° and 40° C," *Analytical Chemistry,* 32:1375.

Sacharow, S. 1971. "Permeability and Food Packaging," *Food Product Development,* 4(5):51–52.

Salame, M. 1986. "Barrier Polymers," *The Wiley Encyclopedia of Packaging Technology,* M. Bakker, ed., New York: John Wiley & Sons, pp. 48–54.

Schultz, T. H., J. C. Miers, H. S. Owens and W. D. Maclay. 1949. "Permeability of Pectinate Films to Water Vapor," *Journal of Physical Colloidal Chemistry,* E104-85:695–696.

Schwartzberg, H. G. 1986. "Modelling of Gas and Vapour Transport through Hydrophilic Films," *Food Packaging and Preservation,* M. Mathlouthi, ed., New York: Elsevier Science Publishing Co., Inc., pp. 115–135.

Smith, S. A. 1986. "Polyethylene, Low Density," *The Wiley Encyclopedia of Packaging Technology,* M. Bakker, ed., New York: John Wiley & Sons, pp. 514–523.

Stannet, V. T., W. J. Koros, D. R. Paul, H. K. Lonsdale and R. W. Baker. 1970. "Recent Advances in Membrane Science and Technology," *Membrane Science and Technology: Industrial, Biological, and Waste Treatment,* J. E. Flinn, ed., New York: Plenum Press, pp. 71–121.

Stoloff, L. 1978. "Calibration of Water Activity Measuring Instruments and Devices: Collaborative Study," *Journal of Official Analytical Chemistry,* 61:1166.

Taylor, C. C. 1986. "Cellophane," *The Wiley Encyclopedia of Packaging Technology,* M. Bakker, ed., New York: John Wiley & Sons, pp. 159–163.

Toas, M. 1989. "Results of the 1985 Round-Robin Test Series Using ASTM E96-80," *Water Vapor Transmission through Building Materials and Systems: Mechanisms and Measurement. ASTM STP 1039,* H. R. Trechsel and M. Bomberg, eds., Philadelphia: American Society for Testing and Materials, pp. 73–90.

Todd, H. R. 1944. "Apparatus for Measuring Gas Transmission through Sheets and Films," *Paper Trade Journal,* 118(10):32–35.

Unknown. 1942. "Moisture-Vapor Test Cabinet," *Modern Packaging,* 16(3):76–82, 100.

Wink, W. A. and G. R. Sears. 1950. "Equilibrium Relative Humidities above Saturated Salt Solutions at Various Temperatures," *TAPPI,* 33(9):96A.

Zimm, B. H. and Lundberg, J. L. 1956. "Sorption of Vapors by High Polymers," *Journal of Physical Chemistry,* (60):425–428.

CHAPTER 8

Application of Coatings[1]

LUCIE AUSTIN GRANT[2]
JACKIE BURNS[3]

INTRODUCTION

THE practice of coating fruits and vegetables dates back to the twelfth and thirteenth centuries, where, in China, commodities were commonly dipped in hot waxes to encourage fermentation (Yehoshua, 1987; Kaplan, 1986). Today, coatings are used for fresh fruits and vegetables to retard moisture loss, improve appearance by imparting shine to the surface, provide a carrier for fungicides or growth regulators, and create a barrier for gas exchange between the commodity and the external atmosphere. In the United States, early coating procedures utilized either dipping, brushing, or individual wrapping in oiled papers to disperse the material over the surface of the commodity. Later, commodities were coated by spraying materials onto rollers or brushes, then allowing the tumbling action of the commodity to evenly spread the coating. Paraffin and beeswax, dispersed in an organic solvent, castor oil, and mineral oil, were used alone or in combination either in a dip, slab-wax process, hot fog, or spray method (Kaplan, 1986; Hardenburg, 1967; Long and Leggo, 1959; Trout et al., 1953; Platenius, 1939; Brooks, 1937; Magness and Diehl, 1924). Current formulations, upon demand of a dynamic fresh fruit and vegetable industry, are applied in a variety of ways and contain a variety of GRAS (generally recognized as safe) ingredients or ingredients approved by the FDA Code of Federal Regulations.

[1]Florida Agricultural Experiment Station Journal Series No. N-00618.
[2]American Machinery Corporation, Orlando, FL 32802.
[3]Citrus Research and Education Center, University of Florida, Institute of Food and Agricultural Sciences, 700 Experiment Station Rd., Lake Alfred, FL 33850.

The coating process for tablets and candies dates back to the eleventh century. Tablets were either hand-rolled or dipped into the coating mixture and allowed to air dry. Later, heat was employed immediately after coating to increase uniformity and hasten drying (Rowell, 1950). Tilted vessels, rotated by hand or by electric motor, heated by open flame and later hot forced air, were used to increase the speed of coating. Coatings were often organic solvent-based. Today, modifications in air-flow design of this basic system are used to adequately dry and polish aqueous-based coatings.

COMMODITIES WHICH ARE COATED

FRUITS

Edible coatings are applied commercially to citrus fruit, apples, and pears to improve gloss, control weight loss, and to carry postharvest fungicides. The first commercial coatings for citrus were applied sixty years ago (Kaplan, 1986). These coatings, largely paraffin-based, were utilized for weight loss control. Later, the desire to impart shine shifted the marketing emphasis from weight loss to the ability of the coating material to impart both weight loss control and shine or gloss. Resin solution water coatings containing shellac and/or alkali-soluble resins, film-forming adjuvants, plasticizers, and emulsifiers are common in the fresh cirtus fruit industry and are used to increase shine and marketability. Resin-based coatings are most often utilized for U.S. domestic markets, whereas the largely shellac-based coatings are used for some export markets. Although both types of coatings impart shine, the durability of shellac-based coatings results in their use when commodities must be stored and shipped for prolonged periods of time. If shellac-coated fruits experience too many changes in temperature and humidity, however, causing them to "sweat," the shellac coatings may whiten, causing the fruit to become unmarketable.

Fungicides are often incorporated into citrus coatings for simplicity and economy. However, effectiveness of decay control is often reduced when fungicides are dissolved into the coating. Two- to threefold increases in fungicide concentration must be used to obtain the same decay control as that of pre-coating aqueous solutions (Eckert and Brown, 1986).

Emulsion coatings containing polyethylene are not as common. Carnauba, and to a lesser extent candelilla and paraffin wax, are used for storage coatings, especially for lemons. In Florida, limes are commonly coated with a mineral-oil emulsion; but as shine becomes more important to marketing, high-shine emulsion coatings are becoming more popular. Coumarone-indene resins or the calcium salts of natural wood rosins dissolved in a blend of petroleum solvents, the so-called solvent waxes (Hall, 1981), continue to be used in the Florida citrus industry. However, their use for export fruit is restricted; and international interest in "natural" products, as well as their high cost, has resulted in a decline in popularity of solvent waxes.

Although apples have been coated commercially for nearly seventy-five years with emulsions containing castor oil, mineral oil, paraffin, and shellac solutions in alcohol (Trout et al., 1953), side-by-side marketing with citrus necessitated the incorporation of high-shine components to apple coatings. As with citrus, apples are usually drenched with postharvest fungicides before entering the packinghouse. However, unlike citrus, apple coatings currently contain very little or no postharvest fungicides. Their main components impart shine with minor emphasis on weight loss control. Because the peel is routinely consumed, only GRAS components with little taste are incorporated into apple coatings. Water-based resin solutions or emulsions containing shellac or carnauba wax mixed with shellac are commonly used, depending on destination. Pears are coated for the same general purposes as citrus and apples; however, the coating (often an apple coating) is diluted twofold. Indeed, the current trend in the pear industry is to not coat at all, but rather to use water for lubrication and ease of transport through the packingline equipment.

Stone fruit, such as peaches, plums, nectarines, and apricots are coated primarily to carry fungicides, but also to control weight loss. The coating formulation may be slightly acidic to avoid peel injury, especially in delicate fruit. For this reason, carnauba wax with no shellac or resin is often used. Emulsions that contain mineral oil and fatty acids (oleic) have also been utilized. Mineral oil–wax emulsions are also being considered since drying after coating is not necessary.

The newest group of fruit to be considered for coating purposes is tropical fruit. Control of weight loss in these commodities is essential, and coatings have been selected for their ability to prevent water loss. An additional benefit attained by coating tropical fruit is the ability to control physiological breakdown in some cases. Mango, papaya, carambola, chayote, and pineapple are coated to some degree. Coatings for tropicals consist largely of emulsions composed of carnauba wax. For these fruit, carnauba controls weight loss and, in addition, drying is not needed. Fruit can be packed and marketed without the danger of loss because of latent damage from excessive drier heat. Interest in the development of edible coatings for tropicals has accelerated, especially those which offer extended shelf life by delaying ripening and that are largely composed of FDA-approved food ingredients (Aylsworth, 1992).

Avocado and melon are coated occasionally as the industry demands to control weight loss. When applied, avocados are coated with either a citrus-type resin solution or emulsions with a mineral oil formulation. Polyethylene or carnauba emulsions are used as coatings for melons.

VEGETABLES

Vegetables that are suitable for coating in general are those bulky organs derived from ovaries or their associated floral parts and root crops. For those organs derived from floral parts, the coatings are very often mineral oil–based and are applied to control water loss, as well as to improve appearance and to reduce abrasive injuries during handling and transport (Hardenburg, 1967; Hartman and Isenberg, 1956). Coatings of mineral oil have the additional advantage of ease of handling. Since their application results in dewatering of the fruit surface, no drying is required before or after the coating process. Tomatoes, cucumbers, bell peppers, eggplant, and squash are commonly coated with mineral oil–based coatings. Carnauba wax is also used for coating vegetables and is usually formulated with emulsifiers and plasticizers. Because of the extensive drying involved, pollution problems, potential fire hazards, and subsequent cost, solvent waxes, once used to a limited extent, are no longer utilized.

Paraffin wax emulsions are frequently used to coat root crops such as rutabaga, turnip, and yucca. In some cases, the stems of sugarcane are also coated. To adhere well to the surface, vegetables to be coated with paraffin waxes must be clean. Because paraffin waxes are applied to vegetable surfaces well above the boiling point of water, any free water on the commodity will form "steam pockets" and will flake during subsequent handling. For this reason, the product must be completely dry before application of paraffin.

TABLETS AND CANDIES

Hard and soft confectioneries, as well as pharmaceutical tablets, are coated to preserve and protect their structure and contents, to increase palatability, to modify the release pattern of a drug in a tablet, and to impart shine to the product (Rowell, 1950). Conventional film coatings used for tablets typically contain polymers, plasticizers, colorants, and solvents. Conventional film coatings are almost exclusively water-based. However, good interaction between the polymer, solvent, and tablet necessitates the use of certain organic solvents on occasion. Pharmaceutical tablets are often coated with modified release film coatings. The coating formulations are designed to remain water-insoluble in the acid environment of the stomach, but to markedly increase in water solubility as the product travels to the more neutral area of the intestinal tract. Confectioneries are often "panned" with chocolate, or panned with high-solids syrups (60–80% solids), sometimes followed by an application of small particle sized granular sugar (Kitt, 1988). Polishing of confectioneries is accomplished by coating or glazing the panned pieces with waxes, oils, and/or shellac dissolved in acceptable solvents for ease of application and drying (Isganitis, 1988).

METHODS OF COATING APPLICATION

FRUITS AND VEGETABLES

Dip Application

Platenius (1939) reported that the Florida citrus industry was the first to realize the value of coating by dip applications for fresh citrus fruit. As an industrial process, coating was accomplished by submerging the commodity into a tank of emulsion. Because of the superior shelf life and gloss of dipped rutabagas from Canada, consumers developed a preference for rutabagas showing glossy appearance; and soon rutabagas, tomatoes, and peppers were being dipped in coating emulsions in this country.

Dipping fruits and vegetables into a tub or tank of the coating material is adequate usually for small quantities of commodities. The produce is washed, dried, then immersed in the dip tank. Time of immersion is not important, but complete wetting of the fruit or vegetable is imperative for good coverage (Long, 1964; Newhall and Grierson, 1956; Van Doren, 1944). The commodity is then either conveyed to a drier where water is removed or allowed to dry under ambient conditions. The continuous dipping of fruits and vegetables into a largely static milieu results in unacceptable buildup of decay organisms, soil, and trash in the dip tank. The dip tanks can be equipped with a porous basket which can be lifted to strain and remove debris. In addition, fruit entering the dip tank must be completely dry to avoid dilution of the resin solution or emulsion coating. For these reasons, other coating methods are more desirable where considerable quantities of fruits and vegetables are to be coated. Still, dipping commodities such as rutabagas, and tropicals such as carambolas and pineapples, continues, especially when thicker applications of the coating are required.

Foam Application

Although largely replaced by other systems, emulsions can be applied with a foam applicator. A foaming agent is added to the coating or compressed air (less than 5 psi or 35 kPa) blown into the applicator tank (Long and Leggo, 1959; Hartman and Isenberg, 1956). The agitated foam is then dropped onto the commodity as it moves over rollers. Cloth flaps or brushes distribute the emulsion over the commodity. Excess coating is removed with squeegees positioned below the rollers and then often recirculated. The foamed emulsion contains very little water and thereby facilitates the drying process. Even distribution of the coating is difficult to achieve because extensive tumbling action is necessary to break the foam and distribute the coating.

Spray Application

Spray application is the conventional method for applying most coatings to fruits and vegetables. Low-pressure spray applicators used in the past delivered coating in excess. Often, recovery wells were utilized and excess coating was recirculated. As with dip application, dilution and contamination were of concern. Later, high-pressure spray applicators, delivering coatings at 60 to 80 psi (414–553 kPa), became available which used much less coating material and gave equal or better coverage, negating the need for recovery wells and recirculation. However, nozzle size was critical since small nozzles often became plugged and large nozzles delivered too much wax. Air-atomizing systems connected to metering pumps have also been used. Air is delivered to the nozzle head at pressures usually at or below 5 psi (35 kPa) and the coating delivered with a metering pump at less than 40 psi (276 kPa). Although nozzle size is not as critical, the air nozzles can be expensive and therefore it is a factor in the selection of an air nozzle system for the application of coatings to produce.

For citrus, apples, and pears, the coating is sprayed onto the fruit which passes under a set of fixed or mobile nozzles. Fruit travels over a slowly rotating bed of brushes. The waxer brush bed is one of the most important pieces of equipment for coating purposes, since application and distribution of the coating occur largely on and across the bed. Straight-cut brushes on the brush bed are more effectively used with round or slightly elliptical commodities, whereas spiral-cut or tumble-trim brush designs are used with small, flat, irregularly shaped produce which requires more tumbling action for good coating coverage. The bed is typically composed of twelve to fourteen brushes, with four to six brushes positioned after the applicator. Too many brushes after the applicator can remove the wax on the commodity surface. With all spray applications where brushes are expected to distribute the coating, all brushes are recommended to be made of a 50% mixture of horsehair and polyethylene bristles. The recommended distance between bristles on each brush is no more than three-eighths in. (0.95 cm). Horsehair, horsehair mixtures, or very soft but durable synthetic types are superior for coating purposes because of their soft texture and high coating absorbency for recommended brush speeds of 100 rpm or less (Hall, 1981; Wardowski et al., 1987). With prolonged use, however, horsehair bristles will lose "body" and fall away from the brush, necessitating more frequent replacement. Brush wear must be checked often, and coatings washed from the brushes with clean water after every application.

Spray nozzles can be selected to deliver full cone, tapered or even-edged flat, or air atomizing spray characteristics. When mounted, attention must be given to operating pressure, distance from the brush bed, and nozzle wear since spray pattern quality and capacity will be directly affected. Full cone spray nozzles are more effectively utilized when a single mobile "wig-wag" nozzle or a single bank of nozzle heads are used to apply coatings, whereas tapered of even-edged flat spray can be chosen with multiple bank stationary nozzle mountings for overlapping or uniform coverage characteristics, respectively. Air atomizing spray nozzles are available that deliver a variety of spray patterns depending on the velocity of the atomizing air. Uniform coating coverage is easier to achieve with spray application but can be adversely affected by wind patterns under and around the nozzles, For this reason, coating equipment is often partially enclosed to minimize the effect of wind currents on coating coverage.

Manifold units consisting of several stationary spray nozzles can be mounted over the brush bed. However, nozzles with lower capacity must be used since several are usually mounted on the manifold and plugging is often a problem (Hall, 1981). Turbulence within the manifold coupled with a drop in pressure along its length result in differences in coating delivery rate across the brush bed. Nozzles of increasing capacity along the manifold and away from the pressure source could be selected to ensure uniform coating delivery. Nozzles which travel along the width of the brush bed are also used to distribute coatings to produce. Beam sliders, swing arms applicators, "lawn sprinklers," and elliptical chain applicators are known in Florida (Hall, 1981). Uniform produce coverage suffers as changes in nozzle direction occur.

Spray applicators deliver coatings over the majority of the commodity surface area. However, good, uniform coverage suffers when commodities of different surface areas are passed over the line in succession and commodity flow and bed fill adjustments are not made in the coating equipment. Good coverage for less tender commodities is also dependent on the tumbling action of the produce over a series of coating-saturated brushes. Because of the sufficient amount of coating available to produce on the brushes, continuous application in some cases is not necessary. In addition, spray application may continue when commodity flow is disrupted by accident or design. Coating can be wasted as a result. For economy and efficiency, programmable spray systems have been developed which control the delivery of coatings as a function of throughput and product type.

A programmable spray system is available in which a high-pressure pump can be programmed to deliver coatings on a percent time-on, time-off basis. Maximum benefit can be attained from this system when the pump is delivering the necessary rate of coating and fruit flow is even and steady. In addition, photosensors have also been installed that "see" produce as it is introduced to the coating equipment, and solenoid valves adjust the rate of coating application from the pumps accordingly. Systems are also available which can sense fruit flow by means of a series of light steel or plastic fingers mounted above the roller bed and positioned before the spray applicator. As the produce passes through the fingers, the system senses the rate of flow, and solenoid values adjust the rate of coating application from the pump. In Israel, air nozzles have been used in concert with a metering pump to deliver the coating (Kaplan, 1986). Another sensing system is available which not only senses the rate of commodity flow, but also compensates and adjusts for inherent differences in commodity flow rate across the width of the brush bed. An additional benefit of this "programmable pulse" system is the use of a pressure pump with conventional spray nozzles, eliminating the need for air in the coating facility and expensive air nozzles. Valves which sense differences in nozzle performance are also used with these systems.

Resin solution and emulsion coatings are pressure-sprayed over rollers onto the surface of root and tender tropical crops and distributed by the spray action itself. Brushes are not used. With some crops, such as tomatoes, cucumbers, squash, limes, and bell peppers, coatings are dripped onto a brush mounted just above the commodity. The produce becomes coated as it passes underneath and touches the overhead brush. The coating is then evenly distributed by the tumbling action of the produce over additional brushes downstream.

Solvent waxes are applied to clean dry citrus fruit in a chamber which is vented from the packinghouse. The coating is applied with nozzles that are fixed and mounted in the center of the chamber (Kaplan, 1986; Hall, 1981; Long, 1964). Fruit is conveyed through the chamber on grooved rollers which effectively turn the fruit at least once as the coating is applied. The solvent wax is directed onto the fruit by fans. The coating is drawn through the conveyed fruit essentially by a partial vacuum created by the venting of the chamber from below. As a result of venting, the solvent evaporates and the solids are further dried by stationary nonheating fans mounted above the conveyor as the fruit exits the chamber. Because solvent waxes create significant air pollution problems, they are becoming unpopular and are even unlawful in California.

Drip Application

The most economical method used today to apply coatings to fruits and vegetables is the drip application method. Different sizes of emitters are available that will deliver a variety of large droplet sizes. Usually mounted on a dual-bank manifold, emitters are spaced one inch apart along the manifold and width of the bed. Irrigation tubing can be used to house the emitters. A metering pump is used to deliver the coating at pressures not greater than 40 psi (276 kPa). Control systems described for spray applications are also feasible for drip-type application.

Besides the economy of drip application, the advantage of this method lies in its ability to deliver the coating either directly to the commodity surface or the brushes themselves. However, because droplet sizes are quite large, good uniform coverage can only be achieved when the commodity has adequate tumbling action over several brushes which are saturated with the coating.

Controlled Drop Application

Controlled drop applicators have been used successfully to coat some produce types. Low-pressure metering pumps are used to deliver the coating to the nozzle housing, where a rotating disc effectively shatters the large drops into smaller ones which are then delivered through a spray nozzle to the commodity. The rotating disc operates typically at speeds of 2000 to 5000 rpm. Lower rotating disc speeds dispense larger drop sizes. Although used successfully in Washington State to coat apples, the citrus industry was slow to utilize this application method because of difficulties in delivering a high solids wax with incorporated fungicide. Now, controlled drop applicators are being used successfully with citrus. Controlled drop applicators can also be used with programmable controllers, which—like spray systems—will vary coating flow to match fruit flow.

TABLETS AND CANDIES

The coating of tablets and candies involves both uniform deposition of the coating material as well as its rapid controlled drying. Both processes occur simultaneously in the coating pan. As the process of coating tablets improved over the years, the pan, once molded into a solid piece opened at one end, is now enclosed and perforated along the side panels for increased ventilation and drying power (Kitt, 1988; Rowell, 1950).

Coatings are delivered to tablets and candies by a pumping system and a spray gun or guns mounted in various positions within the pan. The spray guns atomize the coating solution and can be classified into the air or airless types. In the airless spray gun, coating fluid is delivered under high pressure through a nozzle with a small opening, resulting in atomization. Pressurized pump systems are widely used with airless guns. Changes in fluid delivery rate will change the degree of coating atomization and therefore pumping performance should be closely monitored. In addition, low fluid delivery rates will result in nozzle blockage. With the air spray system, larger nozzle openings are used with pressurized air to achieve atomization. Positive displacement or peristaltic pumps are commonly used with air spray guns. With these pumps, steady flow rate is easier to achieve, while the air stream provides the additional benefit of drying power.

Success in the coating operation is dependent upon drying to produce a uniform tablet or confectionery. In tablet coating, coating solution is constantly being added to the pan during the drying process, but with the coating of confectioneries, this is not necessarily so. Thus, the rate at which the solvent is being introduced into a given sized chamber, the size of the atomized particles, the pan speed, and the distance the spray travels from the nozzle to the tablet bed will influence the rate of drying. In the types of equipment used for tablet coating, forced convection drying is used. Drying can be maximized by safely maximizing the heat difference between the introduced air and the coating solution and by maintaining a low relative humidity of the introduced air. Drying of confectionery polishing solutions can be achieved by directing warm or cool dry air onto the pieces, which will produce a clear polished product (Isganitis, 1988).

Several patterns of air movement through the tablet can be generated by currently available pan designs. Whether air movement originates from the bottom, top, or both areas of the pan or air suspension column, the existing air passes through a series of filters and then is exhausted. In general, friction generated by the movement of product by either forced air or atomized coating solution is cause for concern, especially with weaker candies and tablets (Kitt, 1988). However, waxes deposited on the surface of confectioneries during the polishing process tend to "buff" out better with increased friction (Isganitis, 1988).

KEYS TO SUCCESSFUL COMMODITY COATING

FRUITS AND VEGETABLES

All nonsolvent wax application systems, independent of type, should have a strategy for clean water cleanup after each use. Failure to implement a routine cleanup procedure will result in pump breakdown and clogging of nozzles. Buildup of hardened coatings on brushes, rollers, and side bumpers will result in commodity injury and subsequent decay.

Before arrival to the coating bed, most if not all water must be removed from the fruit surface. Excess water only serves to dilute the coating. Less coating is applied, and foaming may result from the interaction of the coating emulsifiers and water. If fungicides are incorporated in the coating, fungicides are diluted.

Brush wear should be routinely monitored. Incomplete coating coverage, or "missed spots," can often be avoided by replacement of worn brushes on the brush bed and by proper coating bed length to allow adequate tumbling action over the brushes. Incomplete coverage can also be avoided by ensuring that a clean commodity is used for coating purposes. Soil or debris present on the coated commodity surface will be removed by brushes or rollers as it makes its way to the packing table.

Dry the coating as much as possible before packing. Some coating formulations remain "tacky" after drying and commodities must be handled carefully to avoid the creation of missed spots on the finished product. An excellent coating job can be negated by inadequate air flow, temperature, underfill and overfill of the drier.

The use of coatings, although increasingly under scrutiny, has become an important component of the fresh fruit and vegetable industry. The ability to control desiccation, incorporate fungicides for decay control, and to control aspects of product physiology serves to lengthen the market window for commodities with a finite shelf life. Because travel to distant markets requires longer storage and holding periods, greater quality demands are made. These trends in domestic and international marketing underscore the importance of coating in the maintenance of quality in fresh fruits and vegetables.

TABLETS AND CANDIES

Independent of coating type, a clean tablet or confectionery free of dust and debris should be used. If not, the coating will flake away from the piece and result in uneven coating. In addition, as coating is built onto the piece, a rough and uneven product will develop. A dust-free environment is mandatory to achieve a smooth product. Dust, originating from the surface of the tablet in the coating pan, can also arise from excessive drying.

Condensation on the surfaces of products which are not properly temperature-managed will result in localized solubilization of coating and consequently holes in the coating will develop. Inadequate drying after each coat is applied will result in an uneven coat as well as the formation of "doubles" or "clusters," which must be removed before the next process can commence. Recrystallization may also occur, which will result in a lumpy, uneven piece. Products coated with shellac solutions will dull as a result of polymerization reactions and powdering, or unwanted whitening, will develop.

Tumbling action of the product inside the pan is important to ensure an evenly coated and dried product. Care must be taken to avoid unnecessary or excessive tumbling, however. Otherwise the coating will fracture and break away from the product, product dust will develop, and uniformity will suffer.

REFERENCES

Aylsworth, J. D. 1992. "Debate Over Waxing Heats Up," *Amer. Fruit Grower* (May).

Brooks, C. 1937. "Some Effects of Waxing Tomatoes," *Proc. Amer. Soc. Hort. Sci.,* 35:720.

Eckert, J. W. and G. E. Brown. 1986. "Postharvest Citrus Diseases and Their Control," in *Fresh Citrus Fruits,* W. Wardowski, S. Nagy and W. Grierson, eds., Westport, CT: AVI Publishing Co., pp. 315–360.

Hall, D. J. 1981. "Innovations in Citrus Waxing—An Overview," *Proc. Fla State Hort. Soc.,* 94:258–263.

Hardenburg, R. E. 1967. "Wax and Related Coatings for Horticultural Products—A Bibliography," *USDA/ARS publ. #51-15.*

Hartman, J. and F. M. Isenberg. 1956. "Waxing Vegetables," *New York Agric. Ext. Ser. Bull. No. 965.*

Isganitis, D. K. 1988. "Polishing: A Review of Processes and Techniques," *The Manufacturing Confectioner* (October):75–78.

Kaplan, H. J. 1986. "Washing, Waxing, and Color-Adding," in *Fresh Citrus Fruits,* W. F. Wardowski, S. Nagy and W. Grierson, eds., Westport, CT: AVI Publishing Co., pp. 379–396.

Kitt, J. S. 1988. "Panning Problems—Causes and Remidies," *The Manufacturing Confectioner* (October):57–62.

Long, J. K. and D. Leggo. 1959. "Waxing Citrus Fruit," *CSIRO Food Preserv. Quart.,* 19:32–37.

Long, W. G. 1964. "Better Handling of Florida's Fresh Citrus Fruit," *Fla. Agric. Exp. Sta. Bull. No. 681.*

Magness, J. R. and H. C. Diehl. 1924. "Physiological Studies on Apples in Storage," *J. Agric. Res., 27:1–38.*

Newhall, W. F. and W. Grierson, 1956. "A Low-Cost, Self-Polishing, Fungicidal Wax for Citrus Fruit," *Proc. Amer. Soc. Hort. Sci.,* 66:146–154.

Platenius, H. 1939. "Wax Emulsions for Vegetables," *New York Agric. Exp Sta. Bull. No. 723.*

Rowell, T. H. 1950. "The Art of Coating Tablets," industry publication of the F. J. Stokes Co., Chicago, IL, pp. 7–33.

Trout, S. A., E. G. Hall and S. M. Sykes. 1953. "Effects of Skin Coatings on the Behavior of Apples in Storage," *Aust. J. Agric. Res.,* 4:57–81.

Van Doren, A. 1944. "A Report on the Construction and Operation of a Grower-Size Apple Washing Machine," *Proc. Amer. Soc. Hort. Sci.,* 44:183–189.

Wardowski, W. F., W. M. Miller and W. Grierson. 1987. "Packingline Machinery for Florida Packinghouses," *Fla. Coop. Ext. Bull. No. 239.*

Yehoshua, S. B. 1987. "Transpiration," in *Postharvest Physiology of Vegetables,* J. Weichmann, ed., New York, NY: Marcel Dekker, Inc., pp. 113–170.

CHAPTER 9

Edible Coatings and Films Based on Proteins

ARISTIPPOS GENNADIOS[1]
TARA H. McHUGH[2]
CURTIS L. WELLER[3]
JOHN M. KROCHTA[4]

INTRODUCTION

THE film-forming ability of several proteinaceous substances has been utilized in industrial applications for a long time. Ancient Egyptians, Chinese, Greeks, and Romans used casein in glues because of its water resistance. Casein has also functioned in paints, leather finishes, and paper coatings. Corn zein has also been used as a coating former in the manufacture of printing inks, can linings, grease-proof paper, floor coatings, and photographic films. Gelatin was one of the first materials employed in the formation of polymeric capsule walls in microencapsulation.

Additionally, the food industry recognized that proteins possessing film-forming properties could be used for the development of edible, protective food films and coatings, provided that edibility was maintained in every step of the protein-based film or coating preparation. Two current commercial applications of the concept include the extrusion of reconstituted collagen for casings of comminuted meat products, as well as the use of corn zein coatings for nuts and confectionery items.

This chapter provides an overview of edible protein-based films and coatings by compiling the scattered literature on the subject. Information

[1]Department of Biological Systems Engineering, University of Nebraska, Lincoln, NE 68583.
[2]U.S. Dept. of Agriculture, Agricultural Research Service, Western Regional Research Center, Albany, CA 94710.
[3]Departments of Biological Systems Engineering, and Food Science and Technology, University of Nebraska, Lincoln, NE 68583.
[4]Departments of Food Science and Technology, and Biological and Agricultural Engineering, University of California, Davis, CA 95616.

enclosed in research articles and patents relevant to film production methods, property evaluation, and potential applications is presented.

A number of proteins, both of plant and animal origin, that have received attention for production of films and coatings are individually discussed below. These proteins are corn zein, wheat gluten, soy protein, peanut protein, keratin, collagen, gelatin, casein, and milk whey proteins. Other proteins, also known as film formers, that have not received as much consideration and are not reviewed include egg albumen, cottonseed protein, barley hordein, rye secalin, sorghum kafirin, serum albumin, and fish protein.

DEVELOPMENT OF PROTEIN-BASED EDIBLE FILMS AND COATINGS

CORN ZEIN

Protein Nature and Recovery

Corn grains have a protein content in the range of 7% to 11% (Wall and Paulis, 1978). Almost seventy years ago, a classification of plant proteins was proposed by Osborne based on their solubility (Osborne, 1924). He separated them into albumins (water-soluble), globulins (saline-soluble), prolamins (aqueous alcohol-soluble), and glutelins (alkali-soluble and acid-soluble). The work of Osborne provided a solid basis for subsequent study of corn proteins. New approaches and techniques showed that distinction among the four Osborne fractions was not always straightforward and that the fractions are actually composed of mixtures of proteins (Wilson, 1987).

The prolamin fraction of corn is known as zein. Differential solubility in aqueous ethanol yields two zein fractions. About 80% of zein is soluble in 95% ethanol and is characterized as α-zein. The remaining 20%, which is insoluble is 95% ethanol, is characterized as β-zein (Turner et al. 1965). α-Zein consists of monomers and a series of disulfide-linked oligomers of varying molecular weight, whereas β-zein consists of higher molecular weight oligomers (Paulis, 1981).

A fractionation of corn zein into three distinct classes (α, β, and γ) was recently proposed by Esen (1987). According to this nomenclature, α-zein accounts for 75–85% of the total zein and is made up of polypeptides with M_r (relative molecular mass) of 21,000–25,000 and a minor M_r-10,000 polypeptide. β-Zein constitutes 10–15% of the total zein and includes M_r-17,000–18,000 methionine-rich polypeptides. Finally, γ-zein is made up of a proline-rich polypeptide of M_r-27,000 and constitutes 5–10% of the total zein.

Zein is the only corn protein fraction produced commerically in the U.S. Corn gluten obtained as a by-product of corn wet-milling serves as the major source for zein. Zein first became commercially available in 1939 and for many years thereafter as a result of a two-step extraction process. Hot aqueous 86% isopropyl alcohol and hexane were the two successive solvents (Pomes, 1971).

A one-step extraction process was later developed and is still used today. In this process, corn gluten is extracted at 60°C with aqueous 88% isopropyl alcohol containing 0.25% sodium hydroxide. After centrifugation of the extract, zein is precipitated by chilling the supernatant to −15°C. The precipitate, which contains about 30% zein, is dried on a vacuum drum dryer. Additional purification is achieved by redissolving and reprecipitating zein (Reiners et al., 1973). The dried zein powder has a creamy, yellow color.

Zein is insoluble in water (except at pH extremes), as well as in anhydrous alcohols. Typical solvent systems used for zein are aqueous aliphatic alcohol solutions. In general, suitable solvents for zein are either mixtures of water with an organic compound, such as alcohols, acetone, and acetonylacetone, or mixtures of two anhydrous organic compounds, such as alcohols and chlorinated hydrocarbons or glycols (Pomes, 1971).

Insolubility of zein in water is due to its amino acid composition (Table 9.1). A low content of polar amino acids and a high content of nonpolar amino acids, such as leucine, proline, and alanine characterize zein and other seed prolamin proteins (Shewry and Miflin, 1985). In water, the hydrophobic side chain groups of these amino acids tend to associate with one another, preventing protein from dissolving (Wall and Paulis, 1978). Zein is also rich in glutamine, the amide derivative of glutamic acid. The uncharged polar terminal group of glutamine promotes protein association through hydrogen bonding and contributes to insolubility in water as well (Reiners et al., 1973; Wall and Paulis, 1978).

Film-Forming Ability

Interest in the industrial utilization of zein is due mostly to its film-forming abilities when cast from appropriate solvent systems (usually aqueous aliphatic alcohols). Films formed upon solvent evaporation are tough, glossy, scuff-resistant, and grease-resistant (Pomes, 1971). Primarily hydrophobic and hydrogen bonds develop in the film matrix. Disulfide bonds are also present, but in a limited number due to the low content of cystine in commercial zein. Film brittleness necessitates the addition of plasticizers. Suitable plasticizing agents include glycerin, triethylene glycol, fatty acids, glyceryl monoesters, and acetylated monoglycerides (Reiners et al., 1973).

TABLE 9.1. Amino Acid Composition (Moles Amino Acid per 10^5 g Protein) of Corn Zein, Wheat Gliadin, Wheat Glutenin, and Wheat Gluten.

Amino Acid	Corn Zein[1]	Wheat Gliadin[2]	Wheat Glutenin[2]	Wheat Gluten[3]
Lysine	1	5	12–13	9
Histidine	8	14–15	13	15
Arginine	10	15	20	20
Aspartic Acid[4]	41	20	23	22
Threonine	24	18	26	21
Serine	52	38	50	40
Glutamic Acid[5]	166	317	278	290
Proline	94	148	114	137
Glycine	17	25	78	47
Alanine	110	25	34	30
Cystine	—[6]	5	5	7
Valine	31	43	41	45
Methionine	10	12	12	12
Isoleucine	31	37	28	33
Leucine	151	62	57	59
Tyrosine	31	16	23–27	20
Phenylalanine	43	38	27	32
Tryptophan	0	5	8	6

[1]From Reiners et al. (1973).
[2]From Wu and Dimler (1963b).
[3]From Wu and Dimler (1963a).
[4]Includes asparagine.
[5]Includes glutamine.
[6]No value reported.

Water resistance and tensile properties of zein films can be improved by utilizing cross-linking agents. For example, addition of aldehydes, such as formaldehyde, into zein alcoholic solutions, optionally followed by heat curing of cast films, is a known method for obtaining water-resistant zein films (Hansen, 1937; Evans, 1946; Szyperski and Gibbons, 1963). Epoxy resins are also effective in toughening zein films through cross-linking among epoxy groups and basic, phenolic, and aliphatic hydroxyl protein groups (Howland, 1961; Howland and Reiners, 1962). However, edibility of films treated with such cross-linking agents is highly questionable. Films have also been cast from partially deamidated zein solutions (Howland and Reiners, 1962; Wall and Paulis, 1978).

Film Production

Zein films in a free-standing form are required for effective and practical evaluation of barrier and mechanical properties. Several techniques, varying in composition of depositing solutions, substrates for film casting, and drying conditions, were followed in research studies to obtain zein films.

Hansen (1937) developed a formula for preparing film-forming solutions from corn zein, as well as wheat gliadin and barley hordein. Zein was dissolved in a mixture of solvents while also adding a plasticizer and an anti-blushing agent such as an amino acid ester or an organic amino alcohol salt. Addition of the anti-blushing agent prevented blushing or whitening of produced films at high relative humidity environments.

Kanig and Goodman (1962) prepared a zein solution (21.6% w/v) in aqueous ethanol, with oleic acid added as a plasticizer (15.9 v/v). Films from this solution were cast with an applicator on highly polished chromium-plated stainless steel plates. After drying for 6–10 min at 51°C, films were readily peelable from plates.

Takenaka et al. (1967) reported casting zein films by using two different solvents, methanol and 70% acetone. Films were cast from 4% w/v zein solutions in methanol and from 1, 3, and 6% w/v solutions in 70% acetone.

Polyethylene sheets supported by glass plates were used by Mendoza (1975) as a surface for casting zein alcoholic solutions with a thin layer chromatography spreader. Glycerin was selected as a plasticizer and drying was carried out at near ambient temperature.

Dilute zein solutions (1–2% w/w) in aqueous ethanol were cast by Guilbert (1986, 1988). Reportedly, films formed after drying at 35°C for 24 hours were flexible although no plasticizer was added in the preparation formula.

Aydt et al. (1991) obtained zein films by pouring and spreading a commercial zein coating preparation into an aluminum mold. The commercial product used was a zein solution in aqueous ethanol and glycerin with small amounts of antimicrobial (citric acid) and antioxidant agents (BHT and BHA) also present. Films were dried at ambient temperature for 24 hours before being peeled off the aluminum mold.

Park (1991) also cast zein solutions in aqueous ethanol with glycerin and citric acid added as well. Solutions were homogenized and heated to 70°C prior to casting onto glass plates. Heating aided degassing of solutions, preventing bubble formation upon drying.

WHEAT GLUTEN

Protein Nature and Recovery

Wheat proteins account for 8–15% of the dry weight of wheat kernels (Kasarda et al., 1976). Almost 70% of the total protein is contained in the wheat endosperm, the major source of wheat flour (Kasarda et al., 1971).

Fractionation of plant proteins into albumins, globulins, prolamins, and glutelins, as initially proposed by Osborne (1924), has led the way for further classification and characterization studies of wheat proteins.

The terms gliadin and glutenin are traditionally used for the prolamin and the glutelin fractions of wheat protein, respectively (Kasarda et al., 1976). These two are the main wheat endosperm storage proteins, making up about 85% of the wheat flour protein content, while the remaining 15% consists of albumins, globulins, and smaller peptide structures (Holme, 1966). Upon hydration, gliadin and glutenin form a tenacious colloidal complex known as wheat gluten (Pomeranz, 1987). The cohesiveness and elasticity of hydrated gluten provide the dough-forming ability of wheat flour (Krull and Wall, 1969). These properties also generate interest for film formation from gluten protein. The wheat gluten complex and its two major fractions, gliadin and glutenin, are briefly discussed below.

Gluten Fractionation from Wheat Flour

A traditional laboratory method for gluten preparation from wheat flour is the one described by Jones et al. (1959). Lipids are removed first by extraction with n-butyl alcohol. Starch and other soluble nongluten materials are washed away by hand kneading in a stream of dilute sodium chloride. The obtained gluten ball is converted to a powder form after dispersion in dilute acetic acid, centrifugation, and drying. An alternative technique, proposed by Mauritzen and Stewart (1965), involves ultracentrifugation of wheat flour dough.

Wheat gluten is commercially available as "vital wheat gluten." Vital gluten has a protein content of about 75–85% on a dry basis (Kasarda et al., 1971). Most of this protein consists of gliadin and glutenin. Membrane proteins and cell wall proteins are also present, as are small amounts of albumins and globulins that are physically entrapped in the gluten complex. Lipids and carbohydrates are also present at 5–10% and 10–15%, respectively (Kasarda et al., 1971).

Various industrial methods for production of vital wheat gluten from wheat flour are based on hydrating the flour and washing away the starch milk while applying mechanical mixing. Alkali treatment of flour to dissolve gluten and separate it from the alkali-insoluble starch is also used. Centrifugation of dispersed wheat flour slurries is another alternative. An overview of industrial techniques for gluten fractionation from wheat flour is given by Fellers (1973).

Gliadins

Gliadins, as stated earlier, make up the prolamin fraction of wheat endosperm. They are the low molecular weight proteins in the wheat gluten complex, known for their solubility in 70% ethanol. A general nomenclature for wheat gliadins is based on their electrophoretic mobility in aluminum lactate (Woychik et al., 1961). It distinguishes four gliadin groups (α-, β-, γ-, and ω-gliadins). The molecular weights of these groups range from 30,000-40,000 (α-, β-, and γ-gliadins) to 60,000-80,000 (ω-gliadins) (Bietz and Wall, 1980).

The amino acid composition of wheat gliadin (Table 9.1) is characterized by a high content of glutamic acid (38–45% of total amino acid residues), which is present almost entirely as glutamine with very few free ionizable carboxyl groups; a high content of proline (15–30%) and phenylalanine (up to 10%), both nonpolar amino acids; and a low content of lysine and other basic amino acids, such as histidine and arginine (Kasarda et al., 1976; Lasztity, 1984).

The terminal amide group of glutamine promotes extensive hydrogen bonding among the gliadin polypeptide chains. Hydrophobic interactions among nonpolar amino acid residues, as well as the very small number of ionizable basic and acidic groups, are responsible for the limited solubility of gliadin in neutral water (Krull and Wall, 1969). Hydrophobic bonding contributes to the globular conformation of gliadin by twisting and folding gliadin polypeptide chains (Kasarda et al., 1976). Viscosity studies on wheat gliadins verified their globular conformation (Taylor and Cluskey, 1962). Gliadin structure is also affected by development of intramolecular disulfide bonds (Krull and Inglett, 1971; Lasztity, 1986).

Glutenins

Glutenins are the high molecular weight proteins in the wheat gluten complex that are primarily responsible for dough viscoelasticity. It is believed that glutenins result from cross-linking of polypeptide subunits. Molecular weights of glutenins cover a wide range, reaching several million at the high end. According to Jones et al. (1961), the average molecular weight of glutenin is in the order of 2–3 million. The amino acid composition of glutenins has characteristics similar to those of gliadins (Table 9.1). Glutamic acid is present at a high level, mostly as glutamine. The content of proline and other nonpolar amino acids is high. Basic amino acids, such as lysine, are present in small proportions.

It is widely believed that numerous cystine intermolecular disulfide bonds play a major role in formation of the glutenin complex. However, several researchers doubt the high number, and even the existence, of intermolecular disulfide bonds. As a result, various models on the glutenin structure can be found in the literature. A review of these theories has been written by Lasztity (1986). They are briefly summarized below.

Ewart (1979) speculates that glutenin molecules are linear and that polypeptide chains are not highly cross-linked through disulfide bonds. Instead adjacent chains are linked only with a single disulfide bond. An alternative model rejects the existence of any intermolecular disulfide bonds (Kasarda et al., 1976). All bonds of this type in the glutenin structure are considered intramolecular. Protein subunits form aggregates under the influence of strong secondary bonding forces, such as hydrogen, ionic, and hydrophobic interactions.

Bushuk et al. (1980) and Khan and Bushuk (1978, 1979) suggest that protein subunits in glutenin are linked through both intermolecular and intramolecular disulfide bonds, as well as with noncovalent bonds (hydrogen, ionic, hydrophobic bonds). They classify glutenins into acetic acid-soluble (Glutenin I) with molecular weight 68,000 or lower, and acetic acid-insoluble (Glutenin II) with molecular weight above 68,000. Glutenin II subunits are linked together through at least one interpolypeptide disulfide bond and are the major contributors to glutenin elasticity. Glutenin I subunits are linked in the matrix formed by Glutenin II subunits through secondary noncovalent forces.

Structure of Gluten Complex

The gluten complex is believed to be a protein network held in place by extensive covalent and noncovalent bonding. Extended random-coiled glutenin polypeptides provide the frame of this structure. Smaller globular gliadin polypeptides are packed into the network and interact with glutenins. Gliadins bond mostly noncovalently with glutenins, although covalent bonding is possible through disulfide interchange reactions (Bietz and Wall, 1980).

Lipid and carbohydrate components present in vital wheat gluten, are also believed to interact with proteins in the gluten complex. Carbohydrate-protein and lipid-protein interactions in wheat gluten are reviewed by Pomeranz and Chung (1981) and by Lasztity et al. (1987). Free polar lipids (glycolipids and phospholipids) are bound to gliadin proteins by hydrophilic bonds and to glutenin proteins by hydrophobic bonds (Hoseney et al., 1970; Wehrli and Pomeranz, 1970; Pomeranz, 1980). Van der Waals associations between glycolipids and both gliadin and glutenin are also evidenced (Wehrli and Pomeranz, 1970; Pomeranz, 1980).

Non-starchy and non-cellulosic polysaccharides found in gluten are referred to as hemicelluloses or pentosans, even though they contain several other entities (e.g., hexoses, phenolics, proteins) in addition to pentose sugars (Hoseney, 1986). Analysis of sugars contained in pentosans has identified xylose, arabinose, glucose, and smaller amounts of mannose and galactose (Graveland et al., 1982). Covalent binding of carbohydrates to glutenin proteins through hydroxyproline was suggested but was not confirmed (Lasztity, 1986). It is believed that polysaccharide interactions with proteins involve hydrogen bonding between side chains of amino acid residues and hydroxyl or carboxyl polysaccharide groups (Huebner and Wall, 1979; Lasztity et al., 1987). Research findings on interactions of starch with wheat gluten proteins are reported by Dahle (1971) and Hoseney et al. (1971).

Mechanism of Film Formation

Wheat gluten films can be produced by deposition and subsequent drying of wheat gluten dispersions. Aqueous ethanol is the most common solvent employed in film-forming solutions. Akaline or acidic conditions are required for the formation of homogeneous film-forming solutions (Gontard et al., 1992; Gennadios et al., 1993a), while mechanical mixing and heating also aid wheat gluten dispersion. The important role of covalent disulfide bonds in gluten film formation has been noted by various researchers. Intramolecular and intermolecular disulfide bonds in the gluten complex are cleaved and reduced to sulfhydryl groups when dispersing gluten in alkaline environments (Okamoto, 1978). Reduction of disulfide bonds in gluten solutions is promoted by addition of reducing agents, such as mercaptoethanol, sodium sulfite, sodium borohydride, and cysteine (Krull and Wall, 1969; Wall and Beckwith, 1969; Simmonds and Orth, 1973; Okamoto, 1978).

Upon casting of film-forming solutions, disulfide bonds are reformed, linking together polypeptide chains, to yield a film structure. Reoxidation in the air and sulfhydryl-disulfide interchange reactions are the mechanisms that contribute to reformation of disulfide bonds. A number of research papers describe reduction and reoxidation of disulfide bonds (Hlynka, 1949; Beckwith et al., 1965; Beckwith and Wall, 1966; Wall et al., 1968), and sulfhydryl-disulfide interchange reactions (Stewart and Mauritzen, 1966; Redman and Ewart, 1967; McDermott et al., 1969) in wheat gluten. Heating of wheat gluten unfolds protein polypeptides and reveals previously unaccessible sulfhydryl groups (Schofield et al., 1983). Therefore, heating during preparation of film-forming solutions is believed to increase disulfide bond formation upon cooling.

Hydrogen and hydrophobic bonds contribute as well to film structure. Urea, acetamide, and guanidine hydrochloride, all hydrogen bond disrupters, lower the consistency of hydrated wheat gluten confirming the existence of hydrogen bonds (Doguchi and Hlynka, 1967). A similar decrease in consistency is observed by addition of acetone and other organic solvents which possess hydrophobic bond breaking ability.

Addition of plasticizers in wheat gluten film-forming solutions is necessary to induce film flexibility. Films cast without plasticizer are very brittle. This brittleness results from extensive intermolecular associations

(Wall and Beckwith, 1969). Plasticizer molecules mediate between poly-peptide chains, disrupting some of these associations and decreasing the rigidity of the film structure.

Film Production

Reports can be found in the literature describing film formation from wheat gluten–based film-forming solutions. Various methods have been employed to obtain films in a free standing form suitable for property evaluation. A review on the field was recently published by Gennadios and Weller (1990). A summary of research results on wheat gluten film formation is presented below.

Wall and Beckwith (1969) cast and dried films from purified (laboratory prepared) and vital (commercial) wheat gluten, as well as from gliadin and glutenin. Film-forming solutions of 20% protein in 60% ethanol, 20% water, and 20% lactic acid, the latter added as a plasticizer, were prepared and cast. An interesting observation from this work was that purified gluten films had a tensile strength value twice that of films from commercial gluten. Films from purified gluten were also clear, while vital gluten films were translucent and slightly granular. Similarly, Gennadios and Weller (1992) reported that films prepared from a commercial wheat gluten product containing about 82% protein had higher tensile strength (by 120%) than films prepared from another wheat gluten product of about 75% protein content. Such observations indicate that other constituents present in commercial wheat gluten, such as starch, oligosaccharides, and lipids, disrupt the homogeneity of protein film networks. Presumably, starch plays a major role in film weakening by retrograding upon drying and forming crystalline regions within films. Although it is evident that employing wheat gluten of high purity in film formation improves film characteristics, for practical purposes, cost considerations arising from additional purification steps would have to be balanced against improved film properties.

It was reported that films cast from glutenin had substantially higher tensile strength, but lower elongation, than gliadin and whole gluten films, presumably due to the extended random-coiled molecular conformation of glutenin (Wall and Beckwith, 1969; Kolster et al., 1992). Modification of protein by partial conversion of amid groups into methoxyl groups did not affect film strength (Wall and Beckwith, 1969).

Polypeptides obtained from partial hydrolysis of wheat gluten with hydrochloric acid were investigated for their film-forming abilities (Krull and Inglett, 1971; Gutfreund and Yamauchi, 1974). Films cast from aqueous solutions of these polypeptides were very brittle and could not release intact from casting plates. Gluten polypeptides chemically modified by high-pressure reactions with ethylene oxide and ethylenimine were also employed in film formation (Krull and Inglett, 1971; Gutfreund and Yamauchi, 1974). Films cast from mixtures of unmodified polypeptides or whole gluten and polypeptides epoxidized with ethylene oxide were flexible and readily peelable from casting plates without shattering. Film flexibility was due to the plasticizing action of epoxidized polypeptides. A 1:1 ratio of gluten or unmodified polypeptides to epoxidized polypeptides was suggested for film-forming mixtures. Ethylenimine-treated polypeptides also yielded flexible films when mixed with unmodified ones. A 6:1 ratio of unmodified polypeptides to ethylenimine-treated polypeptides was reported as preferable.

Anker et al. (1972) developed a method to produce films from heated alkaline gluten dispersions in alcohol-water mixtures. Monohydroxy aliphatic alcohols with up to three carbon atoms functioned best. An alcohol-water ratio (v/v) between 5:2 and 1:15, and a protein-liquid ratio (w/w) between 1:20 and 1:6 were proposed. Addition of a plasticizer in an amount that corresponded to 20–50% of protein was necessary. Glycerin and diglycerin were the most suitable plasticizers. Alkalinity of the dispersions was provided by an inorganic base, such as ammonium, sodium, and potassium hydroxide. Ammonium hydroxide was preferred since, upon film drying, it evaporated as ammonia resulting in films of netural pH. A pH range of 7–12 was considered appropriate for the dispersions.

The above method, developed by Anker et al. (1972), was adapted by other researchers in order to produce and characterize wheat gluten films (Gennadios et al., 1993a, 1993b, 1993c, 1993d, 1993e; Park and Chinnan, 1990; Aydt et al., 1991; Gennadios and Weller, 1992; Park et al., 1992). Aydt et al. (1991) prepared solutions of vital wheat gluten (15 g) in 95% ethanol (72 mL) and glycerin (6 g). While mixing and warming the solutions, distilled water (48 mL) and 6 N ammonium hydroxide (12 mL) were added. Upon gluten dispersion, which was easily noticeable by a significant decrease in viscosity, solutions were cast on flat glass plates. After frying for one day at ambient temperature, films were peeled from casting plates and tested for various properties.

Park and Chinnan (1990) prepared film-forming solutions with the same component concentration as Aydt et al. (1991), except that glycerin was 20% less. The solutions were homogenized prior to casting. In a variation of this method, Park et al. (1992) and Gennadios et al. (1993b) prepared film-forming solutions by mixing the ingredients without heating. Subsequently, solutions were homogenized, filtered through cheese cloth, conditioned in a water bath at 80°C for 15 min, poured on glass plates, and dried at 55°C for about 30 min. This procedure resulted in films that were smoother in appearance of a greater tensile strength than previous ones.

Gennadios et al. (1993c, 1993d) studied property changes of wheat gluten films by modifying the film-forming solutions employed by Aydt et al. (1991). Modifications included incorporation of acetylated monoglycerides and mineral oil, use of triethylene glycol as a plasticizer in the place of glycerin, and partial substitution of wheat gluten with soy protein isolate, corn zein, and hydrolyzed keratin. Heat curing at 80°C reportedly increased tensile strength and decreased the strain of wheat gluten films (Kolster et al., 1992). Incorporation of glutaraldehyde, a known protein cross-linking agent, into wheat gluten films (4% w/w of protein) increased film tensile strength by 75% (Gennadios and Weller, 1992).

An important study undertaken by Gontard et al. (1992) investigated the effects of gluten concentration, ethanol concentration, and pH of film-forming solutions on various properties of wheat gluten films using response surface methodology. Only acidic conditions (pH between 2 and 6) were studied, employing a volatile acid (acetic acid) to obtain neutral films upon solvent evaporation. Reportedly, films prepared in acidic conditions possessed better sensory and visual properties than films prepared in alkaline conditions by using either sodium hydroxide (yellowish color, alkaline taste, chewy texture) or ammonium hydroxide (ammoniacal odor). However, as concluded in another study, films cast from alkaline solutions (pH 9 to 13) had substantially higher tensile strength values than films cast from acidic (pH 2 to 4) solutions (Gennadios et al., 1993a). Gontard et al. (1992) determined that pH and ethanol concentration of film-forming solutions had strong interactive effects on film opacity, water solubility, and water vapor permeability, whereas mechanical properties were mainly affected by gluten concentration and pH.

Film formation on the surface of heated alkaline wheat gluten solutions was studied by Okamoto (1978). The film formation mechanism involved in this procedure (heat-catalyzed protein polymerization through surface dehydration) is presented in our discussion of soy protein films. Okamoto (1978) suggested a pH range of 10–11 as the optimum for surface film formation from wheat gluten solutions. Blocking of sulfhydryl groups with N-ethylmaleimide decreased film strength by preventing disulfide bond formation, whereas reduction of disulfide bonds by addition of mercaptoethanol yielded films with increased tensile strength. Similarly, higher film tensile strength was also observed when sodium sulfite, another disulfide bond reducing agent, was added in film-forming solutions (Gennadios et al., 1993d). Presumably, reduction increases the number of sulfhydryl groups which become available for disulfide bonding during drying. On the contrary, Wall and Beckwith (1969) reported a decrease in film strength when reducing wheat gluten with sulfite. Nevertheless, it should be noted that all three observations came after using different film formation methods.

SOY PROTEIN

Protein Nature and Recovery

The protein content of soybeans (38–44%) is much higher than the protein content of cereal grains (8–15%) (Snyder and Kwon, 1987). Most of the protein in soybeans can be classified as globulin. A widely used nomenclature system for soy proteins is based on relative sedimentation rates of protein ultracentrifugal fractions. Four such fractions are separable and are designated as 2S, 7S, 11S, and 15S fractions (Wolf and Smith, 1961). The 7S and 11S are the main fractions making up about 37% and 31%, respectively, of the total extractable protein (Wolf et al., 1962).

Protein in the form of meal is one of the typical end products from soybean industrial processing. Protein meal can be further "concentrated" for the production of soy protein concentrates and soy protein isolates. Soy protein concentrates contain at least 70% protein on a dry basis and are produced by extracting and removing soluble carbohydrates and other minor components from the defatted protein meal (Mounts et al., 1987). Aqueous ethanol is the most commonly used extracting medium.

Soy protein isolates contain at least 90% protein on a dry basis. Their production from defatted protein meal involves extraction with dilute alkali and centrifugation. Protein in the extract is precipitated by adjusting the pH to 4.5, recovered by centrifugation, washed, neutralized with food grade alkali, and spray dried (Mounts et al., 1987). The amino acid composition of soy protein isolate is shown in Table 9.2.

The film-forming ability of soy protein has been noted along with a number of other functional properties, such as cohesiveness, adhesiveness, water and fat absorption, emulsification, dough and fiber formation, texturizing capability, and whippability (Wolf and Cowan, 1975). Edible films from soybeans have been traditionally produced in the Orient on the surface of heated soymilk. Protein is the major component of these films, but significant amounts of lipids (primarily) and carbohydrates (secondarily) are also incorporated. Therefore, these films from soymilk are multicomponent films and not strictly soy protein films.

Production of films and coatings from soymilk and soy protein has been reviewed recently by Gennadios and Weller (1991). In the discussion below, after addressing briefly soybean film production in the Orient, the film formation phenomenon from soymilk and from soy protein isolate dispersions will be separately examined.

Soy Film Production in the Orient

Films obtained from soymilk are known as "yuba" in Japan, "tou-fu-pi" in China, "kong kook" in Korea, and "fu chock" in Malaysia (Snyder and

TABLE 9.2. Amino Acid Composition (g Amino Acid per
100 g Protein) of Soy and Peanut Protein[1].

Amino Acid	Soy Protein	Peanut Protein
Lysine	6.4	3.5
Histidine	2.5	2.4
Arginine	7.2	11.2
Aspartic Acid[2]	11.7	11.4
Threonine	3.9	2.6
Serine	5.1	4.8
Glutamic Acid[3]	18.7	18.3
Proline	5.5	4.4
Glycine	4.2	5.6
Alanine	4.3	3.9
Cystine	1.3	1.3
Valine	4.8	4.2
Methionine	1.3	1.2
Isoleucine	4.5	3.4
Leucine	7.8	6.4
Tyrosine	3.1	3.9
Phenylalanine	4.9	5.0
Tryptophan	1.3	1.0

[1]From Boldwell and Hopkins (1985).
[2]Includes asparagine.
[3]Includes glutamine.

Kwon, 1987). They are considered as a specialty item and not as a daily food source, since their production is expensive. Soymilk, necessary to make the films, is prepared from soybeans by subjecting them successively to washing, soaking in water, draining, and wet grinding (Smith and Circle, 1972; Wang, 1981).

The procedure followed to produce soy films from soymilk is described by various authors (Piper and Morse, 1923; Burnett, 1951; Smith and Circle, 1972; Fukushima and Hashimoto, 1980; Guo, 1983; Snyder and Kwon, 1987). Soymilk is heated in open, shallow pans to near boiling point. A creamy yellow film is then formed on top of the milk due to surface dehydration. The film is lifted up with two sticks or by passing a rod underneath it. In this manner films are continuously formed and removed from the surface of soymilk until no further film formation occurs. Small quantities of auramine (a synthetic, yellow dye) are often added to heated soymilk to increase the thickness of formed films (Piper and Morse, 1923; Burnett, 1951). The films, when still wet, are used for wrapping meats and vegetables. Usually they are dried either in air or by slow heating. Upon drying they become brittle and are shaped into sheets, sticks, and small flakes for further use as wrappers in cooking (Snyder and Kwon, 1987).

Film Formation on Surface of Heated Soymilk

Traditional Oriental film production from soymilk created research interest for studying this film-forming process. A number of studies related to film production from heated soymilk and to the mechanism of film formation have been published. The term "soy protein-lipid film" is very common in the literature, since protein and lipids are the two major components of films from soymilk. Other frequently used terms are "yuba," after the Japanese name for these films, and "soymilk skin." Research efforts and findings on film formation from soymilk are presented and discussed below.

Film Production

SOYMILK PREPARATION

Wu and Bates (1972b) proposed an optimal process for soymilk preparation from Bragg soybeans. Specifically, they recommended presoaking soybeans in water for 1 hour at 65°C, followed by draining, grinding with water in a 15:1 water-to-beans ratio, and, finally, passing the mixture through a screw expeller and then a clarifier.

Other researchers ground drained soybeans mixed in a 10:1 water to beans ratio and applied centrifugation to separate soymilk from insoluble solids and impurities (Farnum et al., 1976; Chuah et al., 1983; Sian and Ishak, 1990b).

FILM FORMATION

Heating soymilk up to 85°C in shallow open containers and maintaining it at that temperature with the aid of a water or steam bath is a common method followed for surface film formation (Wu and Bates, 1972a; Jaynes and Chou, 1975; Chuah et al., 1983; Sian and Ishak, 1990a, 1990b). In an optimization study on the processing temperature, Wu and Bates (1972b) observed an increased film formation rate at higher temperatures, such as 95°C, closer to the boiling point. When boiling occurred, film formation was disrupted. Farnum et al. (1976) produced films by heating at 98°C.

Films formed successively on the surface of soymilk were removed until gelation of solutions inhibited further film formation. At this point, dilution of viscous solutions with water sustained film formation up to complete exhaustion of film-forming capability (Wu and Bates, 1973). Simple devices, such as glass or stainless steel rods and L-shaped wires, were employed by researchers for film removal. Films were subsequently dried at ambient or elevated temperatures.

Wu and Bates (1975) attempted to mechanize film formation in a semi-continuous operation. Their system consisted of a flat shallow stainless steel pan with inlet and outlet ports at the ends. Soymilk at a constant temperature (95 ± 5°C) and flow rate (4 L/min) was pumped out from the pan outlet, fed to a steam-heated surface heat exchanger and redistributed to the pan through a T-shaped spreader at the inlet. A plastic screen at the pan inlet prevented foam from forming and floating in the pan, while surface dehydration of films was aided with an electric fan placed 2 m behind the inlet port.

A different method to produce films from soymilk was developed by Jaynes and Chou (1975). They precipitated soymilk protein at its isoelectric point by adjusting the pH to 4.6. This precipitate was centrifuged, collected, neutralized to the original pH of soymilk, and then spread on Teflon coated baking pans. Films were removed from pans after oven-drying at 100°C for 1 hour. Protein content of films obtained in this manner was slightly higher (+4.6% protein on a dry basis) than that of films formed on the surface of heated soymilk in the same study.

CHEMICAL COMPOSITION OF FILMS

Proteins and lipids are the main constituents of soy films, with the rest being carbohydrates and ash. Their composition obviously depends on factors such as soybean variety and quality, and soymilk composition. Average chemical compositions of soy films reported in the literature are shown in Table 9.3.

Another important factor that influences film composition is the production stage at which a film is formed. Wu and Bates (1972a) observed that in successively produced films, protein and lipid content generally decreased, while ash and carbohydrate content increased. Another study by Sian and Ishak (1990b) though, showed that protein content increased in successive films, along with carbohydrate and ash content, while fat

TABLE 9.3. Approximate Chemical Compositions (% Dry Basis) of Soy Protein/Lipid Films.

Reference	Protein	Fat	Carbohydrate	Ash
Burnett (1951)[1]	62	25	—	—
Borgstrom (1968)[1]	52	24	—	—
Wu & Bates (1972a)	55	28	12	2
Jaynes & Chou (1975)	55	34	10	2
Watanabe et al. (1974)	57	26	13	3
Farnum et al. (1976)	46	18	32	4

[1]Only protein and fat contents are reported.

content decreased. This apparent contradiction between the two articles regarding protein content of films can be attributed to differences in time and solution volume for the reported experimental methods. Wu and Bates (1972a) successively removed formed films from 5 L batches of heated soymilk over a period of 6 hours. Sian and Ishak (1990b) heated 650 mL batches of soymilk and reported results for films removed the first 140 min in 20 min intervals. Therefore, data from Sian and Ishak (1990b) corresponded to a small, initial portion of the range of film formation data from Wu and Bates (1972a). Considering, as explained below, that film formation proceeds through protein polymerization, one would expect a depletion of protein concentration in the soymilk after formation of the first few films and a subsequent decrease of protein content in formed films with time.

Mechanism of Film Formation

STRUCTURAL NETWORK OF FILMS

The observation that protein content was higher in films than in the starting soymilk led Wu and Bates (1972a) to characterize film formation as a partial protein concentration and isolation process. The mechanism of the film-forming phenomenon was envisioned by them as an endothermic polymerization of heat-denatured protein accompanied by surface dehydration.

Farnum et al. (1976) described film formation as surface drying of the soymilk emulsion. They proposed that the film structure is a protein matrix formed by heat-catalyzed protein-protein interactions with disulfide, hydrogen, and hydrophobic bonds being the main associative forces in the film network. From protein extractability and electrophoresis data they concluded that 2S and 11S proteins are the major extractable components involved in film formation.

Protein insolubilization through polymerization during heating and subsequent drying of soymilk was evidenced by Fukushima and Van Buren (1970). They showed that the mechanism of polymerization involves intermolecular disulfide and hydrophobic bonds. Heating is essential for formation of these bonds, since it alters the three-dimensional structure of proteins, exposing sulfhydryl groups and hydrophobic side chains. Upon drying, unfolded protein macromolecules approach each other and become linked through disulfide and hydrophobic bonds. The analogy of this case (soymilk heating and drying) and the proposed model (protein polymerization) to film formation on the dehydrated surface of heated soymilk is obvious.

Lipids are believed to be present in films as oil droplets trapped into the protein matrix (Farnum et al., 1976). Electron microscopic observations showed that these oil droplets are less than 0.5 μm in size and are irregularly distributed in the protein matrix (Okamoto, 1978).

PROTEIN-LIPID INTERACTION

Wu and Bates (1972a) examined the manner in which neutral lipids and phospholipids interact with protein in soy films. They produced films from two-component model systems, each system comprised of soy protein isolate and either a neutral lipid (safflower oil) or a phospholipid (soybean lecithin) as the secondary component. The phospholipid showed a higher degree of incorporation into films than the neutral lipid. Their explanation for the high phospholipid incorporation was based on the ability of phospholipids to act as surface active agents. Interaction with protein was characterized by formation of protein-lipid complexes. Nonpolarity of neutral lipids, and therefore limited interaction with protein side chain groups, was considered the reason for lower incorporation of neutral lipids than phospholipids.

According to Karel (1973, 1977), there is no single type of bonding between proteins and lipids. Hydrophobic bonding among nonpolar groups seems to play an important role. Hydrogen bonding and van der Waals forces play a secondary role, while phospholipid interactions with protein may also involve electrostatic bonding.

The ability of soy protein to form films around oil droplets at the interface between oil and water was studied by Flint and Johnson (1981). They suggested that bonding of protein lipophilic groups with oil droplets is involved in the film formation phenomenon.

Watanabe et al. (1975) suggested that the 11S soy protein fraction is more lipophilic than the 7S fraction. In films prepared from homogenized solutions of soybean oil and 7S and 11S soy protein, lipids contained in films from 7S protein were largely extracted, whereas partial extraction was noticed with films from 11S protein.

Farnum et al. (1976) did not observe any preferential uptake of fatty acids during film formation. Lipids seemed to be present in films in the form of triglycerides. Protein to lipid ratios in their films were found to be independent of processing variables and dependent on composition of original soymilk.

Products of lipid peroxidation in protein-lipid systems can react with proteins, leading to protein polymerization and formation of insoluble or sparsely soluble complexes (Roubal and Tappel, 1966; Karel, 1973, 1977). However, the possibility that lipid peroxidation might be of importance in soy protein-lipid film formation was examined and found unlikely (Farnum et al., 1976).

PROTEIN-CARBOHYDRATE INTERACTION

Incorporation of carbohydrates into films from soymilk was investigated by Wu and Bates (1972a) in a two-component model system comprised of soy protein isolate and a carbohydrate (sucrose). Film yields were lower than those of the soy protein isolate and lipid (both neutral and phospholipid) model systems. Poor carbohydrate incorporation was attributed to the hydrophilic nature of sucrose, which kept it dissolved in the solution.

Sucrose, stachyose, and raffinose were identified as the major low molecular carbohydrates present in films from soymilk (Iwane et al., 1986). Farnum et al. (1976) suggested that protein-carbohydrate associative forces in the form of ionic or hydrogen bonds are developed in the film structure. However, occurrence of ionic bonding seems unlikely because these three carbohydrates present in films are not ionic in nature.

Factors Affecting Film Formation

COMPONENT CONCENTRATION IN SOYMILK

Protein is the vital component in soymilk for film formation. Wu and Bates (1972a) showed that model lipid-carbohydrate systems, without protein, resulted in no film formation. Reportedly, a protein concentration in soymilk of 4–5% maximized protein incorporation into films without slowing down the film formation rate (Wu and Bates, 1973). They also recommended that the concentration of total solids be less than 9%. Higher concentrations caused gelation and thickening of the system upon heating, with a detrimental effect on film formation. Protein-lipid ratios below 1.0 were also detrimental to film formation and yielded very oily films (Wu and Bates, 1973).

pH OF SOYMILK

The natural pH of soymilk is about 6.7. However, film formation is favored in more alkaline soymilk solutions. At pH values below natural, the isoelectric range of protein fractions in soymilk is reached, negatively affecting film formation due to protein coagulation. On the other hand, in extremely alkaline environments, negative charges along the protein chains result in repulsive forces that inhibit formation of the protein matrix.

Sian and Ishak (1990a) reported that film formation was possible only within pH 1.5–2.5 and pH 6.3–12.3. Wu and Bates (1972b) observed maximization of both film yield and protein incorporation into films at pH 9.0.

They suggested an optimum soymilk pH range for film formation between 7.0 and 8.0, because film darkening occurred above pH 8.0.

Sian and Ishak (1990a, 1990b) also noticed maximization of film yield at pH 9.0, which they attributed to high protein solubility providing more protein molecules for film formation. They proposed a soymilk pH range of 7.5–9.0 for high film yield, better protein incorporation, and acceptable boiling resistance.

Film Formation from Soy Protein Isolates

Film Production

Films can be produced from soy protein isolates by heating aqueous soy protein isolate dispersions to form surface films or by depositing and drying prepared soy protein–based, film-forming solutions.

SURFACE FILM FORMATION

Wu and Bates (1972a) prepared films from soy protein in a method similar to the one they used for film formation from soymilk. Solutions of soy protein isolates were heated in open, flat, shallow, stainless steel pans at 85°C. Films formed on the surface were removed with glass rods and dried.

The effect of concentration and pH of aqueous soy protein isolate dispersions on film characteristics was investigated (Wu and Bates, 1973). Optimal concentrations, offering the best compromise between film strength, protein incorporation efficiency, and formation rate, were 4.3% at pH 8.5 or 5.3% at pH 9.5. Higher concentrations disrupted film formation, due to gelation of the solutions. Circle et al. (1964) found that thickening and gelation occurred in aqueous soy protein isolate dispersions at concentrations above 7% when heated above 65°C.

Okamoto (1978) examined film formation on the surface of aqueous solutions of 5% soy protein isolate heated at 80°C in open shallow pans. Films from 11S and 7S soy protein fractions were also produced in the same study.

FILM FORMATION BY DEPOSITION OF FILM-FORMING SOLUTIONS

Guilbert (1986, 1988) produced films by deposition of aqueous solutions containing 10% soy protein isolate and 5% glycerin on a weight basis. Glycerin was added as a plasticizer. The deposited film-forming solutions were dried at 35°C for 24 hours.

Gennadios et al. (1993c) prepared a protein-based film-forming solution with the protein source being 70% vital wheat gluten and 30% soy protein isolate. Mixtures of these proteins with 95% ethanol, distilled water, and 6 *N* ammonium hydroxide were dispersed by heating and stirring. Glycerin was added as well to induce film plasticization. The dispersions were passed through a homogenizer and then cast on clean, level glass plates with a thin-layer chromatography spreader bar. Films were peeled off the glass surface after drying for 15 hours in an air-circulating oven set at 32 ± 2°C.

Films from alkaline aqueous solutions of soy protein isolate (5% w/v) and glycerin (3% w/v) were produced by Brandenburg et al. (1993). Film-forming solutions were heated under stirring at 60°C for 10 min, cast on Teflon coated glass plates, and dried at ambient temperature for 15 hours. Use of alkali-treated soy protein isolate yielded films with smoother appearance and fewer insoluble particles, as evidenced with scanning electron microscopy (Brandenburg et al., 1993).

Mechanism of Film Formation

Film formation from aqueous soy protein isolate dispersions is believed to proceed through protein polymerization and solvent evaporation at the interface between film and air. Protein molecules in films are associated through disulfide, hydrophobic, and hydrogen bonds. Color reactivity tests have shown that during surface film formation hydrophobic protein chains orient towards the film side facing the atmosphere, while hydrophilic chains orient towards the film side facing the solution (Okamoto, 1978).

The orientation of protein macromolecules at the air/water interface is explained by Cheesman and Davies (1954) in their review of protein monolayers at interfaces. When protein is in solution, hydrophobic groups are oriented towards the interior of protein molecules, away from water. At the air/water interface these hydrophobic groups extend out of the water into the air interacting with each other, while hydrophilic groups remain submerged. In another review of protein monolayers at interfaces, Mac-Ritchie (1978) discusses interfacial coagulation which might be involved in soy protein surface film formation. When protein concentration in the interface monolayer exceeds a limit, the protein coagulates forming a three-dimensional coagulum at the interface from the two-dimensional monolayer.

Studies have revealed that the 11S fraction of soy protein possesses the capability of forming polymers. Polymerization has been inhibited by reagents that cleave disulfide bonds, such as sodium sulfide, sodium sulfite, mercaptoethanol, sodium cyanide, hydrogen peroxide, and cysteine (Briggs and Wolf, 1957; Circle et al., 1964). These observations indicate that polymerization takes place through disulfide bonds.

Sulfhydryl groups contained in 11S protein are responsible for the formation of disulfide linkages. This was concluded after no polymerization occurred when sulhydryl groups were blocked with reagents such as N-ethylmaleimide and iodoacetamide (Briggs and Wolf, 1957; Wolf and Tamura, 1969).

Both air oxidation and sulfhydryl-disulfide interchange are mechanisms involved in formation of intermolecular disulfide bonds from sulfhydryl groups. The chain-type sulfhydryl-disulfide interchange phenomena associated with proteins are covered in detail by Jensen (1959).

Heating favors soy protein polymerization by disrupting protein structure and exposing sulfhydryl and hydrophobic groups. Alkaline conditions also favor polymerization because they too unfold polypeptide chains and promote sulfhydryl-disulfide interchange (Kelley and Pressey, 1966). Okamoto (1978) concluded that heating above 60°C and alkaline conditions (pH range 6.1–10.2) were necessary for film formation on the surface of soy protein isolate solutions. No film formation occurred at pH values above 10.5. In these highly basic environments, proteins were negatively charged, developing repulsive forces.

The effect of pH on film formation from cast aqueous solutions of soy protein isolate (5% w/v) and glycerin (3% w/v) was studied by Gennadios et al. (1993a). Film formation was achieved in the pH ranges of 1–3 and 6–12, with pH from 8 to 10 being the optimum in terms of film tensile strength. When the pH of solutions was adjusted close to the isoelectric point of soy protein (around pH 4.6), protein coagulated and casting of solutions was not possible.

Motoki et al. (1987a) reported formation of a heteropolymer between 11S soy globulin and milk casein, catalyzed by transglutaminase. They attributed this phenomenon to polymerization of 11S protein through development of intermolecular cross-links by transglutaminase. Casein was assumed bound with lysyl residues of 11S globulin.

The 7S fraction of soy protein also undergoes polymerization by disulfide bonding (Wolf and Smith, 1961; Wolf and Cowan, 1975). Disulfide bonds in 7S globulins were studied by Hoshi et al. (1982). Modifying 7S globulin with N-ethylmaleimide, a sulfhydryl-blocking reagent, prevented insolubilization during storage, indicating the vital role of disulfide linkages in 7S polymerization. The same researchers suggested that hydrophobic and hydrogen bonds also contribute to 7S insolubilization.

Differences were observed between films formed at the surface of heated aqueous solution of 11S and 7S soy globulins (Shirai et al., 1974; Okamoto, 1978). Films from the 11S fraction were smooth and opaque, whereas those from the 7S fraction were translucent and creased. Furthermore, films from the 11S fraction had higher tensile strength, presumably due to the lower tendency of the 7S fraction to form disulfide bonds (Saio et al.,

1971). Interaction of 7S and 11S soy proteins through disulfide bonds has also been suggested (Yamagishi et al., 1983).

PEANUT PROTEIN

Protein Nature and Recovery

Whole seed peanuts have a protein content between 22 and 30% (Ahmed and Young, 1982). Most of these proteins can be classified as albumins and globulins, with globulins accounting for most of the total protein. The two main peanut globulin fractions have been isolated and identified as arachin and conarachin (or nonarachin). Conarachin is the more soluble of the two fractions and its content is about a third of that of arachin (McWatters and Cherry, 1982). Arachin is stored in aleurone grains while conarachin is stored in cytoplasm (Cherry, 1983). Both fractions consist of protein components with molecular weights ranging between 20,000 and 84,000 (Basha and Cherry, 1976). The main difference in amino acid composition between arachin and conarachin is their sulfur content. Conarachin contains more cystine and methionine, and it has about three times as much total sulfur as arachin (Woodroof, 1983).

High-protein-content peanut products are commercially produced from defatted peanut meal in the form of concentrates and isolates. Peanut protein concentrates containing 70% or more protein are produced employing various techniques, such as water leaching at the isoelectric pH, aqueous alcohol leaching, air classification, liquid cyclone fractionation, and moist heat denaturation followed by water leaching (Natarjan, 1980).

In producing peanut protein isolates (90% or more protein content), protein is extracted with dilute alkali from defatted peanut meal. After clarification of the extract, protein is precipitated with acid at the isoelectric pH, washed, concentrated, and dried (Natarajan, 1980). The amino acid composition of peanut protein isolate is shown in Table 9.2.

Film Production

Efforts to develop peanut protein-lipid films on the surface of heated peanutmilk, in a manner similar to film formation on soymilk surface, have been reported in the literature (Wu and Bates, 1973; Aboagye and Stanley, 1985). Wu and Bates (1973) used an experimental peanutmilk system prepared from Florunner peanuts containing 1.6% protein, 3.0% lipid, and 5.6% total solids. Peanut protein-lipid films were successively formed and removed from the surface of peanutmilk by heating in stainless steel pans at 95 ± 5°C. Reportedly, these peanut films were weaker than films produced from soymilk. Addition of soy protein isolate in the peanutmilk system improved film strength, formation rate, and protein incorporation efficiency (Wu and Bates, 1973).

Aboagye and Stanley (1985) studied surface film formation on heated peanutmilk prepared from a valentia-type peanut cultivar. They investigated the effect of pH, heating temperature, and protein concentration on film yield, formation rate, and strength. Temperature was the most significant factor affecting film yield and formation rate. The smaller the temperature difference between film formation temperature and boiling point of water, the greater the film yield and formation rate. Reportedly, the film-forming process was enhanced by using alkaline pH conditions and higher protein concentrations (up to 5% w/v). However, elevated pH levels decreased film strength and a pH around 7 was recommended for optimal film strength (Aboagye and Stanley, 1985).

Surface film formation on heated dispersions of peanut protein was reported by Okamoto (1978). A pH value in the range between 8 and 12 was considered suitable to achieve film formation.

KERATIN

Nature and Recovery

Keratin is the term used to denote the insoluble cystine-containing proteins found in epidermal tissues of vertebrates, such as skin, hair, feathers, nails, claws, hoofs, and scales (Fraser et al., 1972). The large amount of cystine present in keratin and the development of disulfide bonds is mainly responsible for keratin's natural insolubility. Extensive cleavage of these disulfide linkages is necessary for extraction and isolation of keratin. Various approaches, such as reduction, oxidation, sulfitolysis, and oxidative sulfitolysis, have been used to extract and dissolve keratin (Crewther et al., 1965).

Keratin can be more easily extracted from avian feathers than from mammalian epidermal appendages. Feather keratin consists of uniformly sized subunits of around 10,000 molecular weight (Woodin, 1954; Harrap and Woods, 1964b). In terms of amino acid composition (Table 9.4), feather keratin is characterized by a high content of proline, serine, and glycine and by a very low content of lysine, histidine, methionine, and tryptophan (Woodin, 1956; Harrap and Woods, 1964a).

Film Production

Anker (1972) developed a method to produce films from alkaline dispersions of keratin in aqueous alcohol. Keratin isolated from poultry feathers was found more suitable for this process. Films were cast and dried from dispersions with a keratin to liquid ratio between 1:3 and 1:20 (w/w), and a water to alcohol ratio between 0.2:1 to 2.7:1 (v/v). Monohydroxy aliphatic alcohols with one to four carbon atoms were preferable. A plasticizer, such as glycerin, was added in amounts up to 40% of the protein weight in order to increase film flexibility. Alkaline conditions were provided with a water-soluble inorganic base, such as ammonium hydroxide, sodium hydroxide, and potassium hydroxide.

Okamoto (1978) studied film formation on the surface of heated dispersions of chicken feather keratin. Dispersions of 2.5% keratin in 50% ethanol containing mercaptoethanol were heated up to 60°C within the pH range 4 to 10 to achieve surface film formation. Formed keratin films were stronger than films from other proteins, such as wheat gluten, wheat gliadin, and soy protein, produced in the same manner during this study. Incorporation into solutions of *N*-ethylmaleimide, a reagent known to disturb disulfide bonds, resulted in weaker films, verifying the important role of disulfide bonds in film formation.

Film formation from a mixed solution of wheat gliadin and chicken feather keratin was also reported by Okamoto (1978). These films did not have improved properties, a suggestion that no interlinkages were developed between the two proteins. In our opinion this could be attributed to the tendency of wheat gliadin to form, almost exclusively, intramolecular and not intermolecular disulfide linkages.

Gennadios et al. (1993d) produced films by heating, casting, and drying dispersions of vital wheat gluten and hydrolyzed keratin in a 6.5:1 ratio. Dispersions were prepared by dissolving the protein material in aqueous ethanol. Ammonium hydroxide was added to provide alkalinity, as well as a plasticizer (glycerin). Film evaluation indicated an improvement in water vapor and oxygen barrier properties over films from wheat gluten as a single protein source. This implied that some cross-linking occurred in the film structure between keratin and, presumably, the glutenin fraction of wheat gluten.

COLLAGEN

Protein Nature and Recovery

Collagen is a fibrous protein. As a major constituent of skin, tendon, and connective tissues, it is one of the most prevalent and widely distributed proteins in the animal kingdom. Collagen comprises about one-third of the total protein mass in the body and about 90–95% (dry basis) of the corium of animal skins and hides (Gustavson, 1956).

Collagen molecules are arranged in fibrils and packed together to form bundles of parallel fibers with cross-sectional diameters of 20–40 microns.

Intramolecular and intermolecular bridges form in the collagen matrix due to covalent cross-linkages. Possible covalent linkages present in the matrix include ester bonds, imide bonds, and peptide bonds involving side-chain carboxyl groups of aspartic acid and glutamic acid and ϵ-amino groups of lysine (Harrington, 1966). Disulfide bonds, although not numerous due to the low cysteine content of collagen, also play a role in the protein structure (Tanzer, 1976).

Collagen fibers swell upon immersion in alkali, acid, or concentrated solutions of neutral salts and nonelectrolytes; are relatively inelastic; are resistant to proteolytic enzymes but are readily attacked by collagenase; shrink upon heating at a specific temperature, depending on the origin of collagen; and are converted to soluble gelatin by prolonged treatment at a temperature above the shrink-causing temperature (Harrington, 1966).

The amino acid composition of collagen (Table 9.4) is characterized by an unusually high content of glycine, proline, and hydroxyproline. Glycine accounts for about one-third of the total amino acid residues in collagen, while proline and hydroxyproline account for about one-fourth of the residues (Harrington and Von Hippel, 1961; Ramachandran and Ramakrishnan, 1976).

Collagen can be dissolved and isolated from its natural sources by mild extraction with dilute acid, dilute alkali, and neutral salt solutions. Sedimentation studies on soluble collagen have revealed two components, termed α (slower sedimenting) and β (faster sedimenting). The α component has a molecular weight around 100,000 and is a mixture of two polypeptide chains (α_1 and α_2) of about the same mass but differing composition. The β component has a molecular weight of around 200,000 and consists of two different types of covalently cross-linked chain pairs α_1-α_1 and α_1-α_2 (Harrington, 1966; Piez et al., 1968).

Film Production

The film-forming ability of collagen has been traditionally utilized in the meat industry for production of edible sausage casings. Collagen casings are manufactured from the regenerated corium layer of food-grade beef hides (Rust, 1987). Patent literature related to use of collagen in manufacturing of sausage casings can be found dating back to the 1930s (Becker, 1936a, 1936b, 1936c, 1938, 1939, 1941; Becker and Weiss, 1937, 1941; Freudenberg and Becker, 1938). Two different methods have been developed for the industrial manufacture of collagen casings (Courts, 1977; Hood, 1987). A method known as "dry process" evolved in Germany and was adapted by European manufacturers. An alternative method, known as "wet process," was established in North America.

TABLE 9.4. Amino Acid Composition (Residues per 100 Residues) of Selected Animal Proteins.

Amino Acid	Fowl Feather Rachis Keratin[1]	Bovine Corium Collagen[2]	Type A Gelatin[3]	Type B Gelatin[4]
Lysine	0.6	2.5	2.7	2.8
Histidine	0.2	0.5	0.4	0.4
Arginine	3.8	4.8	4.9	4.8
Aspartic Acid[5]	5.6	4.7	4.5	4.6
Threonine	4.1	1.7	1.8	1.8
Serine	14.1	3.9	3.5	3.3
Glutamic Acid[6]	6.9	7.2	7.3	7.2
Proline	9.8	12.9	13.2	12.4
Glycine	13.7	33.7	33.0	33.5
Alanine	8.7	10.7	11.2	11.7
Cysteine	7.8	0.0	0.0	0.0
Valine	7.8	2.0	2.6	2.2
Methionine	0.1	0.4	0.4	0.4
Isoleucine	3.2	1.1	1.0	1.1
Leucine	8.3	2.4	2.4	2.4
Tyrosine	1.4	0.5	0.3	0.1
Phenylalanine	3.1	1.3	1.4	1.4
Tryptophan	0.7	—	—	—
Hydroxyproline	—	9.4	9.1	9.3
Hydroxylysine	—	0.5	0.6	0.4

[1]From Fraser et al. (1972).
[2]From Veis (1964), data were rounded to the nearest 0.1.
[3]From Rose (1987), gelatin from acid-pretreated pigskin.
[4]From Rose (1987), gelatin from alkali-pretreated ossein.
[5]Includes asparagine.
[6]Includes glutamine.

The dry process (Hood, 1987) includes:

(1) Alkaline treatment of hide coriums and acidification to pH 3

(2) Shredding of acid-swollen coriums to preserve maximum fiber structure

(3) Mixing of acid-swollen fibers to produce a high-solids dough (i.e., > 12%)

(4) Addition of plasticizing and cross-linking agents

(5) High-pressure pumping and extrusion of dough to form tubular casings

(6) Drying, conditioning, neutralizing, and/or providing additional cross-linking

The wet process (Hood, 1987) consists of:

(1) Acid- or alkaline-dehairing of hides

(2) Decalcification of hide coriums and grinding into small pieces

(3) Mixing of ground collagenous material with acid to produce a swollen slurry (4–5% solids)

(4) Slurry homogenization

(5) Extrusion into tubular casings (8–10% solids)

(6) Washing casings free of salts

(7) Treatment with plasticizing and cross-linking agents

(8) Drying

For a more detailed description, the reader is referred to a number of patent disclosures on this process (Hochstadt and Lieberman, 1960; Lieberman and Oneson, 1961; Lieberman, 1964a, 1964b, 1965; McKnight, 1964a, 1964b; Petrozziello, 1967; Fagan, 1970).

Several techniques have been developed to improve the properties of collagen films and casings. Collagen casings of increased strength were produced by drying them to a moisture content of 15% to 40% in the presence of a polyhydric alcohol, such as sucrose or polyethylene glycol, or a salt, preferably sodium chloride (Kidney, 1970). Treatment with glyceraldehyde, which lacks the toxicity problems associated with other protein cross-linking aldehydes (e.g., formaldehyde, glutaraldehyde), was found effective in promoting cross-linking and increasing the strength and temperature resistance of collagen coatings (Jones and Whitmore, 1972). Addition of lower alkyl diols with four to eight carbon atoms also improved the mechanical properties of collagen casings by reducing internal hydrogen bonding while increasing intermolecular spacing (Lieberman and Gilbert, 1975; Boni, 1988).

Gas permeability properties of formaldehyde cross-linked and chrome tanned collagen films were studied under various relative humidity conditions (Lieberman and Gilbert, 1973). Formaldehyde reacted with collagen by combining free amino groups of basic amino acids, whereas chrome tanning involved reactions between carboxyl groups of amino acids. A significant inverse relationship was found between permeability and crosslink frequency in formaldehyde cross-linked films. This relationship was enhanced at high relative humidities. Chrome tanning spacer effects resulted in increased oxygen permeabilities at low levels of cross-linkage. At high levels of cross-linkage oxygen permeabilities decreased because of reduced mobility of polymer chains (Lieberman and Gilbert, 1973).

Enzymatic treatments were used to produce collagen casings with a more uniform diameter and wall thickness, as well as increased tenderness (Fujii, 1967; Tsuzuki and Lieberman, 1972; Miller, 1983). A proteolytic enzyme, such as papain, bromelain, ficin, trypsin, chymotrypsin, pepsin, fungal protease, and bacterial protease, was added in the collagen mass prior to extrusion in order to partially (or completely) solubilize collagen.

Films extruded from acid dispersions of collagen have also been subjected to irradiation in the presence of a photosensitive dye (Kuntz, 1964). The dye catalyzed protein cross-linking in the collagen gel. Reportedly, such films had higher tensile strength, lower water retention, less swelling in water, and less shrinkage during heating in water, than films formed from nonirradiated dispersions (Kuntz, 1964).

Surface frictional characteristics of sausage collagenous casings were improved by distributing on their external surface an edible (nontoxic), substantially spherical, nonhygroscopic powder, such as vinyl and olefinic polymers, polyesters, polyisocyanates, polyamides, phenol-aldehydes, melamine-aldehydes, and cellulosic derivatives (Burke, 1975).

Bradshaw et al. (1966) prepared casings by mixing together an aqueous slurry of collagen and a solution or suspension of a polysaccharide able to chemically associate with collagen. Suitable polysaccharides were those possessing reactive ionic groups, such as salts of alginic or pectic acids. Mixtures were extruded and hardened by using a precipitating agent for the polysaccharide, such as calcium chloride.

GELATIN

Protein Nature and Recovery

Gelatin is a protein resulting from partial hydrolysis of collagen. The collagenolytic process involves alkali or acid treatment followed by or accompanied with heating in the presence of water (Eastoe and Leach, 1977). Glue, another product derived from breaking down the fibrous structure of collagen, contains greater amounts of impurities than gelatin because collagenous raw materials employed in glue production are less refined (Young, 1967). Gelatin is differentiated from other proteins by the absence of appreciable internal order and the random configuration of polypeptide chains in aqueous solutions (Veis, 1964).

The breakdown of collagen to gelatin proceeds in two steps: (1) thermal denaturation, occurring at approximately 40°C, which cleaves hydrogen and electrostatic bonds, and (2) hydrolytic breakdown of covalent bonds (Eastoe and Leach, 1977). Principal raw materials commercially used for gelatin manufacture are porkskins and bovine hides, skins, and bones. Some initial preparation of these raw materials, such as degreasing, size reduction, and demineralization of bones, is required prior to further processing. Porkskin is the largest supply of raw material for edible gelatin production in the U.S. (Alleavitch et al., 1989).

A pretreatment (curing) is applied prior to gelatin extraction regardless of the collagenous raw material under processing. Curing removes impurities and initiates collagen hydrolysis. Pretreatment that employs acid processing yields Type A (acid-precursor) gelatin with an isoelectric point between pH 7 and 9, whereas alkaline processing (liming) during pretreatment yields Type B (alkali-precursor) gelatin with an isoelectric point between 4.6 and 5.2 (Rose, 1987; Alleavitch et al., 1989). Type A gelatin is mostly made from porkskins, while bones and bovine hides are usually used as raw materials for Type B gelatin (Viro, 1980).

The acid- or alkali-cured material is washed and subjected to a series of extractions ("cooks"). Each successive extraction is performed at a higher temperature and for a shorter time than the previous extraction. Extraction-derived gelatin liquors are purified by bleaching, and centrifugation or filtration. Subsequently, gelatin liquors are concentrated up to 40–50% solids, sterilized, and dried to 10–15% moisture (Hinterwaldner, 1977b; Rose, 1987; Alleavitch et al., 1989). Amino acid compositions of Type A and Type B gelatins are shown in Table 9.4.

Enzymatic processing is an alternative method to acid or liming processing for conversion of collagen to gelatin. Enzymatic pretreatment offers a number of advantages, including shorter processing time and significantly higher yield. The resultant gelatin has higher purity, better physical properties, and a more narrow molecular weight distribution (Hinterwaldner, 1977a).

The molecular weight of gelatin depends on the raw material employed in gelatin production and on curing, extracting, and handling conditions. As a result, gelatin molecular weights cover a broad range from about 3000 up to 200,000 (Young, 1967).

Commercial value of gelatin increases with the strength of gels formed under standard conditions. Gelatin gel strength is measured with the Bloom gelometer and expressed in "grams Bloom" or "Bloom." Commercial gelatin preparations are available with gel rigidity in the range between 80 and 320 g Bloom (Wainewright, 1977).

Film Production

Gelatin has been known to form clear, flexible, strong, and oxygen-impermeable films when cast from aqueous solutions containing a plasticizer, such as glycerin or sorbitol. Film-forming applications of gelatin in the pharmaceutical and food industry include microencapsulation and manufacture of tablet and capsule coatings.

Gelatin films in a free-standing form were prepared by Guilbert (1986, 1988) by casting aqueous film-forming solutions of 20% gelatin (250 Bloom) and 0–10% glycerin. Bradbury and Martin (1952) prepared gelatin films by drying 5% solutions of gelatin in water. Films dried at 20°C had higher tensile strength than films dried at 60°C. The observed tensile strength difference was attributed to the higher degree of crystallization in films dried at the lower temperature (Bradbury and Martin, 1952). Besides the degree of crystallinity, the existence of gelatin in two different forms interchangeable with temperature was suggested as another reason for differences in gelatin film properties with drying temperature (Robinson, 1953). Presumably, at temperatures higher than 35°C, gelatin exists as single molecules in a configuration that cannot form interchain hydrogen bonds, whereas, at lower temperatures, gelatin has a configuration (collagen-fold configuration) capable of forming interchain hydrogen bonds (Robinson, 1953).

MILK PROTEINS

Milk is secreted from the mammary glands of all mammal species to provide adequate nutrition and immunological protection for their young (Jenness, 1988). The Western breed of dairy cows, *Bos taurus,* is the major source of milk for most industrialized societies (Brunner, 1977). Milk has a protein content of about 33 g/L (Dalgleish, 1989). Whey proteins and casein are the main milk protein fractions and their recovery from whole milk is discussed later. Amino acid compositions of various milk proteins are shown in Table 9.5. High nutritional quality, water solubility, and emulsification capability of milk proteins make their use in edible film formation very attractive.

Approximately 9×10^8 kg of dairy powder ingredients are used annually in the U.S. (Kinsella, 1984). Common forms of dairy powders, such as caseinates, low lactose-low mineral whey, whey concentrate powders, dry sweet whey, whole dry milk, and nonfat dry milk, present potential as edible film-forming ingredients. Recovery of each of the aforementioned dairy powders differs. For example, whole dry milk is best produced by spray-drying milk. End product properties can be altered by heat treatments and homogenization prior to spray-drying (Kinsella, 1984). Film-forming characteristics of cow's milk are discussed below. The provided information is expected to apply, at least partially, to milk from other sources.

Film Formation from Total Milk Proteins

Much literature exists on the formation of problem-causing milk films on processing equipment surfaces (Mabesa et al., 1979, 1980). Such information, although not directly related to edible film research, is a useful indication of milk's potential as a film former. Stainless steel plates were dipped in fresh raw milk and the residual film dried at 37°C and various relative humidities for 30 minutes. Amounts of protein left on stainless steel plates after rinsing were found to increase when dry films were exposed to 100% relative humidity. The optimum temperature for film for-

TABLE 9.5. Amino Acid Composition (% Residues/Mol Protein) of the Major Proteins from Bovine Milk.

Amino Acid	Whey Protein[1] Isolate	β-Lactoglobulin[2] A	Caseinate[3]	α_{s1}-B	Casein[4] β-A1	x-A
Lysine	8.3	9.2	6.9	7.0	5.3	5.3
Histidine	7.8	1.2	2.6	2.5	2.9	1.8
Arginine	1.9	1.8	2.8	3.0	1.9	3.0
Aspartic Acid[5]	10.0	9.9	6.5	7.5	4.3	7.1
Threonine	4.7	4.9	4.5	2.5	4.3	8.9
Serine	7.9	4.3	6.4	8.0	7.6	7.7
Glutamic Acid[6]	13.3	15.4	18.6	19.6	18.7	16.0
Proline	4.9	4.9	10.9	8.5	16.3	11.8
Glycine	2.6	1.8	2.9	4.5	2.4	1.2
Alanine	6.6	8.6	4.4	4.5	2.4	8.3
Cysteine/2	2.4	3.1	0.3	0.0	0.0	1.2
Valine	4.0	6.2	7.3	5.5	9.1	6.5
Methionine	1.6	2.5	2.5	2.5	2.9	1.2
Isoleucine	4.7	6.2	5.2	5.5	4.8	7.1
Leucine	12.3	13.6	9.5	8.5	10.5	4.7
Tyrosine	2.5	2.5	4.0	5.0	1.9	5.3
Phenylalanine	2.5	2.5	4.0	4.0	4.3	2.4
Tryptophan	1.2	1.2	0.8	1.0	0.5	0.6

[1]From Bio-Isolates, Le Sueur Isolates, Le Sueur, MN (1992).
[2]From Kinsella and Whitehead (1989).
[3]From New Zealand Milk Products, Inc., Santa Rosa, CA (1992).
[4]From Brunner (1977).
[5]Includes asparagine.
[6]Includes glutamine.

mation was found to be 37°C. Scanning electron microscopy (SEM) was employed to determine the mechanism by which film solubility decreased and to identify film components. SEM revealed splotches of fat on surfaces and showed a more aggregated and porous appearance for humidified films over those that were dried only. Furthermore, the amount of granulated lactose increased among humidified samples. Electrophoresis studies identified that α- and β-caseins resisted rinsing more in films exposed to 100% relative humidity than in dry films (Mabesa et al., 1979, 1980). These results indicate the possible effects of high relative humidity exposure on milk protein food coatings formed at elevated temperatures, such as decreased solubility and increased granulation of lactose.

Further evidence for formation of milk protein edible films can be found in literature concerning films formed at air-water interfaces of reconstituted milk powders (Leach et al., 1966). When whole milk powder is added to water, small amounts remain insoluble and form a film at the air-water interface impeding further wetting without agitation. Physical and chemical properties of films formed at air-water interfaces during reconstitution of whole milk powder were studied in an effort to eliminate film formation (Leach et al., 1966). Formed films were multimolecular in depth, as shown by film balance studies, and compressible, yet nonelastic. A high concentration of eighteen-carbon saturated fatty acids in films was detected with gas-liquid chromatography. Gel electrophoresis revealed numerous sugar-protein complexes. Increased water or powder temperature prior to forming the reconstituted powder solution reduced film strength. Milk powders were rendered self-dispersing by the presence of liquid fat (Leach et al., 1966). Once again, these observations can be applied to the development of edible films formed from whole milk proteins, as can the possible effects that various types of lipids and temperatures may have on film properties.

In the only available study on films from total milk proteins, Wu and Bates (1973) studied successive film formation on the surface of heated (90 ± 5°C) nonfat dry milk (NFDM) solutions. Maximum NFDM film strength was reached within 15 min of heating prior to film formation. Protein incorporation into films was maximized when starting NFDM solutions had protein concentration of 4–5%, total solids concentration of less than 9%, and pH of 8.0–9.5. NFDM films were stronger than films formed on the surface of soymilk under the same experimental conditions (Wu and Bates, 1973).

Casein

Protein Nature and Recovery

Caseins are the major proteins in milk, representing 80% of the total milk proteins (Dalgleish, 1989). The total concentration of casein in milk ranges between 2.5–3.2% (Brunner, 1977). Caseins are predominantly phosphoproteins that precipitate at pH 4.6 and 20°C. Four principal components, α_{s1}-, α_{s2}-, β-, and $\varkappa$-caseins and a minor component, γ-casein, are identified. Simple Mendelian genetics results in various protein variants or polymorphs for each of the above fractions among different dairy breeds (Brunner, 1977). However, these genetic variants do not appear to significantly affect protein properties (Dalgleish, 1989).

The amino acid composition of various casein components (Table 9.5) is characterized by low levels of cysteine. Consequently, few disulfide cross-linkages are present and the casein molecule assumes an open, random coil form. The primary amino acid sequences of caseins are, perhaps, more important than their amino acid composition, in that these sequences often dictate protein functionality. Caseins show a lack of homologous amino acid sequences with the exception of the sequence -SerP-SerP-SerP-Glu-Glu-, which is found once in α_{s2}- and β-casein and twice in α_{s2}-casein (Dalgleish, 1989). Primary amino acid sequences often dictate possible interactions by cross-linking, due to proximity of functional amino acids.

The α_{s1}- and α_{s2}- caseins have an approximate molecular weight of 23,500, an isoelectric point at pH 5.1, and rather even distributions of hydrophilic residues (Brunner, 1977). They also are more phosphorylated than other caseins and more calcium-sensitive than β-casein (Dalgleish, 1989). Proline is a known disrupter of secondary protein structures. Therefore, the low proline content in α-casein allows for formation of α-helical and β-sheet structures (Brunner, 1977).

The α-casein fraction contains more charged residues and fewer hydrophobic residues than the β-casein fraction, resulting in increased solubility in the absence of calcium (Dalgleish, 1989). From the above characteristics, one would expect α-casein to form flexible films with rather high water vapor permeability and low gas permeability because of the low hydrophobicity and the flexible, random coil structure, resulting in increased intermolecular hydrogen bonding, of α-casein.

β-Caseins, which account for 30–35% of the total casein content in milk, have a molecular weight of 24,000 and an isoelectric point at approximately pH 5.3 (Brunner, 1977). Furthermore, the β-casein fraction has a high hydrophobicity (1334 cal/residue) that could be utilized for forming edible films of low water vapor permeability. The last fifty residues of the N-terminal end are charged and, therefore, hydrophilic, whereas the rest of the molecule is highly hydrophobic (Dalgleish, 1989). This amphiphilic structure allows β-casein to function well as a surfactant. β-Caseins exhibit temperature-, concentration-, and pH-dependent association-dissociation equilibria. At neutral pH values and high temperatures, β-caseins associate into threadlike polymers (Dalgleish, 1989).

$\varkappa$-Casein, making up approximately 15% of the total casein fraction, has a molecular weight of 19,000, an isoelectric point between pH 3.7 and pH

4.2, and is calcium-insensitive (Brunner, 1977). In the presence of calcium, $\varkappa$-casein associates with α_{s1}- and β-caseins to form thermo-dynamically stable micelles (Brunner, 1977). The known ability of $\varkappa$-casein to polymerize through disulfide bonds is dramatically affected by ionic strength (Dalgleish, 1989). Addition of sulfhydryl-blocking agents to $\varkappa$-casein results in an equilibrium between thirty-molecule polymers and monomers of $\varkappa$-casein. $\varkappa$-Caseins are also known to exist as mixed species, containing zero to five carbohydrate moieties. The complete role of these carbohydrate moieties is yet undetermined. However, they are known to increase the hydrophilicity and negative charge of the protein. Apolar residues in $\varkappa$-casein are concentrated at the N-terminal region, whereas the charged portion is located near the C-terminal, resulting in increased amphiphilic character and surface activity. The overall hydrophobicity of $\varkappa$-casein is between that of α- and β-caseins (Dalgleish, 1989).

The proteolytic enzyme rennin plays an important role in $\varkappa$-casein functionality. Rennin cleaves casein between Phe105-Met106, releasing the soluble C-terminal glycopeptide and the insoluble N-terminal para-$\varkappa$-casein. Formation of para-$\varkappa$-casein leads to aggregation of casein micelles due to hydrophobic interactions (Dalgleish, 1989). This process is important in the production of many types of cheese. These reactions could be utilized to form edible films of intermediate water vapor permeability, due to their moderate average hydrophobicities. The potential for disulfide polymerization reactions and their effect on edible film properties should also be investigated.

Association of casein fractions in the form of highly hydrated micelles and the implications of these associations need to be discussed. About 80–90% of caseins in milk exist as colloidally dispersed micelles ranging in size from 10 to 250 nm (Brunner, 1977; Kinsella, 1984). Each micelle can be subdivided into a number of submicelles. A negative correlation between $\varkappa$-casein concentration and micelle size which follows the surface-to-volume ratio of particles indicates that $\varkappa$-casein is located on the surface of micelles, imparting a negative charge which exerts a stabilizing influence. The inner structure of micelles is not completely known. They contain α- and β-caseins in addition to calcium, colloidal calcium phosphate, and citrate. Calcium acts by reducing electrostatic repulsions of casein phosphate groups, consequently facilitating hydrophobic interactions and formation of micelles (Kinsella, 1984). Calcium phosphate appears to play both a structural and a stabilizing role in the micelle, although its exact function remains unknown (Dalgleish, 1989).

In the future, as micellar interactions of caseins become more defined, researchers should attempt to exploit these properties for formation of effective coatings and films. Surfactant capabilities, hydrophobic characteristics, and polymerization potential of these individual casein components should also be explored in an attempt to better characterize relationships between fundamental characteristics and functional properties of proteins. No research has been performed in this area due to the high cost of individual casein fractions. Instead, commercially available caseinates have been explored. Unfortunately, little fundamental information concerning the properties of caseinates is available in the literature. As fractionation processes improve, these options should be explored in an attempt to develop a data base correlating protein chemical and physical characteristics with edible film functional properties. This will enable development of predictive models for protein functionality in edible films.

Numerous procedures are available for the separation of whole casein from whey in skim milk (McKenzie, 1971b). The most common method is precipitation of casein at its isoelectric pH. Several different temperatures can be employed. At 2°C, minimum solubility occurs at pH 4.3, whereas at 20°C and 30°C, minimum solubility occurs at pH 4.6. Skim milk is commonly adjusted to pH 4.6 at room temperature using 1 M hydrochloric acid (Whitney, 1988). Under these conditions, caseins precipitate, whereas whey proteins remain soluble. Therefore, separation can be achieved. Whole casein can also be obtained by either high speed centrifugation in the presence or absence of calcium at various temperatures or by ammonium sulfate precipitation (McKenzie, 1971b).

Once whole casein has been separated from whey, various methods can be used for fractionation of individual caseins, several of which are based on differential solubility (Thompson, 1971). Hipp et al. (1952) developed a procedure to separate α-, β-, and γ-casein. The first step takes advantage of their differential solubilities in 50% alcohol in the presence of ammonium acetate. By varying pH, temperature, and ionic strength, caseins precipitate in the order of α-, β-, and γ-casein. The α_{s1}- and $\varkappa$-casein can be separated with calcium chloride precipitation at 37°C by removing calcium with oxalate and reprecipitating with 0.25 M calcium chloride at 37°C and pH 7 (Waugh et al., 1962). The $\varkappa$-caseins can be subsequently removed from the supernatant by precipitation with sodium sulphate followed by reprecipitation with 50% ethanol in ammonium acetate. Other purification methods of casein fractions include partition methods, electrophoretic methods, and chromatographic methods (Whitney, 1988).

Commercially available caseins are commonly found as rennet caseins, dry casein, or caseinates. Rennet caseins are produced by renneting milk, draining, milling, washing, and drying the curd. Micelle structure is retained in this process. Dry casein is manufactured by adjusting the pH of skim milk to 4.6 and heating to 46°C to obtain a curd. This disrupts the casein micelle by solubilization of calcium phosphate, releasing individual casein molecules. The curd is then drained, washed (to remove lactose), milled, and dried. Commercial caseinates are produced by adjusting acid-coagulated casein to pH 6.7 using calcium or sodium hydroxide. The alkaline treatment solubilizes the casein by altering the net charge which overcomes hydrophobic interactions. The viscous suspension is subsequently pasteurized and spray-dried. Sodium and calcium caseinates are the most common derivatives, although magnesium and potassium caseinates are also available (Kinsella, 1984). The above commercially available casein compounds have been more extensively investigated than purified casein fractions in terms of edible film-forming potential.

Caseinates are soluble above pH 5.5. Furthermore, sodium and potassium caseinates are more soluble and often possess better functional properties than calcium and magnesium derivatives. Calcium caseinate forms dispersions, rather than solutions, which are less viscous than sodium caseinate solutions. Sodium caseinate is very heat stable, and calcium caseinate is heat stable above 4% at 120°C for 15 minutes at pH values greater than neutral (Kinsella, 1984).

Film Production

Caseins have not been as extensively investigated as other protein ingredients for their film-forming potential. Caseins form films from aqueous solutions without further treatment due to their random coil nature and ability to hydrogen bond extensively. It is believed that electrostatic interactions also play an important role in formation of casein-based edible films. The ability to function as a surfactant makes casein a very promising material for formation of emulsion films. Casein films, being transparent, flavorless, and flexible, are attractive for use on food products.

Krochta et al. (1990b) investigated the potential of films formed from casein-lipid emulsions to reduce water vapor transmission rates. Effects of casein pH and calcium cross-linking were also explored. Sodium caseinate solutions were made from 5 g sodium caseinate dissolved in 42.5 g of water, with 2.5 g of glycerin added as a plasticizer. Emulsion films were formed using various lipids, including several acetylated monoglycerides and waxes. Films were cast on plexiglass plates, air dried, and, subsequently dipped for 3 min in buffer solutions followed by soaking in distilled water for 2 min. Then films were dried prior to testing their water barrier properties. Relative water losses from test cells were compared over time.

Several interesting results were reported by Krochta et al. (1990b). Pure caseinate/glycerin films were highly permeable to moisture and very soluble in water. Buffer treatments at the isoelectric point of these films, resulted in decreased film solubility in water. However, no improvement in water vapor resistance was observed. Addition of acetylated monoglycerides to films resulted in a 50% reduction in water loss, while films remained water-soluble. By buffer-treating films, water loss was reduced an additional 50% and films were rendered insoluble. Doubling the amount of acetylated monoglyceride also resulted in a 50% reduction in water loss. Results from wax emulsion films were similar to those from acetylated monoglyceride emulsion films. Beeswax imparted lower water losses than paraffin or carnauba waxes.

A study on water vapor permeability of various types of pure caseinate films found that water vapor permeability of casein materials increased in the following order: magnesium caseinate, calcium caseinate, micellar casein, sodium caseinate, potassium caseinate, and rennet casein (Ho, 1992). Water vapor permeability values at 24°C of magnesium caseinate (0–77% relative humidity gradient) and rennet casein (0–75% relative humidity gradient) films cast from solutions of 6.4 g protein, 1.6 g glycerin, and 92 g deionized water were 43.2 g·mm/kPa·d·m^2 and 55.2 g·mm/kPa·d·m^2, respectively. Film water vapor permeability was determined using the correction factor method described in Chapter 7 of this book.

Ho (1992) also examined the effects of various lipids on water vapor permeability of casein-based edible films at 24°C. Increasing chain length increased the permeability of fatty acid emulsion films. Films with 90% sodium caseinate/10% lauric acid content exhibited water vapor permeability of 12.0 g·mm/kPa·d·m^2 (0–92% relative humidity gradient), whereas 90% sodium caseinate/10% stearic acid films exhibited permeabilities of 26.4 g·mm/kPa·d·m^2 (0–86% relative humidity gradient). Increased fatty acid concentrations decreased water vapor permeability. Finally, in agreement with a previous study by Krochta et al. (1990b), emulsion films with beeswax had lower permeability than films with paraffin or carnauba waxes.

Chaumette (1991) investigated the potential of peroxidase-catalyzed cross-linking of sodium caseinate–acetylated monoglyceride films and its effect on water vapor transmission rates. Free radicals were generated by horseradish peroxidase followed by radical-radical polymerization reactions that cross-linked casein molecules. Tyrosine residue involvement in cross-linking has been well established (Andersen, 1966). Due to the free radical mechanism of these reactions, cross-linking between casein and acetylated monoglycerides is also possible. Solutions of 0.2 M sodium borate buffer (pH 9) containing 1% sodium caseinate, 1% acetylated monoglyceride, 1 mg peroxidase, and 18 mM hydrogen peroxide were prepared. These solutions were incubated at various temperatures for different times between 0 and 30 hours. For film casting, 5% casein and 10% acetylated monoglycerides were added to 90 mL sodium borate buffer along with 76.1 mg peroxidase and 125 μL of hydrogen peroxide or 152.2 mg peroxidase and 250 μL hydrogen peroxide. Results from incubation at 37°C and 55°C and for 5, 16, 26, and 43 hours were compared. Sodium dodecyl sulfate polyacrylamide gel electrophoresis (SDS-PAGE) showed that cross-linking increased with time, temperature, and enzyme concentration. After 23 hours of incubation, all monomers were polymerized. Although it appeared that proteins were cross-linked, resultant films indicated no significant decrease in water vapor transmission rates. Only at low enzyme concentration and 43 hours of incubation did water loss decrease. Interestingly, in many cases water vapor transmission increased significantly (Chaumette, 1991).

Use of buffer solutions required in cross-linking studies complicates film water vapor transmission measurements. Buffers increase electrolyte concentration in films, thereby increasing water vapor transmission. In addition, comparison of water vapor transmission rates should always take into account film thickness variations. Finally, enzyme inactivation prior to film casting may reduce deviations due to variable drying times.

Motoki et al. (1987b) prepared α_{s1}-casein films employing transglutaminase. Transglutaminase is a calcium-dependent enzyme that catalyzes formation of covalent ϵ-(γ-glutamyl)lysyl cross-links. Such films could find use in edible coatings, medical polymers, and enzyme immobilization. Transglutaminase (0.02 units/mg casein) was added into α_{s1}-casein solutions of concentrations 1–5% in 0.1 M tris-HCl buffer (pH 8, 5 mM calcium chloride, 10 mM dithiothreitol, and 0.1% glycerin). Films were obtained by casting these solutions on polymethacrylate boards and drying at 40°C for 5 hours. Highly concentrated protein solutions were firmly gelatinized by transglutaminase, resulting in films of high tensile strength (105 g/cm²) and strain (72%). They were also insoluble in water, mercaptoethanol, urea, and guanidine hydrochloride, even after being heated at 100°C for 10 min. However, films were hydrolyzed by chymotrypsin and, therefore, remained edible (Motoki et al., 1987b). Further work is needed in this area to define the effectiveness of transglutaminase-treated protein films in controlling mass transfer.

Use of transglutaminase-treated α_{s1}-casein films as enzyme immobilization supports was explored (Motoki et al., 1987c). Various enzymes, such as β-glucosidase, α-mannosidase, β-galactosidase, and glucose oxidase, were immobilized by addition to casein solutions prior to transglutaminase-induced gelation. All enzymes were successfully immobilized through entrapment in the protein film lattice, while enzyme reactivity was not lost on repeated usage (Motoki et al., 1987c). Therefore, enzyme immobilization is a possible application of protein films.

Whey Proteins

Protein Nature and Recovery

Whey proteins, accounting for 20% of total milk proteins, are characterized by their solubility at pH 4.6 (Brunner, 1977). Liquid whey, a by-product of cheese manufacture, is produced in large quantities and its annual production is continuously rising. Production of fluid whey in 1985 was estimated at 2×10^{10} kg (Kinsella and Whitehead, 1989). Much of this whey is not utilized, creating serious waste disposal problems. Consequently, significant interest exists in finding new uses for whey proteins, such as in edible films.

Whey protein contains five principal protein types: α-lactalbumin, β-lactoglobulin, bovine serum albumin, immunoglobulins, and proteosepeptones. Various genetic variants of each of these fractions exist among different dairy breeds. However, genetic variants are not expected to affect protein functionality.

β-Lactoglobulin is the major protein in the whey protein fraction. Concentration of β-lactoglobulin in milk is approximately 3.7 g/L (Kinsella and Whitehead, 1989). The amino acid composition of β-lactoglobulin is shown in Table 9.5. The primary amino acid sequence of β-lactoglobulin can be found in Dalgleish (1989). β-Lactoglobulin has a molecular weight of 18,362 for polymorph A and 18,276 for polymorph B. Monomeric β-lactoglobulins contain one free sulfhydryl group and two disulfide groups (Brunner, 1977). Crystallization studies have shown that β-lactoglobulin exists in a globular form, with hydrophobic and sulfhydryl groups located in the interior.

Many studies have been performed on thermodynamic properties of β-lactoglobulin. Near its isoelectric point (pH 5.1–5.6) β-lactoglobulin exists as a dimer. Below pH 3, dimers dissociate into monomers. Dissociation also occurs above pH 8. Between pH 3.8 and 5.1, β-lactoglobulin associates into octamers at low temperatures, due to ionization of carboxyl groups at the interaction sites (Brunner, 1977).

β-Lactoglobulin also undergoes time- and temperature-dependent denaturation reactions at temperatures above 65°C. Denaturation exposes the internal sulfhydryl group in β-lactoglobulin and highly reactive hydrophobic and ϵ-amino groups (Brunner, 1977). In particular, oxidation of free sulfhydryls and disulfide interchange reactions can be induced by heating whey protein solutions, leading to polymerization of β-lactoglobulin molecules. Disulfide interchange reactions also occur with the α-lactalbumin component of whey protein.

α-Lactalbumin, which accounts for 25% of whey protein, is a globular protein having a molecular weight of 14,000 and four disulfide bonds. Calcium actively binds to α-lactalbumin, potentially stabilizing it against denaturation (Kinsella and Whitehead, 1989).

Bovine serum albumin (BSA) is a large globular protein with a molecular weight of 66,000. It contains seventeen disulfide bonds and one free thiol group. BSA has been shown to effectively bind free fatty acids, other lipids, and flavors, which stabilize the molecule against thermal denaturation (Kinsella and Whitehead, 1989).

Immunoglobulins and proteose-peptone fractions represent the remainder of whey proteins. Immunoglobulins, responsible for membrane transport and antigen binding, are thermally unstable. Proteose-peptones are amphiphilic and may affect protein functional properties (Kinsella and Whitehead, 1989).

Although the film-forming ability of individual whey proteins has not been investigated in the past, it is hoped that this potential will be explored in the future as protein separation technology advances. The above described properties of whey proteins can be exploited to form edible films. In particular, the heat denaturation potential of whey protein should be explored for its effects on film barrier and mechanical properties. Protein functionality in films could be modeled by correlating basic functional and chemical characteristics of individual protein fractions with properties of films derived from these fractions.

Currently, individual whey protein fractions are very costly due to the complex nature of their purification processes. Firstly, the casein fraction must be separated from skim milk by acid precipitation. Next, several methods are available for separation of whey proteins from other whey components. For example, whey proteins can be complexed with carboxymethylcellulose (CMC) (Hansen et al., 1971). In this approach, whey is acidified to pH 3.2 and diluted with 0.25 % CMC. Next, the complex is removed by centrifugation. Alternative methods for separation of whey proteins employ ion exhange with sodium hexametaphosphate, ultrafiltration, cellulose acetate membranes, reverse osmosis, and electrodialysis (Whitney, 1988).

Individual whey proteins can be fractionated similarly to caseins by differential solubility and electrophoretic or chromatographic methods. Advantages and limitations of various procedures were discussed (Gordon, 1971; McKenzie, 1971a). In one common differential solubility procedure used for separation of whey proteins, whey is prepared by adding 200 g/L of ammonium sulphate to milk at 40°C and, subsequently, fractionation is achieved (Aschaffenburg and Drewry, 1957).

A simplified procedure for purification of β-lactoglobulin uses 3 % trichloroacetic acid addition to the whey solution, followed by centrifugation at 16,000 g for 15 min (Ebler et al., 1990). Next, extensive dialysis of the supernatant is necessary to remove all lactose and salt from β-lactoglobulin.

Purification procedures for whey protein isolates (WPI) and whey protein concentrates (WPC) have recently been developed. Ultrafiltration techniques are currently employed to isolate WPCs in their nondenatured, functional form. WPCs range from 25 % to 80 % whey protein, whereas WPIs contain 80 % or more protein. WPIs, being highly purified, possess improved functionality over WPCs. High performance hydrophilic ion exchangers are used to purify WPIs.

Film Production

Whey proteins, when appropriately processed, produce transparent, flavorless, and flexible edible films, similar to those produced from caseinates. Wu and Bates (1973) investigated film-forming capabilities of cottage cheese whey and different WPCs. Films were withdrawn from the surface of heated (90 ± 5°C) whey dispersions containing 5% total solids. Sodium and acid WPCs formed better films than calcium WPC and cottage cheese whey.

Major differences exist in the manufacture of casein and whey protein-based films. McHugh et al. (1994) produced whey protein isolate films by heat-treating 8–12% solutions of whey proteins at temperatures between 75°C and 100°C. Heat was necessary for formation of intermolecular disulfide bonds required to produce intact films. Without heat application, films cracked into small crystals upon drying, whereas films obtained from heated solutions were extremely brittle and addition of a food-grade plasticizer was needed to impart film flexibility. In contrast to casein films, whey protein films were water-insoluble due to formation of covalent disulfide bonds. Optimal conditions for production of whey protein-based edible films were heating of 10% whey protein solutions at 90°C for 30 min.

Effects of relative humidity and various plasticizers including glycerin, sorbitol, and polyethylene glycols of different chain lengths, on whey protein edible films were studied by McHugh et al. (1994). In general, plasticizers reduce internal hydrogen bonding in films, thereby increasing film flexibility while increasing water vapor permeability. Sorbitol provided the highest flexibility per unit increase in water vapor permeability among the studied plasticizers. Relative humidity had an exponential effect on water vapor permeability of whey protein films. Plots of permeability versus relative humidity can be used to model water vapor permeability of edible films under variable relative humidity gradients.

McHugh and Krochta (1994b) examined the effects of incorporating various lipids into whey protein-based edible films. Various acetylated monoglycerides, waxes, fatty alcohols, and fatty acids were examined for their effects on the water vapor permeability of films. Beeswax emulsion films exhibited the lowest water vapor permeability. Furthermore, degree of homogenization drastically affected water vapor permeability of emulsion film compositions (McHugh and Krochta, 1994a). As homogenization increased, water vapor permeability decreased, probably because of tortuosity, protein-lipid interfacial effects, or bilayer formation. Water vapor permeability of films decreased with increases in chain length of fatty acids and fatty alcohols. Finally, fatty acid concentration had a profound effect on the water vapor barrier properties of films (McHugh and Krochta, 1994b).

Mechanical properties and water vapor transferability of whey protein films produced by transglutaminase-induced polymerization were examined by Mahmoud and Savello (1992). Reaction mixtures consisted of 5% whey protein in a pH 7.5 buffered solution under reducing conditions in the presence of calcium cations and glycerin. Glycerin concentration of films directly affected film resistance to breakage, moisture content and water vapor transferability. Water vapor transferability was inversely related to film thickness (Mahmoud and Savello, 1992).

PROPERTIES OF PROTEIN-BASED EDIBLE FILMS

A significant amount of information related to development and production of edible films and coatings, either from proteins or from other biological materials, originates from patents assigned to the private sector. Methods of film production are described in patent literature; however, film properties are rarely disclosed. Furthermore, accurate standard methodologies for determination of film permeability properties are often not available, as is discussed in Chapter 7. The result is a lack of comparative film property data, making predictions of film capabilities for potential applications very difficult.

In recent years, several research programs have been initiated by academic and government institutions focusing both on edible film development and film property evaluation. Film property data generated by these sources have appeared in scientific journals, allowing for a better understanding of film capabilities and limitations.

Some of these mechanical property values are presented in this section for selected protein-based films, whereas barrier properties of protein edible films are discussed in detail in Chapter 7 of this book. Comparisons among films using the data given here or from other data in the literature should be made with caution. Reported values are often measured using different procedures, instruments, and testing parameters. Since protein films are generally hydrophilic and susceptible to moisture absorption, film conditioning prior to testing is an important factor. Relative humidity, temperature, and time period of conditioning can have a significant impact on film properties.

MECHANICAL PROPERTIES

Tensile strength and percentage elongation at break, as evaluated with tensile testing, are the two most commonly measured mechanical properties of edible films. Tensile strength expresses the maximum stress developed in a film specimen during tensile testing, while the elongation value represents a film's ability to stretch. The magnitude of these two properties is closely associated with the amount of plasticizer present in a film. Generally, increasing the amount of plasticizer results in films with lower tensile strength and greater elongation.

A number of tensile strength and elongation values for protein-based films are shown in Table 9.6. The reader should be aware when comparing these data that they were generated using different instruments and different testing parameters, such as initial grip separation, cross-head speed, and dimensions of specimens.

Bursting strength, the hydrostatic pressure required to burst a material, was measured for wheat gluten/glycerin (2.5:1) films approximately 100 μm in thickness and conditioned for two days over 50% relative humidity. Values in the range of 75–99 kPa were reported (Aydt et al., 1991; Gennadios et al., 1993c). Propagation tear resistance, the force in grams required to propagate tearing through a specified length of a specimen, was measured in the range of 3–5 g for the same films (Aydt et al., 1991; Gennadios et al., 1993).

APPLICATIONS

MEAT PRODUCTS

Edible coatings comprised of proteins and other nonproteinaceous materials have been investigated for potential applications on fresh meats or value-added processed meat products. Film-forming ability, cohesiveness, and adhesiveness of proteins were utilized in such coatings to meet various objectives, including reduction of moisture migration, retention of natural juices and flavors, enhancement of texture and appearance, and prevention of oil absorption upon cooking.

Films Based on Collagen

Manufacture of collagen casings for packaging of meat products is one of the oldest applications of edible film packaging in the food industry. Artificial casings from collagen and other materials, such as cellulose and thermoplastics, were developed as substitutes for natural casings prepared from the guts of slaughtered meat animals. Natural casings, although still in use, cannot cover all the demands of the meat processing industry. Problems associated with natural casings include limited availability, labor-intensive preparation, high cost, lack of uniformity, susceptibility to breakage during handling and stuffing, and restricted process automation.

Principles of collagen casing production were discussed earlier. The interested reader can find additional information on collagen casings in

TABLE 9.6. Tensile Strength and Elongation at Break of Protein-Based Edible Films and Commonly Used Nonedible Packaging Films.

Reference	Film	Tensile Strength (MPa)	Elongation at Break (%)
Wall & Beckwith (1969)[1]	Wheat Gluten:Lactic Acid (1:1)	0.01	75
Wall & Beckwith (1969)[1]	Wheat Gliadin:Lactic Acid (1:1)	0.01	72
Wall & Beckwith (1969)[1]	Wheat Glutenin:Lactic Acid (1:1)	0.02	63
Watanabe & Okamoto (1976)[2]	5% Wheat Gluten in 3.3 M urea 0.6 M 2-mercaptoethanol added—film formed on heated surface	0.32	—
Gennadios et al. (1993c)[3]	Wheat Gluten:Glycerin (2.5:1)	2.6	276
Gennadios et al. (1993c)[3]	Wheat Gluten:Soy Protein:Glycerin (1.75:0.75:1)	4.4	233
Gennadios et al. (1993b)[4]	Wheat Gluten:Glycerin (2.9:1)	3.9	—
Gennadios et al. (1993b)[4]	Corn Zein:Glycerin (2.4:1)	13.2	—
Aydt et al. (1991)[5]	Corn Zein:Glycerin (3.9:1)	0.4	—
Okamoto (1978)[2]	5% Aqueous solution of 11S soy protein—film formed on heated surface	0.12	—
Okamoto (1978)[2]	5% Aqueous solution of 7S soy protein—film formed on heated surface	0.06	—
Brandenburg et al. (1993)[4]	Soy Protein:Glycerin (1.7:1), pH = 8	4.3	78
Hood (1987)[6]	Collagen:Glycerin (3.4:1) Wet Process Extrusion		
	Longitudinal	8.1	25
	Transverse	10.7	43

(continued)

245

TABLE 9.6. (continued).

Reference	Film	Tensile Strength (MPa)	Elongation at Break (%)
Hood (1987)[6]	Collagen:Sorbitol:Glycerin (8.8:0.8:1)		
	Dry Process Extrusion		
	Longitudinal	9.1	38
	Transverse	8.7	49
Briston (1988)	LDPE[7]	8.6–17.3	500
Briston (1988)	HDPE[7]	17.3–34.6	300
Briston (1988)	PVDC[7]	48.4–138	20–40
Briston (1988)	EVA[7]	13.8	650–800
Briston (1988)	PET[7]	175	70–100
Briston (1988)	OPS[7]	62.2–82.7	20

[1]Measured with Scott tester.
[2]Measured with apparatus described by Watanabe and Okamoto (1976).
[3]Measured with Instron Universal testing instrument. Rectangular specimens 2.54 cm × 10 cm. Initial grip separation 5 cm. Cross-head speed 500 mm/min. Conditioning prior to testing at 50% relative humidity and 23°C for 2 days.
[4]Measured with Instron Universal testing instrument. Rectangular specimens 2.54 cm × 10 cm. Initial grip separation 5 cm. Cross-head speed 500 mm/min. Conditioning prior to testing at 50% relative humidity and 25°C for 2 days.
[5]Measured with Instron Universal testing instrument. Rectangular specimens 2.54 cm × 6 cm. Initial grip separation 4 cm. Cross-head speed 30 mm/min. Conditioning prior to testing at 50% relative humidity and 26°C for over 72 hours.
[6]Measured with Instron Universal testing instrument. Rectangular specimens 3 cm × 10 cm. Cross-head speed 100 mm/min.
[7]Abbreviated terms stand for: LDPE, low density polyethylene; HDPE, high density polyethylene; PVDC, polyvinylidene chloride; EVA, copolymer of ethylene and vinylacetate; PET, polyester (polyethyleneterephthalate); OPS, oriented polystyrene.

related reviews (Courts, 1977; Burke, 1980; Hood, 1987; Rust, 1987) and patent literature (Karmas, 1974).

Enzymatically modified collagen has also been extruded into continuous films suitable for packaging of meat bar snack products (Autry, 1972; Miller, 1972). Acid-swollen collagen was treated with fungal proteases from *Aspergillus oryzae* and *Aspergillus flavus-oryzae*, incubated for 12–48 hours at 20–35°C, extruded on a casting belt, and dried. Wet strength of dried films increased by treatment with gaseous ammonia.

Preparation of casings from mixtures of collagen and other proteins have also been reported. Maser (1987) made sausage casings from blends of collagen (100 parts by weight) and wheat gluten (100 parts or less by weight). It was found that swollen tendon collagen, which does not possess the same cohesive strength as hide corium collagen, could be extruded into adequately strong casings with addition of high bloom gelatin in amounts from 5% to 20% by weight of collagen (Lieberman, 1967). Collagen and casein mixtures have also been extruded to casings (Shank, 1972). These mixtures were less viscous than commonly used acid-swollen collagen slurries and, therefore, required lower pumping and extruding pressures.

Films and Coatings Based on Gelatin

Gelatin coatings carrying antioxidants were effective in reducing rancidification when applied to cut-up turkey meat by spraying or dipping (Klose et al., 1952). Smoke-cured chicken meat coated with an ethylene glycol/gelatin (4:1 on a weight basis) film showed increased moisture retention over a noncoated control (Moorjani et al., 1978).

Whitman and Rosenthal (1971) developed an edible protective coating suitable for primal meat cuts, which usually are not individually packaged with polymeric materials, as well as for other foodstuffs. A hot melt was prepared from a mixture of gelatin and a polyhydric alcohol (e.g., propylene glycol, ethylene glycol, glycerin, sorbitol) and applied on food surfaces either directly or after cooling it down.

Slightly heated aqueous solutions of metal gelatinates, as well as acidic aqueous solutions of gelatin and a meta phosphate polymer, were applied on processed meat products and other food items (Keil et al., 1960; Keil, 1961). Upon drying, transparent coatings formed on food surfaces, offering protection against mold growth, lipid oxidation, and handling abuse.

Spraying aqueous gelatin solutions (10–20% w/w) on battered and breaded meats reduced oil absorption upon frying, due to the grease resistance of gelatin films formed around these meat products (Olson and Zoss, 1985).

A dry coating formulation, containing gelatin, dextrin, starch, and a powdered shortening as major ingredients, was provided for coating poultry meat prior to baking (Jarvis et al., 1983). The coating formed a thin crisp film and imparted the texture, appearance, and taste of roasted and basted poultry upon baking.

An edible wrapping in sheet form was developed using gelatin as the structural matrix (Todd, 1982). Main ingredients of the pliable sheets were gelatin (1–20% w/w), fat (5–25% w/w), a farinaceous material (30–55% w/w), and water (15–40%). The sheets were applied on meats, poultry, and fish prior to baking in order to provide a fat-fried taste, texture, and appearance.

Films and Coatings Based on Other Proteins

Use of proteins other than collagen in manufacture of meat casings has been investigated. Turbak, (1972) produced casings by extruding into a coagulating bath, dispersions of alkaline-swollen film-forming vegetable proteins preferably from corn, wheat, soy, and peanuts. Mixtures of gelatin or corn zein with a hydrophilic substance (7–20% of protein by weight), such as starch, pectin, or alginates, were extruded by Julius (1967).

Wheat gluten, as well as feather keratin, dispersed in aqueous alcohol with the aid of reducing agents was extruded into a gaseous medium to form casings (Mullen, 1971; Schilling and Burchill, 1972).

Egg albumen, wheat gluten, and soy protein concentrate have received consideration as potential predust materials and batters for batter-coated meats and other foods. Such protein materials produced well-adhered crusts when used to predust chicken pieces prior to battering, breading, and deep-frying (Baker et al., 1972). Batter formulations of high egg albumen content were applied on chicken drumsticks and performed equally well to all-flour commercial batters, although partial lack of adhesion was a problem (Baker and Scott-Kline, 1988).

Dehydrated meat was stabilized with protein coatings, preferably made from mixtures of soy protein isolate and egg albumen (Coleman and Creswick, 1966). The process involved mechanical reduction of lean meat into fibers and mixing of fibers in aqueous protein solution. Formed slurries were extruded, cut in pieces, and heated to coagulate the protein. A protective protein matrix formed enclosing meat fibers and aiding flavor retention, while the product possessed good texture and rehydration characteristics.

In a different type of protein film application, still relevant to meat processing, meat curing compositions were coated with aqueous ethanol solutions of a prolamin, such as corn zein, wheat gliadin, and barley hordein (Farkas et al., 1966). Such prolamin coatings prevented curing mixtures from decomposition on storage. Upon preparation of meat curing brines, protein coatings easily dissolved in the sodium chloride solution, releasing their contents.

SEAFOOD

During long-term frozen storage, fish often become rancid due to lipid oxidation and moisture loss. Presently, the most effective protection against frozen storage deterioration is ice glazing, a process whereby frozen fish are coated with sacrificial ice glaze during storage. Although numerous polysaccharide films have been explored as to their effectiveness in retarding lipid oxidation and moisture loss in frozen fish, very few studies have investigated the potential of protein-based films for preservation of frozen fish (Ijichi, 1978).

Hirasa (1991) examined the effect of casein-acetylated monoglyceride emulsion edible coatings on moisture loss and lipid oxidation in frozen fish. Emulsion total solids played an important role in the effectiveness of coatings. Emulsion films had little effect on lipid oxidation; however, moisture loss from salmon fish samples coated with high total solids emulsions was reduced. It was speculated that coatings developed a porous structure as water subliminated at frozen temperatures, leading to a less effective barrier. More research is needed in order to fully define the potential of emulsion films for the long-term preservation of frozen fish. Other potential fish-related applications of protein edible films include application of coatings to breaded frozen fish in order to decrease the amount of moisture loss during storage, while decreasing the amount of lipid absorption during frying. Polysaccharide coatings have been industrially employed for such purposes in the past.

NUTS

The effectiveness of corn zein–based edible coatings to extend the shelf-life of shelled nuts by retarding rancidity, staling, and sogginess, was demonstrated more than thirty years ago. This has led to commercialization of zein for such uses.

Liquid coating formulations consisting of zein, acetylated monoglycerides, and antioxidants in an alcohol solution have been applied on shelled pecans, almonds, walnuts, and peanuts (Cosler, 1958a–c). These storage studies showed that shelf-life of coated samples may be prolonged over several weeks at 100°F and over several months at 70°F. In these coatings, zein provided oxygen and oil resistance, reducing the rates of oxidative rancidity and oil leakage. Acetylated monoglycerides plasticized coatings formed after solvent evaporation and retarded moisture absorption into nuts. Antioxidants commonly used in the food industry, such as citric acid and butylated hydroxyanisole (BHA), were also added to prevent rancidity.

Similar coatings were sprayed on oil roasted peanuts under tumbling in revolving pans (Alikonis and Cosler, 1961). Evaporation of the alcohol solvent was accelerated by air movement. Coated peanuts were subsequently salted and stored in cans at 70 and 100°F. Reportedly, shelf-life of coated peanuts was extended by three to five times at both temperatures. Moreover, salt binding on surface of coated peanuts increased.

Presently, edible corn zein–based coatings for nutmeats, as well as for confections and pharmaceutical products, are commercially available (Andres, 1984). These coatings are mixtures of zein, alcohol, vegetable oils (e.g., partially hydrogenated soybean and cottonseed oils), glycerin, and antioxidants (citric acid, BHA, and BHT). The hydrophobic oils increase moisture resistance of formed coatings, while glycerin functions as a plasticizer. Coatings are applied at an amount of about 1–2% by weight of the products.

Increased retention of salt on surface of dry roasted nuts was also achieved with the use of wheat gluten–based coatings (Noznick and Bundus, 1967). Mixtures of wheat gluten, salt, water, acetic acid, dextrin, and mono-sodium glutamate, were sprayed on hot, dry roasted nuts. Upon drying, salt was bound by coatings formed on nuts. Dextrin supplemented adhesiveness of wheat gluten, while mono-sodium glutamate enhanced flavor. In a variation of the method, the salt was added after application of the coating before the latter became dry.

Film-forming abilities of proteins have also been utilized in color coating unshelled nuts with a colored edible film-forming solution (Johnson, 1969). Edibility was necessary because coatings could come in contact with nutmeats through shell cracks. Food grade pigments were suspended in film-forming solutions of a protein (e.g., casein, albumen, gelatin, zein). Nuts were subsequently coated and air dried. Protein films formed around nutshells retained added colorants. At the same time, the coatings sealed shell openings and prevented diffusion of dye into nutmeats.

CONFECTIONERY PRODUCTS

Alcoholic solutions of corn zein, acetylated monoglycerides, and antioxidants have been effectively used to coat confectionery items. Such items as chocolate pieces, buttercreams, sugar-coated burnt peanuts, and hard gums have been coated in a manner similar to shelled nuts (Cosler, 1957, 1959; Alikonis, 1979). The coatings provided a glossy, decorative finish and extended shelf-life by retarding oxidative rancidity, staling, moisture absorption, and oil leaking. Moreover, the coatings prevented individual candy pieces from sticking together at conditions of increased temperature and humidity. Corn zein coatings also enhanced light stability of colors, increasing opportunities for adding various colorants into candies (Signorino, 1969).

Protein-containing protective coatings developed by Durst (1967), reportedly prolonged shelf-life of coated products, such as brownies, fig bars, chocolate cubes, and donuts. Coating mixtures consisted of a protein, an oil (e.g., cottonseed oil, corn oil, soybean oil, linseed oil, safflower oil), and a plasticizer. Suitable proteins for use in such coatings included egg albumen, gelatin, soy protein, and casein in the form of sodium caseinate.

FRUITS AND VEGETABLES

Protective coatings based on waxes and on sucrose esters of fatty acids have been developed and commercialized for fruits and vegetables. Proteins, however, have not met such success in this type of application. Particularly, use of protein-based coatings on fresh produce of high moisture content has been restricted due to the limited water vapor resistance of such films. Development of composite coatings, combining proteins with hydrophobic materials, presents many opportunities for coating fresh fruits and vegetables. A few research reports and patent disclosures have investigated protein-containing coatings for fruits and vegetables and are discussed below.

Casein coatings did not result in any significant reduction of moisture loss from coated raisins in storage (Watters and Brekke, 1961). Another study revealed that egg albumen and soy protein coatings significantly reduced moisture loss from coated raisins stored with bran flakes (Bolin, 1976). Application of the two protein coatings involved dipping raisins into aqueous protein solutions followed by heat denaturation of proteins.

An industrial procedure for coating fortified rice as it revolves in a tumble unit was described by Mickus (1955). The coating, made of corn zein, stearic acid, and abietic acid (wood resin), retained the enriching vitamins, thiamin and niacin.

The effect of corn zein coatings on tomatoes was investigated by Park (1991). Coating solutions were prepared by dissolving zein, glycerin, and citric acid in 95% ethanol. Coated and uncoated (control) tomatoes were stored at 21°C. Zein films of about 25 μm in thickness formed around tomatoes, reportedly delayed color change, firmness loss, and weight loss. Shelf-life was lengthened by six days as evidenced by sensory evaluation. When thicker zein coatings (about 66 μm) were used, internal oxygen concentration of tomatoes was reduced beyond acceptable limits and anaerobic fermentation occurred (Park, 1991).

A coating method was developed to reduce stickiness of fruits such as dates, figs, raisins, dried prunes, and dried apricots (Gunnerson and Bruno, 1990). Coatings consisted of a plant wax (preferably carnauba wax), a vegetable oil, a protein (preferably corn zein), and a wetting agent able to form a surface emulsion between the wax/oil phase and water. Application of coatings proceeded in two steps. A mixture of the wax, the oil, and the wetting agent was applied first, followed by a solution of the protein with a plasticizer and antioxidants.

Laminate coatings with an amylose ester of a fatty acid as one layer and a protein as the second layer, effectively protected dehydrated fruits and vegetables from moisture absorption and oxidative degradation (Cole, 1969). A solution of the amylose ester in a volatile solvent was applied first, followed by heating to remove the solvent. An aqueous dispersion of a film-forming protein (e.g., corn zein, gelatin, soy protein, casein) was subsequently applied on amylose ester–coated foodstuffs forming a protein coating upon drying.

Gelatin films were utilized to separate fruits (e.g., strawberries) from dairy products (e.g., yogurt) in the same container (Shifrin, 1968). The two products would not be mixed in the stored package, for, otherwise, deterioration and spoilage could occur. Prior to use, the gelatin film could readily be ruptured with a spoon to allow for mixing of fruits with the dairy product.

Several studies have been performed to determine the potential of casein-based emulsion edible films for preservation of fruits and vegetables, such as zucchini, apples, and celery. Caseinate/lipid emulsion films offer advantages over commercial wax edible films in that they can be applied to produce at room temperature. Furthermore, the protein matrix imparts increased film strength and integrity, as well as improved adherence to food surfaces.

Effectiveness of casein and acetylated monoglyceride emulsion coatings at controlling moisture loss and oxidative browning in cut, peeled apple pieces was determined by Krochta et al. (1990a). Coatings from aqueous emulsions of casein with acetylated monoglycerides reduced water loss from apple pieces. Treating these films with a buffer solution at the isoelectric point of casein followed by soaking in water further increased resistance to water loss, as well as formation of insoluble films. Buffers made with ascorbic acid provided antioxidant properties and retarded browning of apple pieces.

Avena-Bustillos et al. (1994) found that calcium caseinate and acetylated monoglyceride emulsion films were effective in reducing water loss from zucchini. Coatings did not significantly change internal carbon dioxide or ethylene concentrations. However, they did decrease internal oxygen concentrations during storage. Using response surface methodology, it was concluded that a 0.8% calcium caseinate and 0.7% acetylated monoglyceride film would be most effective at increasing zucchini shelf-life by increasing water vapor resistance up to 90%.

EGGS

Meyer and Spencer (1973) studied the effect of various coatings, including corn zein and milk casein coatings, on hen's eggs. Shell crushing strength of zein-coated and casein-coated eggs increased by approximately 16% and 8%, respectively, while moisture loss was reduced by approximately 42% and 12%, respectively. Neither protein coating adversely affected egg flavor. In another study of various coatings on hen's eggs, corn zein provided an effective coating against penetration of egg shells by *Pseudomonas fluorescens* and *Salmonella typhimurium* (Tryhnew et al., 1973).

SOLUBLE CONTAINERS

Development of self-supporting, pliable sheets made from edible substances and capable of replacing traditional polymeric materials in routine packaging applications had been a long-standing goal for research related to edible films. Commercial success has so far been very limited. The major problem is producing a film that simultaneously possesses satisfactory mechanical properties, water vapor resistance, and gas barrier properties. Usually one of these three characteristics is lacking. Poor water vapor resistance is the main problem for protein-based films. As a result, use of such films in manufacture of water-soluble containers had been suggested. Two attempts in this direction were reported in the patent literature and are described below.

A method to produce water-soluble containers from casein and vegetable proteins (e.g., soy, peanut, and corn protein) was developed by Georgevits (1967). An acid or a base was added to aqueous protein film-forming solutions (10–40% w/w) to fix the pH at 1.5–3 or 7–10, respectively. A mixture of glycerin and sorbitol was incorporated to induce plasticization and promotion of heat sealability. Foodstuffs such as potato flakes, corn flakes and grains, and skim milk powder, were stored in heat-sealed packages made from cast protein films. When placed in boiling or even cold water or in milk containers they dissolved upon stirring and released their contents. Films that readily disintegrated in boiling water were also cast from acid-swollen collagen treated with proteolytic enzymes, such as ficin, bromelain, and papain (Tsuzuki, 1970). Heat-sealed water-soluble pouches could be made from these collagen films.

ENCAPSULATION AND CONTROLLED RELEASE

Microencapsulation

Microencapsulation is a process of entrapping solid particles, liquid droplets, or gases in thin polymeric coatings (Jackson and Lee, 1991). The outer film (wall) protects the encapsulated material (core) and allows for its controlled release. Application of this technique is widespread in several industrial fields, including the manufacture of foods, pharmaceuticals, cosmetics, and pesticides.

A brief history of the development of microencapsulation, which began in the late 1930s, was compiled by Fanger (1974). As noted in this historical outline, the first microcapsules utilized a protein material, gelatin, as a coating of a colorless dye precursor. Several reviews on theory, applications, and technology of microencapsulation and controlled release of food ingredients are available (Balassa and Fanger, 1971; Bakan, 1973; Karel and Langer, 1988; Karel, 1990; Jackson and Lee, 1991). A recent report on the status of microencapsulation in the food industry, along with a list of available commercial materials, is provided by Dziezak (1988).

Common food ingredients subjected to microencapsulation include acidulants, flavors, colors, leavening agents, salts, enzymes, artificial sweeteners, minerals, and vitamins (Karel, 1990). A number of coating materials are employed in production of microcapsules, such as gums, starch and starch derivatives, cellulosic materials, lipids, proteins, and inorganic materials (Jackson and Lee, 1991). Some microencapsulation processes utilizing protein materials as walls (capsules) are discussed in the following sections.

Gelatin

Gelatin, either by itself or in combination with gums (e.g., gum arabic, acacia), was one of the first materials used in microencapsulation. Oil phases were encapsulated in gelatin or in a gelatin/gum system with a process known as coacervation or phase separation (Vandegaer, 1974). Gutcho (1972, 1976, 1979) provided descriptive information on numerous U.S. patents dealing with microencapsulation techniques utilizing gelatin, most of them referring to nonfood applications. A selected number of patents available for use of gelatin in encapsulation of food ingredients are described here.

Balassa (1970) developed a method for encapsulating materials susceptible to autoxidation, such as aromas, flavors, and vitamins. Each of these materials could form an emulsion with an aqueous colloidal dispersion of gelatin or another protein (e.g., soy protein, casein). Then, as each emulsion was slowly added to a spinning liquid, a slurry was prepared. Each slurry was subsequently dehydrated to form air-impervious capsules.

Alliaceous flavorings, such as garlic, onion, leek, and chive, were encapsulated in gelatin/gum arabic coatings for use in soups or dairy products (Brodnitz and Pascale, 1972). Capsules were formed by spray-drying slurries made of flavors and aqueous solutions of gelatin and gum arabic.

In a method developed by Sato and Kurusu (1974), alcoholic beverages (e.g., wine, whiskey) were mixed with gelatin powder and mixtures were subsequently spray-dried to remove water, while alcohol and other flavoring components were retained within a gelatin coating. The alcohol-containing dried solid product was suitable for use in confectionery items and instant drinks.

A method was provided for encapsulation of the artificial sweetener L-aspartyl-L-phenylalanine methyl ester (Cea et al., 1983). Encapsulation protected the sweetener from breakdown at elevated temperatures and from contact with water. The method operated preferably with a fluidized bed coating process. Gelatin and corn zein were among the materials that could be employed as encapsulants. Fat-soluble vitamins, such as vitamin A, were successfully stabilized by encapsulation in gelatin (Koff and Widmer, 1964; Timreck, 1970; de Man et al., 1986).

Ruminant animal feeds were supplemented with encapsulated polyhalogenated hydrocarbons in an effort to suppress methane production within the rumen and increase production of useful fatty acids (Grass et al., 1972). A coacervation technique was applied utilizing gelatin, gum arabic, and polyethylenemaleic anhydride copolymer as capsule polymeric walls. Formed walls were hardened with addition of glutaraldehyde.

Egg Albumen

A process for encapsulating liquid oils in protein walls was described by Soloway (1964). The water-insoluble organic liquids were dispersed in aqueous protein solutions. Application of heat denatured and coagulated albumen allowed formation of protein coatings around oil droplets.

The aforementioned technique was improved by Kosar and Atkins (1968). In their process, an infrared absorber was included in the oil droplets and dispersions were subjected to infrared radiation. As radiation energy converted to heat energy in the absorber, temperature increased to a level suitable for protein coagulation. This modification offered convenience since conventional heating of dispersions was avoided.

Wheat Gluten

Wheat gluten was used for encapsulation and controlled release of fat-soluble edible food substances, such as flavors, vitamins, and coloring agents (Noznick and Tatter, 1967). An alkaline or acidic dispersion of gluten was prepared and mixed with a vegetable oil or an animal fat to form an oil-in-water emulsion. Preferably, an emulsifying agent (e.g., lecithin, glyceryl esters, glycol esters, polyglycerol esters of fatty acids) was added as well. The oil or fat was the carrier of the fat-soluble material to be encapsulated. Formation of microcapsules was achieved by spray-drying of emulsions.

Corn Zein

Preparation of corn zein microcapsules containing water-immiscible core materials, such as animal or vegetable oils, was described by Brynko and Bakan (1963). Core materials were dispersed in alkaline aqueous zein solutions. Lowering the pH of solutions below 5.5 caused phase separation and formation of zein walls around oil particles. Treatment with a protein cross-linking agent, such as glutaraldehyde, provided additional hardening of zein microcapsules. Deamidated zein could also be used in this method.

Casein

Polyunsaturated fats can be encapsulated and stabilized in formaldehyde-treated casein and fed to ruminants. Microcapsules remain stable at the neutral conditions of the rumen, preventing hydrogenation of polyunsaturated fatty acids, thereby increasing polyunsaturated fat content of milk and meat. This process, which is carried out by spray-drying formaldehyde-treated emulsions of sodium caseinate and vegetable oils, originated in Australia (Scott et al., 1970; Faichney et al., 1972; Scott, 1975; Tracey, 1975; Scott and Hills, 1978; Ashes et al., 1984). Similar research has also been conducted by researchers of the USDA (Plowman et al., 1972; Bitman et al., 1973, 1974; Edmondson et al., 1974). Although experimental findings have been encouraging, the process has not met commercial success due to cost considerations and the need for more research.

Whey Protein

The potential of whey proteins as microencapsulation agents is currently explored in the laboratory of Dr. M. Rosenburg at the University of California, Davis. The ability of whey proteins to encapsulate anhydrous milk fat, vegetable oils, and other volatiles using spray-drying is being evaluated. Although no results have been published yet, whey protein microcapsules appear very promising.

Macrocoating Encapsulation

Besides their use in microencapsulation applications, protein coatings, especially those from gelatin and corn zein, are extensively used in the pharmaceutical industry for production of conventional capsules (macrocapsules, tablets). Microcapsules range in diameter size from several tenths of a μm to a few thousands μm, while capsules have diameters greater than 5000 μm (Jackson and Lee, 1991). Protein shells provide protection and controlled release of contained medicaments, while also masking tastes and odors. A large number of U.S. patents related to protein coatings for medicinal tablets can be found in Gutcho (1972, 1976, 1979).

Control of Surface Additive Concentration

Use of protein-based films for controlled release of preservative agents is another potential application. Alcoholic film-forming solutions of corn zein, glycerin, and an acetylated monoglyceride product were applied by spraying onto the surface of an intermediate moisture cheese analog (Torres and Karel, 1985; Torres et al., 1985). A high concentration of sorbic acid was applied to the outer coating surface. The coating reduced the diffusion rate of the preservative into the food interior and improved surface microbial stability against growth of *Staphylococcus aureus*.

In another study, lactic acid–treated casein films and tannic acid–treated gelatin films were tested for control of sorbic acid concentration on the surface of a model food system (saccharose 44%, agar 1.5%, water 54.5%; 0.95 water activity) (Guilbert, 1988). Casein films retained 30% of sorbic acid at the surface after thirty-five days of storage (at 25°C and 95% RH), while poor retention and swelling were observed with gelatin films. Tannic acid–treated gelatin films showed very good retention of an antioxidant (α-tocopherol) on the surface of the same food model. Guilbert (1988) also reported that casein films containing sorbic acid significantly improved microbial stability of intermediate moisture fruits (apricots and papaya) inoculated with *Aspergillus niger* and *Saccharomyces rouxii*.

FUTURE RESEARCH NEEDS ON PROTEIN-BASED EDIBLE FILMS

Development of edible films and coatings from proteins and other biological materials has recently evolved as an important interdisciplinary research field. Interest in the concept is likely to continue to grow in the near future. Several potential innovative applications of edible films in the areas of food packaging, storage, and preservation have been envisioned by food technologists. Consumer demands for foods of increased quality and longer shelf-life have stimulated interest in this area. Furthermore, use of edible films could aid in reducing the amounts of food packaging materials required from nonrenewable sources. Edible film research may also be promoted by the apparent opportunities for further utilization of agricultural commodities and other biological products in the production of such edible films. A number of issues that should be addressed by researchers involved in production of edible films and coatings are pointed out and discussed below.

MULTICOMPONENT EDIBLE FILMS

Advantages and limitations of protein films are currently well defined. In general, protein films are excellent oxygen and carbon dioxide barriers at low relative humidities, whereas their resistance to water vapor transmission is limited. The good gas barrier properties are drastically reduced in high humidity environments because of protein films' susceptibility to moisture absorption and swelling.

Improvement of resistance to water vapor remains one of the main objectives for protein films. Several chemical and physical treatments (i.e., tanning, treatment with aldehydes, treatment at the protein isoelectric point) show some effectiveness in promoting cross-linking, "hardening" the protein film structure, and improving film barrier and mechanical properties.

A more interesting approach appears to be the use of proteins with other film-forming lipid and polysaccharide materials. By utilizing proteins' chemical, physical, and functional properties, increasingly effective edible films can be produced. For example, films cast from protein-lipid emulsions or bilayer protein-lipid films combine lipid hydrophobicity with the improved mechanical properties and excellent oxygen barrier ability of proteins, producing effective edible coatings.

As noted by Krochta (1992), development of multicomponent edible films resembles the production of multilayer synthetic packaging films where moisture-sensitive materials that are very effective oxygen barriers, such as cellophane, polyamides (PA), and ethylene vinyl alcohol (EVOH) are sandwiched between layers of other moisture-resistant polymers.

This approach has been followed by several researchers for the development of bilayer films from cellulosic derivatives and lipids (Kamper and Fennema, 1984a, 1984b; Greener and Fennema, 1989; Kester and Fennema, 1989a, 1989b; Hagenmaier and Shaw, 1990; Rico-Peña and Torres, 1990, 1991; Vojdani and Torres, 1990).

FILM CHARACTERIZATION

As mentioned earlier when discussing properties of protein-based films, a vast amount of information on the subject is included in patent disclosures albeit quantitative data are lacking. A thorough film characterization is necessary for tailoring protein films to specific applications.

Several variables are involved in film production, such as types of ingredients added in film-forming solutions, pH and temperature of solutions, and drying temperatures of cast films. The effect of each individual factor on film properties has to be quantified, as well as the interactions among these factors. This would allow for optimization of film-producing processes.

PROTEIN NUTRITIONAL QUALITY

A common assumption would be that application of protein edible films to food systems results in a nutritional enhancement of the food. This argument is also supported by the better sensory characteristics of protein films when compared to films from lipid materials (Krochta, 1992). Protein nutritional quality largely depends on the pattern and concentration of essential amino acids provided for synthesis of nitrogen-containing compounds in humans (McBean and Speckmann, 1988). However, one must be aware of two important points; film-forming proteins might be of poor dietary value and protein nutritional quality might be adversely affected during film preparation.

Some proteins used on an experimental or commercial basis for film production are deficient in essential amino acids. Corn zein, for instance, is deficient in lysine and tryptophan (Wright, 1987). Wheat gluten also contains low levels of lysine (Lasztity, 1984). Methionine is a limiting essential amino acid in soy protein and other oilseed proteins (Jansen, 1977).

Processing conditions during film production, such as heat treatments, pH changes, and solvent systems, may have a harmful effect on protein nutritive quality. Based on rat assays, Bates and Wu (1975) determined a protein efficiency ratio (PER) of 1.26 for soy protein–lipid films formed on the surface of heated soymilk. This value was lower than usual PER values of most soy-derived textured vegetable proteins that range between 1.8 and 2.6.

Nutritional quality of edible protein films could be improved by fortification with one of more of the limiting amino acids. Addition of DL-methionine raised PER values of soy protein–lipid films (Bates and Wu, 1975; Chuah et al., 1983). Film production by combination of two or more protein sources could also correct amino acid deficiencies in a similar manner to the way that protein supplementation and complementation are utilized in animal feed preparation.

Protein in cow's milk is both present in significant quantities and of exceptionally high quality. The presence, in varying quantities, of all the amino acids needed by humans results in the high quality of milk proteins. Furthermore, the distribution pattern of amino acids in milk proteins resembles that required by humans (McBean and Speckmann, 1988). These nutritional properties of milk proteins make milk protein edible films potential food nutrient enhancers.

Whey proteins are slightly superior to casein due to the limiting quantities of sulfur-containing amino acids in casein. Both caseins and whey proteins have a surplus of the essential amino acids lysine, threonine, methionine, and isoleucine. Therefore, supplementing vegetable proteins, particularly cereal proteins, which are deficient in these amino acids, with milk proteins is very attractive. Tryptophan is another essential amino acid which is abundant in milk (McBean and Speckmann, 1988).

As mentioned earlier, it is important to be aware of processing conditions that may alter the high nutritional quality of milk proteins. Heat-induced denaturation may diminish or enhance nutritional quality in proteins, depending on the extent of denaturation. For example, studies suggest that ultra-high-temperature treated milk may be more digestible than raw or conventionally pasteurized milk due to increased protein denaturation (McBean and Speckmann, 1988). The biological value and net utilization of milk protein–based edible films should be tested prior to making any nutritional claims concerning such films.

Casein films can be formed from aqueous solutions without application of heat, resulting in optimum nutritional retention. Whey protein films, however, are often formed by heat denaturation. Milk protein emulsion films also often undergo heat treatments. Therefore, tests must be performed before proposing health claims.

ALLERGIES AND INTOLERANCE TO PROTEINS

Researchers involved in developing and applying edible protein-based films and coatings need to consider potential allergenic responses to specific protein sources by some individuals. Allergy to food proteins is linked to several diseases and possibly exacerbates many other clinical conditions (Levinsky, 1982).

The relationship of wheat and other cereal proteins to celiac disease should be particularly mentioned. Celiac disease, also known as non-tropical sprue or gluten-sensitive enteropathy, is an intolerance rather than an allergy. It is characterized by nutrient malabsorption as a consequence of gluten-dependent damage to the mucosa of the small intestine (Skeritt, 1988). The gliadin fraction of gluten has been pointed out as the most toxic fraction (Kasarda, 1978).

Incidence of gluten intolerance is estimated to be 1 in 200–500 individuals in parts of Ireland, Austria, and Scandinavia and 1 in 1000–5000 for the rest of Europe and North America (Skeritt et al., 1990). Individuals diagnosed with celiac disease are required to completely remove wheat, rye, barley, triticale, and possibly oats from their diets. Any processed food product they consume has to be free of the above cereals. Two other cereals, corn and rice, are considered safe (Kasarda, 1978).

Therefore, application of any commercially developed edible films and coatings based on cereal proteins should be accompanied by proper labeling; especially when such films and coatings are applied on fresh fruits and vegetables. This type of use demands great thought, since it could deprive celiac patients of their dietary needs.

Application of milk protein–based edible films presents further challenges to food processors in terms of effectively labeling in order to prevent allergic reactions associated with milk proteins. Cow's milk is the most common allergenic food among young infants (Taylor, 1992), resulting in particularly sensitive issues concerning proper labeling of food films. Both casein and whey fractions contain major allergens in their proteins. The most common allergens are β-lactoglobulin, casein, and α-lactalbumin. However, other cow's milk proteins play important roles in some cases. Many different symptoms are associated with milk protein allergies, including gastrointestinal, respiratory, dermatological, and neurological disorders (Taylor, 1986).

Various process-related factors affect human immune responses to cow's milk proteins. Most importantly to the application of milk protein edible films is the fact that heat treatment of milk proteins diminishes their antigenicity. However, high levels of heat treatment are necessary to obtain such effects, resulting in significant loss of nutritional quality (Taylor, 1986). Reduction of antigenicity can also be achieved through enzymatic hydrolysis reactions (Jost et al., 1987), a process that may one day be included in production of milk protein edible films. All these effects must be considered when applying milk protein edible films to food systems.

Another fairly common adverse reaction is associated with lactose, a carbohydrate contained in considerable amounts in cow's milk. Lactose is broken down to usable monosaccharides in the small intestine by the enzyme lactase (β-galactosidase). Certain individuals are characterized by lactase deficiency, a condition of low intestinal lactase activity (Savaiano and Levitt, 1987). The most common type of lactase deficiency is termed primary adult lactase deficiency and results from a genetically programmed decrease of lactase production following weaning (Friedl, 1981; Lisker, 1984). Primary adult lactase deficiency is widespread affecting approximately 70% of the world's population (Savaiano and Levitt, 1987). Lactase deficiency leads to lactose malabsorption, which is the body's inability to metabolize some or all of the lactose, and may or may not be accompanied by symptoms (Friedl, 1981; Savaiano and Levitt, 1987; Houts, 1988). Symptomatic lactose malabsorption is referred to as lactose intolerance (Friedl, 1981; Savaiano and Levitt, 1987; Houts, 1988). Whey protein isolate preparations contain very low levels of lactose (less than 1%) and, therefore, whey protein edible films are not expected to pose any problems related to lactose intolerance. However, milk proteins, whey protein concentrates, and caseinates all possess elevated levels of lactose. Consequently, use of these materials would necessitate proper labeling of film ingredients on food products in order to prevent potential consumer lactose intolerance related problems.

In conclusion, the potential of protein edible films for regulating mass transfer in food systems offers numerous benefits to both food processors and consumers. Protein edible films have not been extensively used in the food industry in the past. However, potential functional and nutritional properties of such films warrant increased consideration. In the future, a data base with results from many different proteins will likely be developed, correlating protein chemical and physical properties with mass transfer and other edible film characteristics. Such data will allow for the development of predictive models for performance of protein films. This chapter represents the beginning of such a compilation. Furthermore, it is hoped that one day these edible film characteristics will provide a new category for defining protein functional properties. This is a very promising area of scientific research, one that should be pursued in the upcoming years.

REFERENCES

Aboagye, Y. and D. W. Stanley. 1985. "Texturization of Peanut Proteins by Surface Film Formation. 1. Influence of Process Parameters on Film Forming Properties," *Can. Inst. Food Sci. Technol. J.*, 18:12–20.

Ahmed, E. M. and C. T. Young. 1982. "Composition, Quality, and Flavor of Peanuts," in *Peanut Science and Technology*, H. E. Pattee and C. T. Young, eds., Yoakum, TX: American Peanut Research and Education Society, Inc., pp. 655–688.

Alikonis, J. J. 1979. *Candy Technology.* Westport, CT: AVI Publishing Company, Inc.

Alikonis, J. J. and H. B. Cosler. 1961. "Extension of Shelf Life of Raosted and Salted Nuts and Peanuts," *Peanut J. Nut World*, 40(5):16–17.

Alleavitch, J., W. A. Turner and C. A. Finch. 1989. "Gelatin," in *Ullmann's Encyclopedia of Industrial Chemistry, Vol. A12, Fifth Edition*, B. Elvers, S. Hawkins, M. Ravenscroft, J. F. Rounsaville and G. Schultz, eds., New York, NY: VCH Publishers, pp. 307–317.

Anderson, S. O. 1966. "Covalent Crosslinks in a Structural Protein, Resilin," *Acta Physiol. Scand.*, 66(263):1–81.

Andres, C. 1984. "Natural Edible Coating Has Excellent Moisture and Grease Barrier Properties," *Food Process.*, 45(13):48–49.

Anker, C. A. February 15, 1972. U.S. patent 3,642,498.

Anker, C. A., G. A. Foster and M. A. Loader. April 4, 1972. U.S. patent 3,653,925.

Aschaffenburg, R. and J. Drewry. 1957. "Improved Method for the Preparation of Crystalline β-Lactoglobulin and α-Lactalbumin from Cow's Milk," *Biochem. J.*, 65:273–277.

Ashes, J. R., L. J. Cook and G. S. Sidhu. 1984. "Effect of Dietary Cholesterol Protected against Ruminal Hydrogenation on the Plasma Cholesterol and Liver of Sheep," *Lipids*, 19:159–163.

Autry, R. F. May 23, 1972. U.S. patent 3,664,849.

Avena-Bustillos, R. J., J. M. Krochta, M. E. Saltveit, R. J. Rojas-Villegas and J. A. Sauceda-Perez. 1994. "Optimization of Edible Coating Formulations on Zucchini to Reduce Water Loss," *J. Food Eng.*, 21:197–214.

Aydt, T. P., C. L. Weller and R. F. Testin. 1991. "Mechanical and Barrier Properties of Edible Corn and Wheat Protein Films," *Trans. ASAE*, 34:207–211.

Bakan, J. A. 1973. "Microencapsulation of Foods and Related Products," *Food Technol.*, 27(11):34–35, 38–40, 42, 44.

Baker, R. C. and D. Scott-Kline. 1988. "Development of a High Protein Coating Batter Using Egg Albumen," *Poult. Sci.*, 67:557–564.

Baker, R. C., J. M. Darfler and D. V. Vadehra. 1972. "Prebrowned Fried Chicken. 2. Evaluation of Predust Materials," *Poult. Sci.*, 51:1220–1222.

Balassa, L. L. February 17, 1970. U.S. patent 3,495,988.

Balassa, L. L. and G. O. Fanger. 1971. "Microencapsulation in the Food Industry," *CRC Crit. Rev. Food Technol.*, 2:245–265.

Basha, S. M. M. and J. P. Cherry. 1976. "Composition, Solubility, and Gel Electrophoretic Properties of Proteins Isolated from Florunner (*Arachis hypogaea* L.) Peanut Seeds," *J. Agric. Food Chem.*, 24:359–365.

Bates, R. P. and L. C. Wu. 1975. "Protein Quality of Soy Protein-Lipid Films (Yuba) and Derived Fractions," *J. Food Sci.*, 40:425–426.

Becker, O. W. July 7, 1936a. U.S. patent 2,046,541.

Becker, O. W. October 6, 1936b. U.S. patent 2,056,595.

Becker, O. W. October 6, 1936c. U.S. patent 2,056,596.

Becker, O. W. April 26, 1938. U.S. patent 2,115,607.

Becker, O. W. June 13, 1939. U.S. patent 2,161,908.

Becker, O. W. December 23, 1941. U.S. patent 2,267,488.

Becker, O. W. and E. Weiss. December 21, 1937. U.S. patent 2,103,138.

Becker, O. W. and E. Weiss. September 16, 1941. U.S. patent 2,256,040.

Beckwith, A. C. and J. S. Wall. 1966. "Reduction and Reoxidation of Wheat Glutenin," *Biochim. Biophys. Acta,* 130:155–162.

Beckwith, A. C., J. S. Wall and R. W. Jordan. 1965. "Reversible Reduction and Reoxidation of the Disulfide Bonds in Wheat Gliadin," *Arch. Biochem. Biophys.,* 112:16–24.

Bietz, J. A. and J. S. Wall. 1980. "Identity of High Molecular Weight Gliadin and Ethanol-Soluble Glutenin Subunits of Wheat: Relation to Gluten Structure," *Cereal Chem.,* 57:415–421.

Bio-Isolates. 1992. LeSueur Isolates, 719 North Main, Le Sueur, MN.

Bitman, J., T. R. Wrenn, L. P. Dryden, L. F. Edmondson and R. A. Yoncoskie. 1974. "Encapsulated Vegetable Fats in Cattle Feeds," in *Microencapsulation: Processes and Applications,* J. E. Vandegaer, ed., New York, NY: Plenum Press, pp. 57–69.

Bitman, J., L. P. Dryden, H. K. Goering, T. R. Wrenn, R. A. Yoncoskie and L. F. Edmondson. 1973. "Efficiency of Transfer of Polyunsaturated Fats into Milk," *JAOCS,* 50:93–98.

Boldwell, C. E. and D. T. Hopkins. 1985. "Nutritional Characteristics of Oilseed Proteins," in *New Protein Foods, Vol. 2.,* A. M. Altschul and H. L. Wilcke, eds., Orlando, FL: Academic Press, Inc.

Bolin, H. R. 1976. "Texture and Crystallization Control in Raisins," *J. Food Sci.,* 41:1316–1319.

Boni, K. A. December 27, 1988. U.S. patent 4,794,006.

Borgstrom, G. 1968. *Principles of Food Science.* New York, NY: Macmillan Company.

Bradbury, E. and C. Martin. 1952. "The Effect of the Temperature of Preparation on the Mechanical Properties and Structure of Gelatin Films," *Proc. Roy. Soc. A,* 214:183–192.

Bradshaw, N. J., C. J. Schram and A. Courts. September 1, 1966. British patent 1,040,770.

Brandenburg, A. H., C. L. Weller and R. F. Testin. 1993. "Edible Films and Coatings from Soy Protein," *J. Food Sci.,* 58:1086–1089.

Briggs, D. R. and W. J. Wolf. 1957. "Studies on the Cold-Insoluble Fraction of the Water Extractable Soybean Proteins. I. Polymerization of the 11S Component through Reactions of Sulfhydryl Groups to Form Disulfide Bonds," *Arch. Biochem. Biophys.,* 72:127–144.

Briston, J. H. 1988. *Plastic Films, Third Edition.* New York, NY: John Wiley & Sons, Inc.

Brodnitz, M. H. and J. V. Pascale. March 7, 1972. U.S. patent 3,647,481.

Brunner, J. R. 1977. "Milk Proteins," in *Food Proteins,* J. R. Whitaker and S. R. Tannenbaum, eds., Westport, CT: AVI Publishers, Inc., pp. 175–208.

Brynko, C. and J. A. Bakan. December 31, 1963. U.S. patent 3,116,206.

Burke, N. I. November 4, 1975. U.S. patent 3,917,855.

Burke, N. I. 1980. "Use of Corium Layer in Edible Casings," *J. Am. Leather Chem. Assoc.,* 75:459–463.

Burnett, R. S. 1951. "Soybean Protein Food Products," in *Soybeans and Soybean Products,* K. S. Markley, ed., New York, NY: Interscience Publishers, Inc., pp. 949–1002.

Bushuk, W., K. Khan and G. McMaster. 1980. "Functional Glutenin: A Complex of Covalently and Non-Covalently Linked Components," *Ann. Technol. Agric.,* 29:279–294.

Cea, T., J. D. Posta and M. Glass. May 17, 1983. U.S. patent 4,384,004.

Chaumette, G. 1991. "Peroxidase-Catalyzed Crosslinking of a Sodium Caseinate-Myvacet Film and Effect on Water Vapor Transmission," United States Department of Agriculture, Western Regional Research Center, Albany, CA.

Cheesman, D. F. and J. T. Davies. 1954. "Physicochemical and Biological Aspects of Proteins at Interfaces," *Adv. Protein Chem.*, 9:439–501.

Cherry, J. P. 1983. "Peanut Protein Properties, Processes, and Products," in *Peanuts: Production, Processing, Products, Third Edition,* J. G. Woodroff, ed., Westport, CT: The AVI Publishing Company, Inc., pp. 337–359.

Chuah, E. C., A. Z. Idrus, C. L. Lim and C. C. Seow. 1983. "Development of an Improved Soya Protein-Lipid Film," *J. Food Technol.*, 18:619–627.

Circle, S. J., E. W. Meyer and R. W. Whitney. 1964. "Rheology of Soy Protein Dispersions. Effect of Heat and Other Factors on Gelation," *Cereal Chem.*, 41:157–172.

Cole, M. S. November 18, 1969. U.S. patent 3,479,191.

Coleman, R. J. and N. S, Creswick. May 31, 1966. U.S. patent 3,253,931.

Cosler, H. B. May 7, 1957. U.S. patent 2,791,509.

Cosler, H. B. 1958a. "Prevention of Staleness and Rancidity in Nut Meats and Peanuts," *Manuf. Confect.*, 37(8):15–18, 39–40.

Cosler, H. B. 1958b. "Prevention of Staleness and Rancidity in Nut Meats and Peanuts," *Candy Ind. Confect. J.*, 3:17–18, 22, 35.

Cosler, H. B. 1958c. "Prevention of Staleness, Rancidity in Nut Meats and Peanuts," *Peanut J. Nut World*, 37(11):10–11, 15.

Cosler, H. B. 1959. "A New Edible, Nutritive, Protective Glaze for Confections and Nuts," *Manuf. Confect.*, 39(5):21–23.

Courts, A. 1977. "Uses of Collagen in Edible Products," in *The Science and Technology of Gelatin,* A. G. Ward and A. Courts, eds., New York, NY: Academic Press, Inc., pp. 395–412.

Crewther, W. G., R. D. B. Fraser, F. G. Lennox and H. Lindley. 1965. "The Chemistry of Keratins," *Adv. Protein Chem.*, 20:191–346.

Dahle, L. K. 1971. "Wheat Protein-Starch Interaction. I. Some Starch-Binding Effects of Wheat-Flour Proteins," *Cereal Chem.*, 48:706–714.

Dalgleish, D. G. 1989. "Milk Proteins—Chemistry and Physics," in *Food Proteins,* J. E. Kinsella and W. G. Soucie, eds., Champagne, IL: American Oil Chemists Society, pp. 155–178.

de Man, J. M., L. de Man and T. Wygerde. 1986. "Stability of Vitamin A Beadlets in Nonfat Dry Milk," *Milchwissenschaft,* 41:468–469.

Doguchi, M. and I. Hlynka. 1967. "Some Rheological Properties of Crude Gluten Mixed in the Farinograph," *Cereal Chem.*, 44:561–575.

Durst, J. R. June 6, 1967. U.S. patent 3,323,922.

Dziezak, J. D. 1988. "Microencapsulation and Encapsulated Ingredients," *Food Technol.*, 42(4):136–140, 142–143, 146–148, 151–153.

Eastoe, J. E. and A. A. Leach. 1977. "Chemical Constitution of Gelatin," in *The Science and Technology of Gelatin,* A. G. Ward and A. Courts, eds., New York, NY: Academic Press, Inc., pp. 73–107.

Ebler, S., L. G. Phillips and J. E. Kinsella. 1990. "Purification of β-Lactoglobulin: Isolation of Genetic Variants and Influence of Purification Method on Secondary Structure," *Milchwissenschaft,* 45(11):694–698.

Edmondson, L. F., R. A. Yoncoskie, N. H. Rainey and F. W. Douglas. 1974. "Feeding

Encapsulated Oils to Increase the Polyunsaturation in Milk and Meat Fat," *JAOCS,* 51:72–76.

Esen, A. 1987. "A Proposed Nomenclature for the Alcohol-Soluble Proteins (Zeins) of Maize (*Zea mays* L.)," *J. Cereal Sci.,* 5:117–128.

Evans, C. D. June 18, 1946. U.S. patent 2,402,128.

Ewart, J. A. D. 1979. "Glutenin Structure," *J. Sci. Food Agric.,* 30:482–492.

Fagan, P. V. October, 1970. U.S. patent 3,535,125.

Faichney, G. J., H. L. Davies, T. W. Scott and L. J. Cook. 1972. "The Incorporation of Linoleic Acid into the Tissues of Growing Steers Offered a Dietary Supplement of Formaldehyde-Treated Casein-Safflower Oil," *Austr. J. Biol. Sci.,* 25:205–212.

Fanger, G. O. 1974. "Microencapsulation: A Brief History and Introduction," in *Microencapsulation: Processes and Applications,* J. E. Vandegaer, ed., New York, NY: Plenum Press, pp. 1–20.

Farkas, C. R., M. Vogel and I. F. Levy. April 12, 1966. U.S. patent 3,245,808.

Farnum, C., D. W. Stanley and J. I. Gray. 1976. "Protein-Lipid Interactions in Soy Films," *Can. Inst. Food Sci. Technol. J.,* 9:201–206.

Fellers, D. A. 1973. "Fractionation of Wheat into Major Components," in *Industrial Uses of Cereals,* Y. Pomeranz, ed., St. Paul, MN: American Association of Cereal Chemists, Inc., pp. 207–228.

Flint, F. O. and R. F. P. Johnson. 1981. "A Study of Film Formation by Soy Protein Isolate," *J. Food Sci.,* 46:1351–1353.

Fraser, R. D. B., T. P. MacRae and G. E. Rogers. 1972. *Keratins. Their Composition, Structure and Biosynthesis.* Springfield, IL: Charles C. Thomas, Publisher.

Freudenberg, H. and O. W. Becker. April 12, 1938. U.S. patent 2,114,220.

Friedl, J. 1981. Lactase Deficiency: Distribution, Associated Problems, and Implications for Nutritional Policy," *Ecol. Food Nutr.,* 11:37–48.

Fujii, T. April 18, 1967. U.S. patent 3,314,861.

Fukushima, D. and H. Hashimoto. 1980. "Oriental Soybean Foods," in *World Soybean Research Conference II: Proceedings,* March 26.29, 1979, F. T. Gorbin, ed., Boulder, CO: Westview Press, Inc., pp. 729–743.

Fukushima, D. and J. Van Buren. 1970. "Mechanisms of Protein Insolubilization during the Drying of Soy Milk. Role of Disulfide and Hydrophobic Bonds," *Cereal Chem.,* 47:687–696.

Gennadios, A. and C. L. Weller. 1990. "Edible Films and Coatings from Wheat and Corn Proteins," *Food Technol.,* 44(10):63–69.

Gennadios, A. and C. L. Weller. 1991. "Edible Films and Coatings from Soymilk and Soy Protein," *Cereal Foods World,* 36:1004–1009.

Gennadios, A. and C. L. Weller. 1992. "Tensile Strength Increase of Wheat Gluten Films," *ASAE Paper No. 92-6517,* presented at the *1992 International Winter Meeting,* American Society of Agricultural Engineers, December 15–18, Nashville, TN.

Gennadios, A., A. H. Brandenburg, C. L. Weller and R. F. Testin. 1993a. "Effect of pH on Properties of Wheat Gluten and Soy Protein Isolate Films," *J. Agric. Food Chem.,* 41:1835–1839.

Gennadios, A., H. J. Park and C. L. Weller. 1993b. "Relative Humidity and Temperature Effects on Tensile Strength of Edible Proteins and Cellulose Ether Films," *Trans. ASAE,* 36:1867–1872.

Gennadios, A., C. L. Weller and R. F. Testin. 1993c. "Property Modification of Edible Wheat Gluten Films," *Trans. ASAE,* 36:465–470.

Gennadios, A., C. L. Weller and R. F. Testin. 1993d. "Modification of Physical and Barrier Properties of Edible Wheat Gluten-Based Films," *Cereal Chem.,* 70:426–429.

Gennadios, A., C. L. Weller and R. F. Testin. 1993e. "Temperature Effect on Oxygen Permeability of Edible Protein-Based Films," *J. Food Sci.,* 58:212–214, 219.

Georgevits, L. E. March 21, 1967. U.S. patent 3,310,446.

Gontard, N., S. Guilbert and J. L. Cuq. 1992. "Edible Wheat Gluten Films: Influence of the Main Process Variables on Film Properties Using Response Surface Methodology," *J. Food Sci.,* 57:190–195, 199.

Gordon, W. G. 1971. "α-Lactalbumin," in *Milk Proteins, Vol. II,* H. A. McKenzie, ed., New York, NY: Academic Press, Inc., pp. 331–365.

Grass, G. M., Jr., R. C. Parish and J. E. Trei. May 2, 1972. U.S. patent 3,660,562.

Graveland, A., P. Bosvelt, W. J. Lichtendonk, H. H. E. Moonen and A. Scheepstra. 1982. "Extraction and Fractionation of Wheat Flour Proteins," *J. Sci. Food Agric.,* 33:1117–1128.

Greener, I. K. and O. Fennema. 1989. "Barrier Properties and Surface Characteristics of Edible, Bilayer Films," *J. Food Sci.,* 54:1393–1399.

Guilbert, S. 1986. "Technology and Application of Edible Protective Films," in *Food Packaging and Preservation. Theory and Practice,* M. Mathlouthi, ed., Essex, England: Elsevier Applied Science Publishers Ltd., pp. 371–394.

Guilbert, S. 1988. "Use of Superficial Edible Layer to Protect Intermediate Moisture Foods: Application to the Protection of Tropical Fruits Dehydrated by Osmosis," in *Food Preservation by Moisture Control,* C. C. Seow, ed., Essex, England: Elsevier Applied Science Publishers Ltd., pp. 199–219.

Gunnerson, R. E. and R. C. Bruno. August 7, 1990. U.S. patent 4,946,694.

Guo, X. A. 1983. "Research on Heat Denaturation of Soy Protein after Solvent Extraction, and Traditional Chinese Soy Foods," in *Soybean Research in China and the United States. Proceedings of the First China/USA Soybean Symposium and Working Group Meeting,* July 26–30, 1982, B. J. Irwin, J. B. Sinclair and J. L. Wang, eds., Urbana-Champaigne, IL: Board of Trustees of the University of Illinois, pp. 64–66.

Gustavson, K. H. 1956. *The Chemistry and Reactivity of Collagen.* New York, NY: Academic Press, Inc.

Gutcho, M. H. 1972. *Capsule Technology and Microencapsulation.* Park Ridge, NJ: Noyes Data Corporation.

Gutcho, M. H. 1976. *Microcapsules and Microencapsulation Techniques.* Park Ridge, NJ: Noyes Data Corporation.

Gutcho, M. H. 1979. *Microcapsules and Other Capsules.* Park Ridge, NJ: Noyes Data Corporation.

Gutfreund, K. and T. Yamauchi. 1974. "Potentially Useful Products Derived from Gluten Hydrolyzate," in *Proceedings of the 8th National Conference on Wheat Utilization Research,* October 10–12, 1973, Denver, CO, D. A. Fellers, ed., Albany, CA: U.S. Dept of Agriculture, pp. 32–38.

Hagenmaier, R. D. and P. E. Shaw. 1990. "Moisture Permeability of Edible Films Made with Fatty Acid and (Hydroxypropyl)methylcellulose,"*J. Agric. Food Chem.,* 38:1799–1803.

Hansen, D. W. December 21, 1937. U.S. patent 2,102,623.

Hansen, P. M. T., J. Hidalgo and I. Gould. 1971. "Reclamation of Whey Protein with Carboxymethylcellulose," *J. Dairy Sci.*, 54:830–834.

Harrap, B. S. and E. F. Woods. 1964a. "Soluble Derivatives of Feather Keratin I. Isolation, Fractionation and Amino Composition," *Biochem. J.*, 99:8–18.

Harrap, B. S. and E. F. Woods. 1964b. "Soluble Derivatives of Feather Keratin II. Molecular Weight and Conformation," *Biochem. J.*, 92:19–26.

Harrington, W. F. 1966. "Collagen," in *Encyclopedia of Polymer Science and Technology: Plastics, Resins, Rubbers, Fibers, Vol. 4*, H. F. Mark, N. G. Gaylord and N. M. Bikales, eds., New York, NY: Interscience Publishers, pp. 1–16.

Harrington, W. F. and P. H. Von Hippel. 1961. "The Structure of Collagen and Gelatin," *Adv. Protein Chem.*, 16:1–138.

Hinterwaldner, R. 1977a. "Raw Materials," in *The Science and Technology of Gelatin*, A. G. Ward and A. Courts, eds., New York, NY: Academic Press, Inc., pp. 295–314.

Hinterwaldner, R. 1977b. "Technology of Gelatin Manufacture," in *The Science and Technology of Gelatin*, A. G. Ward and A. Courts, eds., New York, NY: Academic Press, Inc., pp. 315–364.

Hipp, N. J., M. L. Groves, J. H. Custer and T. L. McMeekin. 1952. "Separation of α-, β-, and γ-Casein," *J. Dairy Sci.*, 35:272–281.

Hirasa, K. 1991. "Moisture Loss and Lipid Oxidation in Frozen Fish—Effect of a Casein-Acetylated Monoglyceride Edible Coating," M. S. thesis, University of California, Davis.

Hlynka, I. 1949. "Effect of Bisulfite, Acetaldehyde, and Similar Reagents on the Physical Properties of Dough and Gluten," *Cereal Chem.*, 26:307–316.

Ho, B. 1992. "Water Vapor Permeabilities and Structural Characteristics of Casein Films and Casein-Lipid Emulsion Films," M.S. thesis, University of California, Davis.

Hochstadt, H. R. and E. R. Lieberman. January 5, 1960. U.S. patent 2,920,000.

Holme, J. 1966. "A Review of Wheat Flour Proteins and Their Functional Properties," *Baker's Dig.*, 40(6):38–42, 78.

Hood, L. L. 1987. "Collagen in Sausage Casings," *Adv. Meat Res.*, 4:109–129.

Hoseney, R. C. 1986. *Principles of Cereal Science and Technology.* St. Paul, MN: American Association of Cereal Chemists, Inc.

Hoseney, R. C., K. F. Fenney and Y. Pomeranz. 1970. "Functional (Breadmaking) and Biochemical Properties of Wheat Flour Components. VI. Gliadin-Lipid-Glutenin Interaction in Wheat Gluten," *Cereal Chem.*, 47:135–140.

Hoseney, R. C., K. F. Finney, Y. Pomeranz and M. D. Shogren. 1971. "Functional (Breadmaking) and Biochemical Properties of Wheat Flour Components. VIII. Starch," *Cereal Chem.*, 48:191–201.

Hoshi, Y., F. Yamauchi and K. Shibasaki. 1982. "On the Role of Disulfide Bonds in Polymerization of Soybean 7S Globulin during Storage," *Agric. Biol. Chem.*, 46:2803–2807.

Houts, S. S. 1988. "Lactose Intolerance," *Food Technol.*, 42(3):110–113.

Howland, D. W. November 28, 1961. U.S. patent 3,010,917.

Howland, D. W. and R. A, Reiners. 1962. "Preparation and Properties of Epoxy-Cured Zein Coatings," *Paint Varnish Prod.*, 52(2):31–34, 38.

Huebner, F. R. and J. S. Wall. 1979. "Polysaccharide Interactions with Wheat Proteins and Flour Doughs," *Cereal Chem.*, 56:68–73.

Ijichi, K. E. 1978. "Evaluation of an Alginate Coating during Frozen Storage of Red Snapper and Silver Salmon," M.S. thesis, University of California, Davis.

Iwane, A., T. Yasui and C. Tsutsumi. 1986. "Changes in Low Molecular Weight Carbohydrates in Yuba (Soymilk Skin) during Yuba-Film Formation," *Nippon Shokuhin Kogyo Gakkaishi*, 33:783–785.

Jackson, L. S. and K. Lee. 1991. "Microencapsulation and the Food Industry," *Lebensm. Wiss. Technol.*, 24:289–297.

Jansen, G. R. 1977. "Amino Acid Fortification," in *Evaluation of Proteins for Humans*, C. E. Bodwell, ed., Westport, CT: The AVI Publishing Company, Inc.

Jarvis, G. W., S. A. Angalet and W. J. Horan. January 4, 1983. U.S. patent 4,367,242.

Jaynes, H. O. and W. N. Chou. 1975. "New Method to Produce Soy Protein-Lipid Films," *Food Prod. Dev.*, 9(4):86, 90.

Jenness, R. 1988. "Composition of Milk," in *Fundamentals of Dairy Chemistry*, N. P. Wong, R. Jenness, M. Keeney and E. H. Marth, eds., New York, NY: Van Nostrand Reinhold, pp. 1–38.

Jensen, E. V. 1959. "Sulfhydryl-Disulfide Interchange," *Science*, 130:1319–1323.

Johnson, R. A. December 23, 1969. U.S. patent 3,485,640.

Jones, H. W. and R. A. Whitmore. September 26, 1972. U.S. patent 3,694,234.

Jones, R. W., N. W. Taylor and F. R. Senti. 1959. "Electrophoresis and Fractionation of Wheat Gluten," *Arch. Biochem. Biophys.*, 84:363–376.

Jones, R. W., G. E. Babcock, N. W. Taylor and F. R. Senti. 1961. "Molecular Weights of Wheat Gluten Fractions," *Arch. Biochem. Biophys.*, 94:484–488.

Jost, R., J. C. Monti and J. J. Pahud. 1987. "Whey Protein Allergenicity and Its Reduction by Technological Means," *Food Technol.*, 41(10):118–121.

Julius, A. July 4, 1967. U.S. patent 3,329,509.

Kamper, S. L. and O. Fennema. 1984a. "Water Vapor Permeability of Edible Bilayer Films," *J. Food Sci.*, 49:1478–1481, 1485.

Kamper, S. L. and O. Fennema. 1984b. "Water Vapor Permeability of an Edible, Fatty Acid, Bilayer Film," *J. Food Sci.*, 49:1482–1485.

Kanig, J. L. and H. Goodman. 1962. "Evaluative Procedures for Film-Forming Materials Used in Pharmaceutical Applications," *J. Pharm. Sci.*, 51(1):77–83.

Karel, M. 1973. "Protein-Lipid Interactions," *J. Food Sci.*, 38:756–763.

Karel, M. 1977. "Interaction of Food Proteins with Water and with Lipids, and Some Effects of These Interactions on Functional Properties," in *Biotechnological Applications of Proteins and Enzymes*, Z. Bohak and N. Sharon, eds., New York, NY: Academic Press, Inc., pp. 317–338.

Karel, M. 1990. "Encapsulation and Controlled Release of Food Components," in *Biotechnology and Food Process Engineering*, H. G. Schwartzberg and M. A. Rao, eds., New York, NY: Marcel Dekker, pp. 277–294.

Karel, M. and R. Langer. 1988. "Controlled Release of Food Additives," in *Flavor Encapsulation*, S. J. Risch and G. A. Reineccius, eds., Washington, DC: American Chemical Society, pp. 177–191.

Karmas, E. 1974. *Sausage Casing Technology*. Park Ridge, NJ: Noyes Data Corporation.

Kasarda, D. D. 1978. "The Relationship of Wheat Proteins to Celiac Disease," *Cereal Foods World,* 23:240–244, 262.

Kasarda, D. D., J. E. Bernardin and C. C. Nimmo. 1976. "Wheat Proteins," in *Advances in Cereal Science and Technology, Vol. 1,* Y. Pomeranz, ed., St. Paul, MN: American Association of Cereal Chemists, Inc., pp. 158–236.

Kasarda, D. D., C. C. Nimmo and G. O. Kohler. 1971. "Proteins and the Amino Acid Composition of Wheat Fractions," in *Wheat Chemistry and Technology, Second Edition,* Y. Pomeranz, ed., St. Paul, MN: American Association of Cereal Chemists, Inc., pp. 227–299.

Keil, H. L. February 14, 1961. U.S. patent 2,971,849.

Keil, H. L., R. F. Hagen and R. W. Flaws. September 20, 1960. U.S. patent, 2,953,462.

Kelley, J. J. and R. Pressey. 1966. "Studies with Soybean Protein and Fiber Formation," *Cereal Chem.,* 43:195–206.

Kester, J. J. and O. Fennema. 1989a. "An Edible Film of Lipids and Cellulose Ethers: Barrier Properties to Moisture Vapor Transmission and Structural Evaluation," *J. Food Sci.,* 54:1383–1389.

Kester, J. J. and O. Fennema. 1989b. "An Edible Film of Lipids and Cellulose Ethers: Performance in a Model Frozen-Food System," *J. Food Sci.,* 54:1390–1392, 1406.

Khan, K. and W. Bushuk. 1978. "Glutenin: Structure and Functionality in Breadmaking," *Baker's Dig.,* 52(2):14–17, 18–20.

Khan, L. and W. Bushuk. 1979. "Studies of Glutenin. Comparison by Sodium Dodecyl Sulfate-Polyacrylamide Gel Electrophoresis of Unreduced and Reduced Glutenin from Various Isolation and Purification Procedures," *Cereal Chem.,* 56:63–73.

Kidney, A. J. April 7, 1970. U.S. patent 3,505,084.

Kinsella, J. E. 1984. "Milk Proteins: Physicochemical and Functional Properties," *CRC Crit. Rev. Food Sci. Nutr.,* 21:197–262.

Kinsella, J. E. and D. M. Whitehead. 1989. "Proteins in Whey: Chemical, Physical, and Functional Properties," *Adv. Food Nutr. Res.,* 33:343–438.

Klose, A. A., E. P. Macchi and H. L. Hanson. 1952. "Use of Antioxidants in the Frozen Storage of Turkeys," *Food Technol.,* 6:308–311.

Koff, A. and P. F. Widmer. August 4, 1964. U.S. patent 3,143,475.

Kolster, P., H. J. Kuiper and J. M. Vereijken. 1992. "Non-Food Applications of Wheat Gluten," presented at the *1992 Annual Meeting of the American Association of Cereal Chemists,* September 20–23, Minneapolis, MN.

Kosar, J. and G. M. Atkins, Jr. October 15, 1968. U.S. patent 3,406,119.

Krochta, J. M. 1992. "Control of Mass Transfer in Foods with Edible Coatings and Films," in *Advances in Food Engineering,* R. P. Singh and M. A. Wirakartakusumah, eds., Boca Raton, FL: CRC Press, Inc., pp. 517–538.

Krochta, J. M., J. S. Hudson and R. J. Avena-Bustillos. 1990a. "Casein-Acetylated Monoglyceride Coatings for Sliced Apple Products," presented at the *1990 Annual Meeting of the Institute of Food Technologists,* June 16–20, Anaheim, CA.

Krochta, J. M., A. E. Pavlath and N. Goodman. 1990b. "Edible Films from Casein-Lipid Emulsions for Lightly-Processed Fruits and Vegetables," in *Engineering and Food, Vol. 2, Preservation Processes and Related Techniques,* W. E. Spiess and H. Schubert, eds., New York, NY: Elsevier Science Publishers, pp. 329–340.

Krull, L. H. and G. E. Inglett. "Industrial Uses of Gluten," *Cereal Sci. Today,* 16:232–236, 261.

Krull, L. H. and J. S. Wall. 1969. "Relationship of Amino Acid Composition and Wheat Protein Properties," *Baker's Dig.*, 43(4):30–34, 36, 38–39.

Kuntz, E. October 13, 1964. U.S. patent 3,152,976.

Lasztity, R. 1984. *The Chemistry of Cereal Proteins.* Boca Raton, FL: CRC Press, Inc.

Lasztity, R. 1986. "Recent Results in the Investigation of the Structure of the Gluten Complex," *Die Nahrung,* 30:235–244.

Lasztity, R., F. Bekes, F. Orsi, I. Smied and M. Ember-Karpati. 1987. "Protein-Lipid and Protein-Carbohydrate Interactions in the Gluten Complex," in *Gluten Proteins,* R. Lasztity and F. Bekes, eds., Singapore: World Scientific Publishing Co., Inc., pp. 343–363.

Leach, B. E., N. H. Rainey and M. J. Pallansch. 1966. "Physical and Chemical Analysis of Film Formed at Air-Water Interface on Reconstitution of Whole Milk Powders," *J. Dairy Sci.,* 49(12):1465–1468.

Levinsky, R. J. 1982. "Allergy Aspects of Food Proteins," in *Food Proteins,* P. F. Fox and J. J. Condon, eds., New York, NY: Elsevier Science Publishing Co., Inc., pp. 133–143.

Lieberman, E. R. March 3, 1964a. U.S. patent 3,123,482.

Lieberman, E. R. March 3, 1964b. U.S. patent 3,123,653.

Lieberman, E. R. December 7, 1965. U.S. patent 3,221,372.

Lieberman, E. R. October 10, 1967. U.S. patent 3,346,402.

Lieberman, E. R. and S. G. Gilbert. 1973. "Gas Permeation of Collagen Films as Affected by Cross-Linkage, Moisture, and Plasticizer Content," *J. Polymer Sci.,* 41:33–43.

Lieberman, E. R. and I. B. Oneson. December 19, 1961. U.S. patent 3,014,024.

Lisker, R. 1984. "Lactase Deficiency," in *Genetic Factors in Nutrition,* A. Velázquez and H. Bourges, eds., Orlando FL: Academic Press, Inc.

Mabesa, R. C., R. T. Marshall and M. E. Anderson. 1979. "Factors Influencing the Tenacity of Dried Milk Films Exposed to High Humidity," *J. Food Prot.,* 42:631–637.

Mabesa, R. C., R. T. Marshall and M. E. Anderson. 1980. "Milk Films Exposed to High Humidity: Studies with Electron Microscopy and Electrophoresis," *J. Food Prot.,* 43:29–35.

MacRitchie, F. 1978. "Proteins at Interfaces," *Adv. Protein Chem.,* 32:283–326.

Mahmoud, R. and P. A. Savello. 1992. "Mechanical Properties of and Water Vapor Transferability through Whey Protein Films," *J. Dairy Sci.,* 75:942–946.

Maser, F. April 9, 1987. European patent 0 244 661 A2.

Mauritzen, C. M. and P. R. Stewart. 1965. "The Ultracentrifugation of Doughs Made from Wheat Flour," *Aust. J. Biol. Sci.,* 18:173–189.

McBean, L. D. and E. W. Speckmann. 1988. "Nutritive Value of Dairy Foods," in *Fundamentals of Dairy Chemistry,* N. P. Wong. ed., New York, NY: Van Nostrand Reinhold, pp. 343–408.

McDermott, E. E., D. J. Stevens and J. Pace. 1969. "Modification of Flour Proteins by Disulphide Interchange Reactions," *J. Sci. Food Agric.,* 20:213–217.

McHugh, T. H. and J. M. Krochta. 1994a. "Dispersed Phase Particle Size Effects on Water Vapor Permeability of Whey Protein–Beeswax Edible Emulsion Films," *J. Food Proc. Pres.* (in press).

McHugh, T. H. and J. M. Krochta. 1994b. "Water Vapor Permeability Properties of Edible Whey Protein–Lipid Emulsion Films," *J. Amer. Oil Chem. Soc.,* 71: 307–312.

McHugh, T. H., J. F. Aujard and J. M. Krochta. 1994. "Plasticized Whey Protein Edible Films: Water Vapor Permeability Properties," *J. Food Sci.*, 59.

McKenzie, H. A. 1971a. "β-Lactoglobulins," in *Milk Proteins, Vol. II*, H. A. McKenzie, ed., New York, NY: Academic Press, Inc., pp. 257–330.

McKenzie, H. A. 1971b. "Whole Casein: Isolation, Properties, and Zone Electrophoresis," in *Milk Proteins, Vol. II*, H. A. McKenzie, ed., New York, NY: Academic Press, Inc., pp. 87–116.

McKnight, J. T. March 3, 1864a. U.S. patent 3,123,483.

McKnight, J. T. October 6, 1964b. U.S. patent 3,151,990.

McWatters, K. H. and J. P. Cherry. 1982. "Potential Food Uses of Peanut Seed Proteins," in *Peanut Science and Technology*, H. E. Pattee and C. T. Young, eds., Yoakum, TX: American Peanut Research and Education Society, Inc., pp. 689–736.

Mendoza, M. 1975. "Preparation and Physical Properties of Zein Based Films," M.S. thesis, University of Massachusetts, Amherst.

Meyer, R. and J. V. Spencer. 1973. "The Effect of Various Coatings on Shell Strength and Egg Quality," *Poult. Sci.*, 52:703–711.

Mickus, R. R. 1955. "Seals Enriching Additives on White Rice," *Food Eng.*, 27(11):91–94, 160.

Miller, A. T. May 23, 1972. U.S. patent 3,664,844.

Miller, A. T. June 14, 1983. U.S. patent 4,388,331.

Moorjani, M. N., K. C. M. Raja, P. Puttarajappa, V. S. Khabade, N. S. Mahendrakar and M. Mahadevaswamy. 1978. "Studies on Curing and Smoking Poultry Meat," *Ind. J. Poult. Sci.*, 13(1):52–57.

Motoki, M., N. Nio and K. Takinami. 1987a. "Functional Properties of Heterologous Polymer Prepared by Transglutaminase between Milk Casein and Soybean Globulin," *Agric. Biol. Chem.*, 51:237–239.

Motoki, M., H. Aso, K. Seguro and N. Nio. 1987b. "α_{s1}-Casein Film Prepared Using Transglutaminase," *Agric. Biol. Chem.*, 51:993–996.

Motoki, M., H. Aso, K. Seguro and N. Nio. 1987c. "Immobilization of Enzymes in Protein Films Prepared Using Transglutaminase," *Agric. Biol. Chem.*, 51:997–1002.

Mounts, T. L., W. J. Wolf and W. H. Martinez. 1987. "Processing and Utilization," in *Soybeans: Improvement, Production, and Uses, Second Edition*, J. R. Wilcox, ed., Madison, WI: American Society of Agronomy, Inc., Crop Science Society of America, Inc., and Soil Science Society of America, Inc., pp. 820–866.

Mullen, J. D. October 26, 1971. U.S. patent 3,615,715.

Natarajan, K. R. 1980. "Peanut Protein Ingredients: Preparation, Properties, and Food Uses," *Adv. Food Res.*, 26:215–273.

New Zealand Milk Products, Inc. 1992. 3637 Westwind Blvd., Santa Rosa, CA.

Noznick, P. P. and R. H. Bundus. 1967. April 18, 1967. U.S. patent 3,314,800.

Noznick, P. P. and C. W. Tatter. November 7, 1967. U.S. patent 3,351,531.

Okamoto, S. 1978. "Factors Affecting Protein Film Formation," *Cereal Foods World*, 23:256–262.

Olson, S. and R. Zoss. April 16, 1985. U.S. patent 4,511,583.

Osborne, T. B. 1924. *The Vegetable Proteins*. London, England: Longmans, Green, and Co.

Park, H. J. 1991. "Edible Coatings for Fruits and Vegetables: Determination of Gas Diffusivities, Prediction of Internal Gas Composition and Effects of the Coating on Shelf Life," Ph.D. dissertation, University of Georgia.

Park, H. J. and M. S. Chinnan, 1990. "Properties of Edible Coatings for Fruits and Vegetables," *ASAE Paper No. 90–6510,* presented at the *1990 International Winter Meeting,* American Society of Agricultural Engineers, December 18-21, 1990, Chicago, IL.

Park, H. J., C. L. Weller, P. J. Vergano and R. F. Testin. 1992. "Factors Affecting Barrier and Mechanical Properties of Protein-Based Edible, Degradable Films," presented at the *1992 Annual Meeting of the Institute of Food Technologists,* June 20-24, New Orleans, LA.

Paulis, J. W. 1981. "Disulfide Structures of Zein Proteins from Corn Endosperm," *Cereal Chem.,* 58:542-546.

Petrozziello, M. A. November 7, 1967. U.S. patent 3,351,338.

Piez, K. A., P. Bornstein and A. H. Kang. 1968. "The Chemistry and Biosynthesis of Interchain Crosslinks in Collagen," in *Symposium on Fibrous Proteins,* W. G. Crewther, ed., New York, NY: Plenum Press, pp. 205-211.

Piper, C. V. and W. J. Morse. 1923. *The Soybean.* New York, NY: McGraw-Hill Book Company, Inc.

Plowman, R. D., J. Bitman, C. H. Gordon, L. P. Dryden, H. K. Goering and T. R. Wrenn. 1972. "Milk Fat with Increased Polyunsaturated Fatty Acids," *J. Dairy Sci.,* 55:204-207.

Pomeranz, Y. 1980. "Interaction between Lipids and Gluten Proteins," *Ann. Technol. Agric.,* 29:385-397.

Pomeranz, Y. 1987. *Modern Cereal Science and Technology,* New York, NY: VCH Publishers, Inc.

Pomeranz, Y. and O. K. Chung. 1981. "Starch-Lipid-Protein Interactions in Cereal Systems," in *Utilization of Protein Resources,* D. W. Stanley, E. D. Murray and D. H. Lees, eds., Westport, CT: Food & Nutrition Press, Inc., pp. 129-157.

Pomes, A. F. 1971. "Zein," in *Encyclopedia of Polymer Science and Technology: Plastics, Resins, Rubbers, Fibers, Vol. 15,* H. F. Mark, N. G. Gaylord and N. M. Bikales, eds., New York, NY: Interscience Publishers, pp. 125-132.

Ramachandran. G. N. and C. Ramakrishnan. 1976. "Molecular Structure," in *Biochemistry of Collagen,* G. N. Ramachandran and A. H. Reddi, eds., New York, NY: Plenum Press, pp. 45-84.

Redman, D. G. and J. A. D. Ewart. 1967. "Disulphide Interchange in Dough Proteins," *J. Sci. Food Agric.,* 18:15-18.

Reiners, R. A., J. S. Wall and G. E. Inglett. 1973. "Corn Proteins: Potential for Their Industrial Use," in *Industrial Uses of Cereals,* Y. Pomeranz, ed., St. Paul, MN: American Association of Cereal Chemists, Inc., pp. 285-302.

Rico-Peña, D. C. and J. A. Torres. 1990. "Oxygen Transmission Rate of an Edible Methylcellulose-Palmitic Acid Film," *J. Food Process. Eng.,* 13:125-133.

Rico-Peña, D. C. and J. A. Torres. 1991. "Sorbic Acid and Potassium Sorbate Permeability of an Edible Methylcellulose-Palmitic Acid Film: Water Activity and pH Effects," *J. Food Sci.,* 56:497-499.

Robinson, C. 1953. "The Hot and Cold Forms of Gelatin," in *Nature and Structure of Collagen,* J. T. Randell, ed., New York, NY: Academic Press, Inc., pp. 96-105.

Rose, P. I. 1987. "Gelatin," in *Encyclopedia of Polymer Science and Engineering, Vol.*

7, *Second Edition*, H. F. Mark, N. M. Bikales, C. G. Overberger and G. Menges, eds., New York, NY: John Wiley & Sons, Inc., pp. 488–513.

Roubal, W. T. and A. L. Tappel. 1966. "Polymerization of Proteins Induced by Free-Radical Lipid Peroxidation," *Arch. Biochem. Biophys.*, 113:150–155.

Roy, P. K., A. H. R. Feldbrugge and D. P. Bresnahan. March 23, 1982. U.S. patent 4,321,280.

Rust, R. E. 1987. "Sausage Products," in *The Science of Meat and Meat Products, Third Edition*, J. F. Price and B. S. Schweigert, eds., Westport, CT: Food and Nutrition Press, Inc., pp. 457–485.

Saio, K., M. Kajikawa and T. Watanabe. 1971. "Food Processing Characteristics of Soybean Proteins. Part II. Effect of Sulfhydryl Groups on Physical Properties of Tofu-Gel," *Agric. Biol. Chem.*, 35:890–898.

Sato, J. and T. Kurusu. January 15, 1974. U.S. patent 3,786,159.

Savaiano, D. A. and M. D. Levitt. 1987. "Milk Intolerance and Microbe-Containing Dairy Foods," *J. Dairy Sci.*, 70:397–406.

Schilling, E. D. and P. I. Burchill. July 4, 1972. U.S. patent 3,674,506.

Schofield, J. D., R. C. Bottomley, M. F. Timms and M. R. Booth. 1983. "The Effect of Heat on Wheat Gluten and the Involvement of Sulphydryl-Disulphide Interchange Reactions," *J. Cereal Sci.*, 1:241–253.

Scott, T. W. 1975. "Polyunsaturated Meat and Dairy Products," *Cereal Foods World*, 20:72–76.

Scott, T. W. and G. D. L. Hills. February 14, 1978. U.S. patent 4,073,960.

Scott, T. W., L. J. Cook, K. A. Ferguson, I. W. McDonald, R. A. Buchanan and G. D. L. Hills. 1970. "Production of Polyunsaturated Milk Fat in Domestic Ruminants," *Austr. J. Sci.*, 32:291–293.

Shank, J. L. October 3, 1972. U.S. patent 3,695,902.

Shewry, P. R. and B. J. Miflin. 1985. "Seed Storage Proteins of Economically Important Cereals," in *Advances in Cereal Science and Technology, Vol. 7*, Y. Pomeranz, ed., St. Paul, MN: American Association of Cereal Chemists, Inc., pp. 1–83.

Shifrin, G. A. November 19, 1968. U.S. patent 3,411,921.

Shirai, M., K. Watanabe and S. Okamoto. 1974. "Contribution of 11S and 7S Globulins of Soybean Protein to the Formation and Properties of Yuba-Film," *Nippon Shokuhin Kogyo Gakkaishi*, 21:324–328.

Sian, N. K. and S. Ishak. 1990a. "Effect of pH on Formation, Proximate Composition and Rehydration Capacity of Winged Bean and Soybean Protein-Lipid Film," *J. Food Sci.*, 55:261–262.

Sian, N. K. and S. Ishak. 1990b. "Effect of pH on Yield, Chemical Composition and Boiling Resistance of Soybean Protein-Lipid Film," *Cereal Foods World*, 35:748, 750, 752.

Signorino, C. A. 1969. "Non-Candy Coatings Enhance Products in Many Ways," *Candy Industry*, 133(9):5, 24.

Simmonds, D. H. and R. A. Orth. 1973. "Structure and Composition of Cereal Proteins as Related to Their Potential Industrial Utilization," in *Industrial Uses of Cereals*, Y. Pomeranz, ed., St. Paul, MN: American Association of Cereal Chemists, Inc., pp. 51–120.

Skerritt, J. H. 1988. "Immunochemistry of Cereal Grain Storage Proteins," in *Advances in Cereal Science and Technology, Vol. 9*, Y. Pomeranz, ed., St. Paul, MN: American Association of Cereal Chemists, Inc., pp. 263–338.

Skeritt, J. H., J. M. Devery and A. S. Hill. 1990. "Gluten Intolerance: Chemistry, Celiac-Toxicity, and Detection of Prolamins in Foods," *Cereal Foods World*, 36:638–639, 641–644.

Smith, A. K. and S. J. Circle. 1972. "Protein Products as Food Ingredients," in *Soybeans: Chemistry and Technology, Vol. 1*, A. K. Smith and S. J. Circle, eds., Revised Second Printing, Westport, CT: AVI Publishing Company, Inc., pp. 339–388.

Snyder, H. E. and T. W. Kwon. 1987. *Soybean Utilization*, New York, NY: Van Nostrand Reinhold Company, Inc.

Soloway, S. June 16, 1964. U.S. patent 3,137,631.

Stewart, P. R. and C. M. Mauritzen. 1966. "The Incorporation of [35S]Cysteine into the Proteins of Dough by Disulphide-Sulphide Interchange," *Aust. J. Biol. Sci.*, 19:1125–1137.

Szyperski, R. T. and J. P. Gibbons. 1963. "Zein Systems Developed for Heat Cured Coatings," *Paint Varnish Prod.*, 53(8):65.

Takenaka, H., H. Ito, H. Asano and H. Hattori. 1967. "On Some Physical Properties of Film Forming Materials," *Gifu Yakka Daigaku*, 17:142–146.

Tanzer, M. L. 1976. "Cross-Linking," in *Biochemistry of Collagen*, G. N. Ramachandran and A. H. Reddi, eds., New York, NY: Plenum Press, pp. 137–162.

Taylor, S. L. 1986. "Immunologic and Allergic Properties of Cows' Milk Proteins in Humans," *J. of Food Prot.*, 49:239–250.

Taylor, S. L. 1992. "Chemistry and Detection of Food Allergens," *Food Technol.*, 46(5):146, 148–152.

Taylor, N. W. and J. E. Cluskey. 1962. "Wheat Gluten and Its Glutenin Component: Viscosity, Diffusion and Sedimentation Studies," *Arch. Biochem. Biophys.*, 97:399–405.

Thompson, M. P. 1971. "α_s- and β-Caseins," in *Milk Proteins, Vol. II*, H. A. McKenzie, ed., New York, NY: Academic Press, Inc., pp. 117–173.

Timreck, A. E. September 1, 1970. U.S. patent 3,526,682.

Todd, J. M. October 26, 1982. U.S. patent 4,356,202.

Torres, J. A. and M. Karel. 1985. "Microbial Stabilization of Intermediate Moisture Food Surfaces. III. Effects of Surface Preservative Concentration and Surface pH Control on Microbial Stability of an Intermediate Moisture Cheese Analog," *J. Food Process. Preserv.*, 9:107–119.

Torres, J. A., M. Motoki and M. Karel. 1985. "Microbial Stabilization of Intermediate Moisture Food Surfaces. I. Control of Surface Preservative Concentration," *J. Food Process. Preserv.*, 9:75–92.

Tracey, M. V. 1975. "Altering Fatty Acid Composition of Ruminant Products," *Cereal Foods World*, 20:77–80, 90–100.

Tryhnew, L. J., K. W. B. Gunaratne and J. V. Spencer. 1973. "Effect of Selected Coating Materials on the Bacterial Penetration of the Avian Egg Shell," *J. Milk Food Technol.*, 36:272–275.

Tsuzuki, T. September 22, 1970. U.S. patent 3,529,530.

Tsuzuki, T. and E. R. Lieberman. August 1, 1972. U.S. patent 3,681,093.

Turbak, A. F. August 8, 1972. U.S. patent 3,682,661.

Turner, J. E., J. A. Boundy and R. J. Dimler. 1965. "Zein: A Heterogeneous Protein Containing Disulfide-Linked Aggregates," *Cereal Chem.*, 42:452–461.

Vandegaer, J. E. 1974. "Encapsulation by Coacervation," in *Microencapsulation: Processes and Applications*, J. E. Vandegaer, ed., New York, NY: Plenum Press, pp. 21–37.

Veis, A. 1964. *The Macromolecular Chemistry of Gelatin*. New York, NY: Academic Press, Inc.

Viro. F. 1980. "Gelatin," in *Encyclopedia of Chemical Technology, Vol. 11, Third Edition*, G. J. Bushey, C. I. Eastman, A. Klingsberg and L. Spiro, eds., New York, NY: John Wiley & Sons, Inc., pp. 711–719.

Vojdani, F. and J. A. Torres. 1990. "Potassium Sorbate Permeability of Methylcellulose and Hydroxypropyl Methylcellulose Coatings: Effect of Fatty Acids," *J. Food Sci.*, 55:841–846.

Wainewright, F. W. 1977. "Physical Tests for Gelatin and Gelatin Products," in *The Science and Technology of Gelatin*, A. G. Ward and A. Courts, eds., New York, NY: Academic Press, Inc., pp. 507–534.

Wall, J. S. and A. C. Beckwith. 1969. "Relationship between Structure & Rheological Properties of Gluten Proteins," *Cereal Sci. Today*, 14(1):16–18, 20–21.

Wall, J. S. and J. W. Paulis. 1978. "Corn and Sorghum Grain Proteins," in *Advances in Cereal Science and Technology, Vol. 2*, Y. Pomeranz, ed., St. Paul, MN: American Association of Cereal Chemists, Inc., pp. 135–219.

Wall, J. S., M. Friedman, L. H. Krull, J. F. Cavins and A. C. Beckwith. 1968. "Chemical Modification of Wheat Gluten Proteins and Related Model Systems," *J. Polymer Sci., Part C*, 24:147–161.

Wang, H. L. 1981. "Oriental Soybean Foods," *Food Devel.*, 15(5):29–34.

Watanabe, K. and S. Okamoto. 1976. "Formation of Yuba-Like Films and Their Physical Properties," *New Food Industry*, 18(4):65–77.

Watanabe, T., H. Ebine and M. Okada. 1974. "New Protein Food Technologies in Japan," in *New Protein Foods, Vol. 1A*, A. M. Altschul, ed., New York, NY: Academic Press, Inc., pp. 414–453.

Watanabe, K., T. Watanabe and S. Okamoto. 1975. "On the Contributions of Lipid to the Properties of Yuba-Film," *Nippon Shokuhin Kogyo Gakkaishi*, 22:143–147.

Watters, G. G. and J. E. Brekke. 1961. "Stabilized Raisins for Dry Cereal Products," *Food Technol.*, 15:236–238.

Waugh, D. F., M. L. Ludwig, J. M. Gillespie, B. Melton, M. Foley and E. S. Kleiner. 1962. "The α_s-Caseins of Bovine Milk," *J. Amer. Chem. Soc.*, 84:4929–4938.

Wehrli, H. P. and Y. Pomeranz. 1970. "A Note on the Interaction between Glycolipids and Wheat Flour Macromolecules," *Cereal Chem.*, 47:160–166.

Whitman, G. R. and H. Rosenthal. January 19, 1971. U.S. patent 3,556,814.

Wilson, C. M. 1987. "Proteins of the Kernel," in *Corn: Chemistry and Technology*, S. A. Watson and P. E. Ramstad, eds., St. Paul, MN: American Association of Cereal Chemists, Inc., pp. 273–310.

Wolf, W. J. and J. C. Cowan. 1975. *Soybeans as a Food Source*. Cleveland, OH: CRC Press, Inc.

Wolf, W. J. and A. K. Smith. 1961. "Food Uses and Properties of Soybean Protein. II. Physical and Chemical Properties of Soybean Protein," *Food Technol.*, 15(5):12–13, 16, 18, 21, 23, 26, 28, 31, 33.

Wolf, W. J. and T. Tamura. 1969. "Heat Denaturation of Soybean 11S Protein," *Cereal Chem.*, 46:331–344.

Wolf, W. J., G. E. Babcock and A. K. Smith. 1962. "Purification and Stability Studies of the 11S Component of Soybean Proteins," *Arch. Biochem. Biophys.*, 99:265–274.

Woodin, A. M. 1954. "Molecular Size, Shape and Aggregation of Soluble Feather Keratin," *Biochem. J.*, 57:99–109.

Woodin, A. M. 1956. "Structure and Composition of Soluble Feather Keratin," *Biochem. J.*, 63:576–581.

Woodroof, J. G. 1983. "Composition and Nutritive Value of Peanuts," in *Peanuts: Production, Processing, Products, Third Edition,* J. G. Woodroof, ed., Westport, CT: The AVI Publishing Company, Inc., pp. 165–179.

Woychik, J. H., J. A. Boundy and R. J. Dimler. 1961. "Starch Gel Electrophoresis of Wheat Gluten Proteins with Concentrated Urea," *Arch. Biochem. Biophys.*, 94:477–482.

Wright, K. N. 1987. "Nutritional Properties and Feeding Value of Corn and Its By-Products," in *Corn: Chemistry and Technology,* S. A. Watson and P. E. Ramstad, eds., St. Paul, MN: American Association of Cereal Chemists, Inc., pp. 447–478.

Wu, L. C. and R. P. Bates. 1972a. "Soy Protein-Lipid Films. 1. Studies on the Film Formation Phenomenon," *J. Food Sci.*, 37:36–39.

Wu, L. C. and R. P. Bates. 1972b. "Soy Protein-Lipid Films. 2. Optimization of Film Formation," *J. Food Sci.*, 37:40–44.

Wu, L. C. and R. P. Bates. 1973. "Influence of Ingredients upon Edible Protein-Lipid Film Characteristics," *J. Food Sci.*, 38:783–787.

Wu, L. C. and R. P. Bates. 1975. "Protein-Lipid Films as Meat Substitutes," *J. Food Sci.*, 40:160–163.

Wu, Y. V. and R. J. Dimler. 1963a. "Hydrogen-Ion Equilibria of Wheat Gluten," *Arch. Biochem. Biophys.*, 102:230–237.

Wu, Y. V. and R. J. Dimler. 1963b. "Hydrogen Ion Equilibria of Wheat Glutenin and Gliadin," *Arch. Biochem. Biophys.*, 103:310–318.

Yamagishi, T., A. Miyakawa, N. Noda and F. Yamauchi. 1983. "Isolation and Electrophoretic Analysis of Heat-Induced Products of Mixed Soybean 7S and 11S Globulins," *Agric. Biol. Chem.*, 47:1229–1237.

Young, H. H. 1967. "Gelatin," in *Encylcopedia of Polymer Science and Technology: Plastics, Resins, Rubbers, Fibers, Vol. 7,* H. F. Mark, N. G. Gaylord and N. M. Bikales, eds., New York, NY: Interscience Publishers, pp. 446–460.

CHAPTER 10

Edible Coatings from Lipids and Resins

ERNESTO HERNANDEZ[1]

INTRODUCTION

F OOD products are commonly coated to extend shelf-life and to improve appearance. Coatings can be used to encapsulate food, candy,
medication, vitamins, or flavoring agents for convenient ingestion, and to
protect flavor. In the case of fruits and vegetables, coatings are used to prevent weight loss (retain moisture), slow down aerobic respiration, and
improve appearance by providing gloss. Lipids and resins are useful in
reducing surface abrasion during handling operations, while at the same
time sealing tiny scratches that may nevertheless occur on the surface of
fruits and vegetables. Edible coatings can also be used as carriers of fungicides, antioxidants, antimicrobials, coloring agents, flavoring agents, and
growth regulators.

Lipid compounds include neutral lipids of glycerides which are esters of
glycerol and fatty acids, and the waxes which are esters of long-chain
monohydric alcohols and fatty acids. From this group, acetylated
monoglycerides, natural waxes, and surfactants are commonly utilized in
edible coatings (Kester and Fennema, 1986). Resins are a group of acidic
substances that are usually secreted by special plant cells into long resin
ducts or canals in response to injury or infection in many trees and shrubs.

One reason lipids and resins are commonly added to food coatings is to
impart hydrophobicity. Extensive reviews are available in the literature on
lipid- and resin-based edible films in general (Kester and Fennema, 1986;
Guilbert, 1986; Sward, 1972). Waxes such as carnauba, beeswax, paraffin,

---

[1]Food Protein R&D Center, Texas A&M University, College Station, TX 77843.

rice bran, and candelilla have been used in combination with other lipids, resins, or polysaccharides to coat fresh fruits and vegetables such as citrus and apples (Dalal et al., 1971; Durand et al., 1984; Ahmad et al., 1979; Paredes-Lopez et al., 1974; Lakshminarayana et al., 1974; Hitz and Haut, 1942; Claypool, 1939), although of these waxes, only carnauba is still commonly used today.

This chapter discusses properties of lipids and resins as they are used in food coatings, including how the composition of films affects the permeability of the coating as it is applied on the food item. The effect of film composition on its stretchability and flexibility is also discussed.

LIPIDS AND RESINS USED IN EDIBLE COATINGS

WAXES AND OILS

Paraffin Wax

Paraffin wax is derived from the wax distillate fraction of crude petroleum. It is composed of hydrocarbon fractions of generic formula C_nH_{2n+2} ranging from eighteen to thirty-two carbon units (Bennett, 1975). Synthetic paraffin wax is allowed for food use in the United States. It consists of a mixture of solid hydrocarbons resulting from the catalytic polymerization of ethylene. Both natural and synthetic paraffins are refined to meet FDA specifications for ultraviolet absorbance (21 CFR, Code of Federal Regulations, 184.1973). Synthetic paraffin wax should not have an average molecular weight lower than 500 or higher than 1200. Additional physical properties are listed in Table 10.1. Paraffin waxes are permitted for use as protective coatings for raw fruit and vegetables and for cheese. They can also be used in chewing gum base, as a defoamer, and as a component in microencapsulation of spice-flavoring substances.

Carnauba Wax

Carnauba wax is an exudate of palm tree leaves from the Tree of Life (*Copernica cerifera*), found mostly in Brazil. It has the highest melting point and specific gravity of commonly found natural waxes and is added to other waxes to increase melting point, hardness, toughness, and luster (Table 10.1). Refined carnauba wax consists mostly of saturated wax acid esters with twenty-four to thirty-two hydrocarbons and saturated long-chain monofunctional alcohols (Bennett, 1975). Carnauba wax is considered a GRAS (generally recognized as safe) substance and is permitted for use in coatings for fresh fruits and vegetables, in chewing gum, in con-

TABLE 10.1. Properties of Some Waxes.

Wax	Melting Point (°C)	Sp. Gr. g/ml	Sapon. Number	Refr. Index	Iodine Number	Hardness (cm × 10⁻²)[a]
Paraffin[b]	50.0–51.1	0.880–0.915	0	1.410–1.450	0	31.3
Carnauba[b]	82.5–86.0	0.996–0.998	78–88	1.463 (60°)	7–14	4.7
Beeswax[c]	61.0–65.0	0.950–0.960	88–102	1.439–1.445 (75°)	8–11	2.0–4.0
Candelilla[b]	65.0–68.8	0.982–0.993	46–66	1.455	15–30	4.7
Polyethylene[b]	97.0–115.0	0.922	>0.1	—	—	2.0–4.0[d]

[a]At 150 g/60 sec/25°C.
[b]Bennett (1975).
[c]Tulloch (1970).
[d]100 g/5 sec/25°C.

281

fections, and in sauces with no limitations other than good manufacturing practice (Table 10.2). There are several grades of carnauba wax available (from crude to refined), but coating companies are reluctant to indicate which grades are most used in edible coatings.

Beeswax

Also known as white wax, beeswax is secreted by honey bees for comb building. The wax is harvested by centrifuging the honey from the wax combs, and then melted with hot water, steam, or solar heating. The wax is subsequently refined with diatomaceous earth and activated carbon, and finally bleached with permanganates or bichromates. Beeswax consists mostly of monofunctional alcohols C_{24}–C_{33}, hydrocarbons C_{25}–C_{33}, and long-chain acids C_{24}–C_{34} (Tulloch, 1970). This wax is very plastic at room temperature, but becomes brittle at colder temperatures. It is soluble in most other waxes and oils. Some of its properties are listed in Table 10.1. Beeswax is considered a GRAS substance and is allowed for direct use in food with some limitations (Table 10.2). Maximum allowed levels are 0.065% in chewing gum, 0.005% in confections and frostings, 0.04% in hard candy, 0.1% in soft candy, and 0.002% in all other categories.

Candelilla Wax

Candelilla wax is an exudate of the candelilla plant (*Euphorbia cerifera, E. antisphylitica, Pedilanthus parvonis, P. aphyllus*), a reed-like plant that grows mostly in Mexico and southern Texas. The wax is recovered by immersing the plant in boiling water, after which the wax is skimmed off the surface, refined, and bleached. Its degree of hardness is intermediate between beeswax and carnauba. This wax sets very slowly, taking several days to reach maximum hardness. Some of its properties are listed in Table 10.1. Candelilla wax is considered a GRAS substance, and is allowed for certain food uses including chewing gum and hard candy with no limitations other than good manufacturing practices (CFR, 184.1978) (Table 10.2).

Polyethylene Wax

Oxidized polyethylene is defined as the basic resin produced by the mild air oxidation of polyethylene, a petroleum by-product. It should have a minimum number average molecular weight of 1200, as determined by high-temperature vapor pressure osmometry, a maximum of 5% total oxygen by weight, and an acid value of 9–19 according to the Federal Code of Regulations (Table 10.2). Several grades of polyethylene wax are

TABLE 10.2. Lipids, Lipid Emulsifiers, Resins and Rosins Commonly Used in Food Coatings with Hydrophilic-Lipophilic Balance (HLB) and Required HLB Values for Emulsification.

Component	HLB[a, b]	Required HLB[a, b]	Uses/Regulatory Status[c] (21 CFR)
Acetylated monoglycerides	1.5	—	Coating agent, emulsifier (172.828)
Acetylated sucrose diester	1.0	—	
Ammonium lauryl sulfate	31	—	Emulsifier, scald agent (172.822)
Beeswax	3.5	9	Coating agent, candy glaze (GRAS, 184.1973)
Butyl stearate	—	11	Plasticizer (181.27)
Carnauba wax	—	15	Coating agent, confections, fruit and fruit juices (GRAS, 184.1978)
Castor oil	—	8	Coating agent, anticaking (172.876)
Chlorinated paraffin	—	12–14	Coating agent, chewing gum (172.615, 175.105, 175.250)
Cocoa butter	—	6	Coating, confections (GRAS)
Corn oil	—	8	Coating and emulsifying agent, texturizer (GRAS)
Coumarone indene resin	—	—	Coating agent (172.215, 175.300)
Ethylene glycol monostearate	2.9	—	Emulsifier, confections (73.1)
Glycerol monostearate	3.8	—	Coating agent, emulsifier (GRAS, 184.1324)
Isopropyl palmitate	12	—	
Lard	—	5	Coating agent, emulsifier (GRAS)
Lauric acid	12.8	6	Defoaming agent, lubricant (172.860)
Lauryl alcohol	9.5	14	Intermediate, flavoring agent (172.515)
Mineral oil	—	10	Coating agent, confections (172.878)
Oleic acid	1	17	Emulsifier, binder, lubricant (172.862)

(continued)

283

TABLE 10.2. (continued).

Component	HLB[a, b]	Required HLB[a, b]	Uses/Regulatory Status[c] (21 CFR)
Oxidized polyethylene	—	—	Coating agent, defoaming agent (172.260, 175.300, 176.200)
Palm oil	—	7	Coating agent, emulsifier, texturizer (GRAS, 184.1585)
Paraffin wax	—	10	Coating agent, lubricant, chewing gum (172.615, 178.3800)
Potassium oleate	20	—	Emulsifier, stabilizer (172.863)
Propylene glycol monostearate	—	3.4	Emulsifier, stabilizer (172.856)
Shellac resin	—	—	Coating component (175.300)
Sodium alkyl sulfate	40	—	Emulsifier, scald agent (172.210)
Sodium oleate	18	—	Emulsifier, anticaking agent (172.863)
Sorbitan monostearate	4.7	—	Emulsifier, defoamer, stabilizer (172.515, 172.842)
Soybean oil	—	6	Coating agent, texturizer, formulation aid (GRAS)
Soya lecithin	8	—	Emulsifier, antioxidant (GRAS, 184.1400)
Stearyl alcohol	4.9	15	Intermediate (172.864)
Triethanolamine oleate soap	12	—	
Tallow	—	6	Coating agent, texturizer (GRAS)
Wood rosin	—	—	Coating component (175.300)

[a]Griffin (1979).
[b]Shinoda and Kunieda (1984).
[c]CFR (Code of Federal Regulations).

284

available to obtain desired emulsion properties. These waxes differ in molecular weight (affecting viscosity), density (affecting hardness), and softening point (Eastman Chemical Inc., 1990). This wax, although less popular of late, is used to make emulsion coatings. Methods of emulsification include wax to water emulsification where the wax is combined with a fatty acid (food grade oleic acid) and heated. Morpholine is then added and the mixture is combined with hot water, agitated, and cooled.

In the pressure emulsification method, all ingredients except dilution water are combined, heated, pressurized, and agitated. Hot water is then added to the mixture along with further agitation. Sometimes an alkali-soluble resin such as shellac resin is added at 20% solids to create a more uniform and continuous coating. Oxidized polyethylene is allowed for use as a protective coating or component of protective coatings for some fresh produce where, generally, the peel is not normally ingested (avocado, banana, coconut, citrus, mango, melon, pineapple, papaya, pumpkin, etc.) and certain nuts in their shells. Polyethylene emulsions can be applied at the 10–15% total solids level (Wineman, 1984; Allied Signal, Inc., 1986; Eastman Chemical Products, 1984). Polyethylene wax can also be combined with paraffin wax to help reduce moisture loss but at a slight sacrifice in gloss or appearance.

Mineral Oil

White mineral oil consists of a mixture of liquid paraffinic and naphthenic hydrocarbons and is allowed for use as a food release agent and as a protective coating agent for fruits and vegetables in an amount not to exceed good manufacturing practices (Table 10.2).

FATTY ACIDS AND MONOGLYCERIDES

Fatty acids and polyglycerides and their derivatives are used primarily as emulsifiers and dispersing agents. Most fatty acids derived from vegetable oils are considered GRAS (Table 10.2) and are commonly used in combination with glycerides as emulsifiers. Mono- and diglycerides are the most commonly used food emulsifiers and are usually prepared by transesterification of glycerol and triglycerol. Common sources of triglycerides and their derivatives are vegetable oils such as soybean (Erickson et al., 1980). Their emulsifying properties are mentioned below.

Long-chain alcohols such as stearyl alcohol ($C_{18}H_{38}O$) are commonly used as additives in edible coatings due to their high melting point and hydrophobic characteristics. They are usually extracted from sperm whale oil and are allowed for use as emulsifiers in foods, pharmaceuticals, and cosmetics (Bennet, 1975). Long-chain fatty acids such as stearic and palmitic acids are also commonly used in edible coatings for their higher melting points and hydrophobicity (Hagenmaier and Shaw, 1990).

EMULSIONS

Macro- and Microemulsions

Emulsion waxes are true wax emulsions or wax suspensions. These types of coatings provide good moisture barriers but they impart little shine to the fruit. Most research on preparation of wax emulsions, particularly formulations of edible films, is proprietary and little information on composition and methods of preparation is available. Waxes are dispersed in water as macroemulsions, particle size range 2×10^3–10^5 Å, or as microemulsions, particle size 1000–2000 Å. Melted wax disperses similarly to conventional oils in oil-in-water (o/w) macroemulsions (Hernandez and Baker, 1991). The wax globules obtained through high-pressure homogenization show similar average size and size distribution profiles (i.e., normal and lognormal) as conventional o/w emulsions. One important difference between the high-pressure homogenization and conventional methods, however, is that at room temperatures, the dispersed phase for the wax suspension in the high-pressure homogenization method is in solid state. This has the advantage that little or no flocculation of the suspended globules takes place, and if the globules are small enough, no settling or phase separation occurs. Wax solutions and emulsions can be applied by foaming, spraying, dipping, and brushing. Foaming is recommended as the most suitable because it leaves a thin film and prevents waste; however, spraying and brushing are the most popular methods because they are easier and more convenient to operate (Akamine et al., 1975).

Waxes, such as carnauba and beeswax, can also be dispersed as microemulsions (Prince, 1977). These waxes are relatively easy to microemulsify because of their high content of hydroxyl and ester groups. Procedures for the formulation of wax microemulsions require the selection of suitable emulsifiers that are compatible with the dispersed and continuous phases. Two emulsifiers are normally used when preparing a microemulsion, a surfactant that is partially soluble in the continuous and dispersed phases and a co-surfactant which is usually an alcohol. The use of microemulsified waxes in food coatings is desirable because the small droplet size enables the wax to coalesce into a uniform film. As the coating dries, the surface of the product is glossy because it reflects light, giving the coated product a shiny finish.

The most common method of preparing o/w or wax-in-water microemulsions is the so-called inversion process (Prince, 1977). Usually

100% of the emulsifier is added to the phase to be dispersed and water is then gradually added until the mixture progresses from a viscoelastic gel stage to a clear fluid, o/w microemulsion, resulting in a translucent suspension. The formation of a microemulsion depends more on the interaction between the dispersed phase and the emulsifiers than on the mechanical method used to disperse the wax. The opposite is true for macroemulsions, where the mechanical method of dispersion, i.e., high-pressure homogenization, colloid mill, or high-speed stirring, will determine the average particle size and particle size distribution of suspended globules in the emulsion.

Commonly Used Emulsifiers

The choice of emulsifiers and co-surfactants is very wide and new ones are being developed at a fast pace. The selection of emulsifiers for formulation of microemulsions requires extensive experimental work. However, this selection can be greatly facilitated using two criteria: hydrophilic-lipophilic balance (HLB) and phase inversion temperature (PIT). HLB assigns a value of 1 to 40 to emulsifiers depending on the hydrophobic and hydrophilic portion of the surfactant. Very hydrophilic surfactants, such as sodium lauryl sulfate, have been assigned a value of 40. HLB values can be calculated from the molecular structure, but since many surfactants are ionic and have a complex behavior, it is usually done empirically (Prince, 1977). HLB values can be calculated for some types of nonionic surfactants from the saponification number of the ester and the acid number of the fatty acid. A major disadvantage of HLB is that it makes no allowance for changes in surfactant behavior due to temperature, pH, and the presence of other electrolytes in the continuous phase (Shinoda and Kunieda, 1984).

In general, surfactants with low HLB values are recommended for water-in-oil (w/o) emulsification and surfactants with high HLB values are recommended for o/w emulsification. PIT system selects emulsifiers based on the temperature at which an o/w emulsion turns into a w/o emulsion. The oil/water interfacial tension is considered minimum at PIT, and the emulsion prepared at this temperature will produce the smallest suspended particles. HLB and PIT systems are often used in combination, when factors such as hydrophobicity of dispersed phase, temperature, and concentration of components are to be considered. Some commonly used lipid emulsifiers are listed in Table 10.2 along with other components of wax formulations. Even though this is not a comprehensive list, it illustrates the wide range of HLB values of available lipid emulsifiers. Table 10.2 also lists some reported HLB values required for the emulsification of some wax components.

There is a wide variety of emulsifiers available commercially that are suitable for the manufacture of wax emulsions and microemulsions. Many food emulsifiers are derivatives of glycerols and fatty acids. Some examples include the polyglycerols-polystearates with a wide range of HLB values (2.5–11.0) and the TWEEN and polysorbate series (poly-oxyethylene sorbitan derivatives) with high HLB values (10.5–15.0). Lecithin, a phospholipid in which one fatty acid radical has been replaced with phosphoric acid, which in turn is esterified with choline (phosphati-dylcholine) is widely used as an emulsifier.

Most natural waxes, such as beeswax, carnauba, and candelilla, possess emulsifying properties since they are composed of long-chain alcohols and long-chain esters. However, not much information is available on their hydrophilic-lipophilic properties in either solid or liquid state.

RESINS AND ROSINS

Shellac

Shellac resin is a secretion by the insect *Laccifer lacca* and is mostly produced in central India. This resin is composed of a complex mixture of aliphatic alicyclic hydroxy acid polymers, i.e., aleuritic and shelloic acids (Griffin, 1979). This resin is soluble in alcohols and in alkaline solutions. It is also compatible with most waxes, resulting in improved moisture-barrier properties and increased gloss for coated products. Some of its properties are listed in Table 10.3. Since shellac is not a GRAS substance, it is only permitted as an indirect food additive in food coatings and adhesives (21 CFR 175.300). A petition has been submitted by its manufacturers for an upgrade to GRAS status (*Federal Register,* 54, 142).

Other natural resins permitted for use in food coatings include copal, damar, and elemi. They are used mostly in coatings for the pharmaceutical industry but little work has been reported on their uses in food.

Wood Rosin

Rosins are obtained from the oleoresins of pine trees, either as an exudate or as tall oil, a by-product from the wood pulp industry. Rosin is the residue left after distillation of volatiles from the crude resin. Wood rosin is approximately 90% abietic acid ($C_{20}H_{32}O_2$) and its isomers, and 10% dehydroabietic acid ($C_{20}H_{28}O_2$). Some of its properties are listed in Table 10.3. Wood resin can be modified by hydrogenation, polymerization, isomerization, and decarboxylation to make it less susceptible to oxidation and discoloration, and to improve its thermoplasticity. Drying oils, such as some vegetable oils, may be esterified with some glycol derivatives (e.g.,

TABLE 10.3. Properties of Some Resins[b, c].

Resin	Melting Point (°C)	Sp. Gr. (gr/ml)	Sapon. Number	Iodine Number	Acid Number
Shellac	115–120	1.035–1.1400	230–262	10–18	73–95
Wood Rosin	82[a]	1.07—1.09	168	—	162
Wood Rosin (Hydrogenated)	75[a]	1.045	—	214–218[b]	160
Copal	119	1.072	—	—	139

[a]Softening point.
[b]Sward (1972).
[c]Martin (1982).

289

butylene, ethylene, polyethylene, and polypropylene), to form resinous and polymeric coating components. Rosin and its derivatives are widely utilized in coatings for citrus and other fruits (Sward, 1972).

Coumarone Indene

Coumarone indene resin is a petroleum and/or coal tar by-product. It is 100% aromatic in content and exhibits excellent resistance to alkalis, dilute acids, and moisture. Several grades are available, varying in melting point, molecular weight, and viscosity. It is approved for use in coatings (Table 10.2) and is often a component of the so-called solvent wax for citrus (Neville Chemical Co., 1988).

SOLVENT WAXES

Solvent waxes consist mostly of resins, with little or no actual wax, and small amounts of petroleum solvent. Other solvents such as ethanol and isopropanol can also be used. Use of solvents permits waxes and resins to blend together, producing coatings that have good moisture-barrier properties and glossy finish. The most commonly used resin is coumarone indene. Solvent waxes are easy to apply to fruit and vegetables because of their low viscosity and rapid drying rates. However, they are diminishing in use (Kaplan, 1986) as they are more expensive than water-based waxes, and are more dangerous to handle on production lines (due to air pollution from solvent vaporization). Also, the solvent sometimes migrates into the fruit or vegetable, affecting their flavor and wholesomeness.

Resin solution waxes have one or more of the following ingredients: shellac, modified protein, natural gums, lipids, morpholine, and sucrose esters of fatty acids. These components are either dissolved or suspended in alkaline solutions (Hall, 1981; Drake et al., 1987). These types of coatings also contain little or no wax.

CHARACTERISTICS OF LIPID- AND RESIN-BASED COATINGS

COHESION AND FLEXIBILITY

Lipid- and resin-based edible films should possess both cohesiveness and flexibility. These characteristics are determined by the molecular weight of both the hydrophobic and hydrophilic phases, as well as their branching and polarity. Molecules with low polarity and high linearity tend to produce films with high degrees of cohesiveness and rigidity. On the other hand, polar molecules with a high degree of branching will promote cross-linking and formation of crystals. Lipid- and resin-based films should have good flexibility without breaking. Plasticizers are added to reduce brittleness and increase flexibility and tear resistance of edible films. This is particularly important when the product is to be stored at low temperature (Kester and Fennema, 1989c). Acetoglycerides, plasticizers sometimes used in coating formulations, have the ability to crystallize into a relatively stable α-polymorphic form. Other plasticizers commonly used are fatty acids, monoglycerides, phospholipids, and ester derivatives of glycerol (Kester and Fennema, 1989c). However, plasticizers tend to extend, dilute, and soften the structure of the film, thus increasing the permeability of the film to moisture and gases. Hence, careful design and evaluation of formulations for edible films is important.

Lipids in edible films are often supported on a polymer structure matrix, usually a polysaccharide. Derivatives of cellulose are the most common polysaccharides used as supporting matrices in composite lipid edible films. Hydroxypropyl methylcellulose is used for its solubility in ethanol, particularly when alcohol-soluble waxes are used as coating agents. Ethyl cellulose is frequently used to add toughness to the film. Further discussion of polysaccharides in edible films can be found in Chapter 11. With wax coatings, it is desirable that the crystallites in the wax film be in a closely packed arrangement, oriented parallel to the polysaccharide or support base. It is also important to minimize the "cooling" shock phenomenon. This occurs as the wax on the surface of the product cools, resulting in contraction, and leading to formation of small unoriented wax crystals which generate fissures and defects on the product surface (Kester and Fennema, 1989b). The ability of a hydrophobic substance to retard moisture transfer depends on the homogeneity of its final repartition in the supporting matrix and/or on the surface (Martin-Polo et al., 1992a). Orientation, length, and size and shape of the aliphatic chains may explain the differences observed in permeability of lipid-containing films (Martin-Polo et al., 1992b). In other words, any conditions that produce larger and closer packed plate crystals oriented parallel to the plane of the support base will tend to reduce the transfer of moisture.

The surfaces of films prepared with molten wax are smooth and uniform, whereas films prepared with alcohol solutions or lipid emulsions tend to be irregular, and in some cases contain globules formed in the emulsion (Greener and Fennema, 1989).

PERMEABILITY

Transport of gases and water vapor through the coating depends on the gas concentration at each side of the film and it partially obeys Fick's law of mass diffusion

$$q = D(C_1 - C_2)/t \qquad (10.1)$$

(providing the coating is intact, i.e, no holes), where q is the amount of gas diffusing through a unit area of film per unit time, D is diffusion constant, C_1 is the concentration of gas at the high-pressure side, C_2 is the concentration of gas at lower-pressure side, and t is film thickness.

Expressed in terms of gas pressures and in accordance to Henry's law,

$$q = DS(p_1 - p_2)/t \qquad (10.2)$$

where S is the solubility coefficient for the gas in the film, and p_1 and p_2 are the pressures of the gases at each side of the film coating. The product DS is usually expressed as P or film permeability constant.

In most cases, the permeability increases as water is absorbed in the film (i.e., when subjected to high-humidity environments) and Fick's diffusion equation is not strictly obeyed. Long and Thompson (1955) calculated an integral diffusivity (D) of some polymer films by taking into account sorption and desorption curves of the gases absorbed in the polymers,

$$D = \pi/32(K_a^2 - K_d^2) \qquad (10.3)$$

where K_a and K_d are the initial slopes of the sorption and desorption curves of vapor absorbed into the film.

D, S and therefore P are also temperature dependent,

$$S = S_o \exp(-H/RT) \qquad (10.4)$$

and

$$D = D_o \exp(-E_d/RT) \qquad (10.5)$$

where H is the heat of solution, E_d is the activation energy for the diffusion process, and S_o and D_o are solubility and diffusion constants that are dependent on polymer structure (Kumins, 1965).

In actual food coatings, the transport of gases and vapor through the film becomes dependent on factors such as coating formulation, the temperature of application, drying temperature, storage time, type of product to which the coating is applied, uniformity of the film, relative humidity gradients, etc. (Guilbert, 1986).

Some methods to test edible films have been reported for pharmaceutical and food use (Hagenmaier and Shaw, 1991; Kamper and Fennema, 1984b; Kanig and Goodman, 1962), but given the complexity of films and the number of factors involved, there is still a need for a reliable method to quantitatively test edible films. So far, most of the data reported on edible coatings is difficult to reproduce and has mainly qualitative value. Also, the testing of coatings on artificial support systems may yield different permeability data compared to testing of films on food products, especially respiring fruit and vegetable tissues.

In general, permeability of films decreases as the concentration of the hydrophobic phase increases. However, excessive addition of lipids may result in collapse of the film due to loss of cohesiveness as the matrix provided by the saccharide or resin becomes too weak. The presence of hydrophilic components such as polysaccharides and plasticizers in the film increases moisture sorption and swelling, resulting in distortion of the film (Greener and Fennema, 1989).

Waxes and solid fat films tend to form platelets as they crystallize due to side-by-side association of the lipid dimers, which causes the crystals to grow perpendicularly. If the plate crystals grow perpendicularly to the direction of the gas flow, they will be better barriers. However, if the platelets grow in the direction of gas flow, the permeability of the film will increase. The addition of fatty acids and other plasticizers helps to reduce the formation of these platelets due to better thermogelation with polysaccharides such as methylcellulose. This results in more entrapment of fatty acids (Kester and Fennema, 1989b). Carnauba wax and beeswax tend to produce smaller platelets and are favored as the hydrophobic component of edible film.

When edible films are prepared as w/o and o/w emulsions, the film cast on the product is likely to develop voids through which gases and moisture will migrate (Ukai et al., 1976). When the film is so tightly cast that no appreciable amount of void space develops, moisture transfer takes place mainly through the hydrophilic portion of the film. In reality, both the presence of voids and the ratio of hydrophilic/hydrophobic materials will determine the permeability of the film to moisture and gases. Therefore, the size of the suspended globules in the emulsion plays an important role in the migration of water vapor and gases through the film. It has been reported that certain fatty acids result in lower water vapor transfer rates in conjunction with certain polysaccharide film formers than do others. For example, lauric acid decreased the permeability of a chitosan film while other fatty acids (methyl laurate, acetylated monoglyceride, propylene glycol stearate, palmitic acid, octanoic acid, and butyric acid) did not have the same effect (Wong et al., 1992). This was due to the altered microstructure of the composite film whose surface appeared in electron micrographs as a sheet-like structure, stacked in layers. None of the other composite films (chitosan + the other fatty acids mentioned above) showed this characteristic.

Permeability to Gases

Films prepared with shellac and other polar resins have lower permeability to gases such as oxygen (O_2), carbon dioxide (CO_2), and ethylene than films prepared with waxes. In general, coating polymers with polar groups (i.e., ester and hydroxyl groups) tends to result in lower permeability to O_2 (Hagenmaier and Shaw, 1991). Therefore, if flow of the above gases through the film is desired, the fruit coating should have low concentrations of shellac and rosins. More research is needed in this area to determine proper levels of resins that are compatible with fruit respiration. This is especially important for coated fruits and vegetables where the product's respiration can deplete internal O_2 concentrations (causing anaerobic conditions) and/or result in a buildup of CO_2 and ethylene levels (promoting CO_2 injury and senescence, respectively). Either condition can reduce the quality and shelf-life of fresh produce.

Stearyl alcohols have been reported to be more resistant to oxygen transmission than tristearin, beeswax, and acetylated monoglycerides by 39, 43, and 61%, respectively (Kester and Fennema, 1989b). In general, permeability to oxygen increased as the degree of unsaturation increased and as the chain length decreased. Long-chain alcohols, such as stearyl alcohol, and long-chain fatty acids, such as stearic acid, form hydrogen bonding during crystallization, resulting in higher melting points and crystals that are usually in an orthorhombic arrangement. Beeswax also seems to crystallize in an orthorhombic packing pattern, which is denser than the hexagonal arrangement and more effective as a gas and vapor barrier.

The influence of temperature on gas transfer across lipids generally obeys the Arrhenius equation. Lebovitz (1966) reported that the energy of activation (E_a) for most permeating processes across a film ranges between 6–16 Kcal/mole. Kester and Fennema (1989b) calculated an E_a of 7.5 Kcal/mole for transport of gases across tristearin and stearyl alcohol. The E_a for permeability of this composite film was similar to the E_a of cuticular plant tissue.

Permeability to Water Vapor

Waxes are normally more resistant to moisture transport than all the other components in edible film formulations (permeability data for different coating components are given in Chapter 7). The addition of fatty acids, besides acting as a plasticizer, increases the bonding of the wax to the polysaccharide layer in edible films. It has been reported that the tensile strength of films increased with fatty acid concentration (Hagenmaier and Shaw, 1990). In this case, film uniformity was reported to be more important for food protection than film thickness.

Usually, the permeability of lipid films to water vapor will increase with increasing polarity, unsaturation, and branching of the lipids. For films that are partially soluble in water, the diffusivity of vapor increases as moisture is absorbed in the polar portion of the film. This effect is accentuated at high relative humidities. Waxes such as beeswax are very good moisture barriers because of the tight orthorhombic arrangement of the crystals. Waxes are relatively more permeable to water than to oxygen because water is more soluble in the wax. Water permeability of waxes (e.g., beeswax) has been reported to be approximately 10 times smaller than for acetostearin films, 25 times smaller than for common oil films, and 100–200 times less than for protein films such as casein (Kester and Fennema, 1989c). Water vapor transfer rates (WVTR) depend not just on the humidity gradient, but also on the relative humidity level on one side of the film. At equal relative humidity gradients, WVTR will increase at higher relative humidities on one side of the film. This is due to the plasticizing effect of moisture in the hydrophilic portion of the film coating.

Edible films, in general, show an Arrhenius type of behavior at moderate temperatures, namely, the WVTR will increase with the temperature. However, at lower temperatures an increase in WVTR takes place as temperature decreases (Kamper and Fennema, 1984b). This was reported to be due to a shrinking and breaking of the film as the components become more rigid. This factor raises the need to add plasticizers to wax-in-water formulations. This may also be due to an increase in the hydration of the film at lower temperatures, thus facilitating moisture transfer through the film. Hydration of polysaccharides, such as methylcellulose, at $0°C$ is reported to be twice as high as at room temperature. Edible films usually do not perform well at water activity (a_w) higher than 0.9, except for waxes, which have been reported to stay intact at a_w levels higher than 0.95 (Kester and Fennema, 1989a).

In some cases, for coated fruits and vegetables, an increase in WVTR will be due to cracks and fissures in the film (Fox, 1958). Hence, it is important to avoid shock "cooling" as much as possible. This is usually accomplished by drying the fruit at a higher than ambient temperature to allow gradual drying and subsequent cooling of the film to occur.

APPLICATIONS OF LIPID- AND RESIN-BASED COATINGS

FRUITS AND VEGETABLES

Waxes are found naturally on fruit and vegetable surfaces and help prevent moisture loss during dry seasons. Cuticular wax in some fruits occurs as a two-layer structure of "soft" and "hard" waxes. The inner layer is composed of thin platelets while the outer layer is composed of tubular projections. Cuticular waxes consist of long-chain alcohols, aldehydes, esters, fatty acids, and hydrocarbons. This wax layer may be partially removed during washing treatments. In plums, the cuticular wax thickness was calculated to be 3 μm and the amount to be 300 μg/cm^2 (Bain and McBean, 1967). In comparison, Brusewitz and Singh (1985) calculated the thickness of natural wax coatings in oranges to be 1–2 μm.

Edible films have to be tailored to the respiration rates and anatomy of a given product (Claypool and King, 1941). For example, oranges, with their thick peel, are more susceptible to anaerobic conditions (resulting in senescence and decay) if the proportion of waxes and lipids in the film is too high (Cohen et al., 1990; Nisperos-Carriedo et al., 1990; Brusewitz and Singh, 1985). Apples and pears have thinner peels than do citrus fruits, and so are less likely to go anaerobic, making it possible to apply films with a higher wax content. Coatings containing lipid and resin components are more likely to enhance anaerobic conditions in some climacteric fruit that undergoes high rates of respiration, especially when such fruit is subjected to elevated temperatures. Fungicides, such as imazalil, can be incorporated into wax emulsions to prevent stem-end rot (Brown, 1984). Application of fungicide with wax coatings can also prevent the migration of the fungicide into the fruit.

Effectiveness of edible coatings, particularly in fruit and vegetables, depends on how well the fruit is washed before coating, the formulation of the wax, and the drying rate. Drying rates are fairly independent of air speed within a certain range (2–4 m/s). However, at air velocities above 4 m/s, significantly higher drying rates have been reported (Brusewitz and Singh, 1985).

Typical problems when coating fruit and vegetables are "dusting," "powdering," "excessive tackiness," or "rewetting." Dusting is the improper adherence of the wax on the fruit surface as a result of premature drying of wax droplets as they are sprayed in the conveyor belt before the droplets settle on the fruit surface. This can also occur due to improper coating formulation. Powdering is the breakdown of the coating on the fruit surface. This results from improper washing and drying of the fruit. When the wax is applied over a dirty surface, the wax subsequently peels off. Powdering also results from excessive shrinkage of the fruit due to water loss and surface shriveling, which causes some of the coating to peel off. Excessive tackiness makes material handling operations difficult, which can result in fruit build-up on packing lines and poor belt-and-roll sizing (Guilbert, 1986). Rewetting of the waxed fruit surface is caused by fruit sweating and breakdown during shipping, storage, and retail display. This can be due to mishandling of the fruit before or after waxing, to temperature changes, or to waxing too ripe or soft fruit that cannot withstand waxing and packaging without bruising.

Usual problems encountered when preparing edible films include phase separation of emulsion components, distortion of film during application or casting, and cracking or pinhole formation on the product surface. It is desirable that the film be resistant to impact and stable to temperature changes or atmosphere stresses such as high relative humidity and high a_w. It is important to optimize coating thickness to be compatible with fruit respiration (Brusewitz and Singh, 1985). Albrigo and Ismail (1983) found that waxed oranges average 0.6% weight loss per week, and waxing reduced weight loss by approximately 20% compared to unwaxed fruit. The film thickness in commercially coated oranges has been reported to be between 2–5 μm (Brusewitz and Singh, 1985). Excessive coating on the fruit can affect the normal maturation process, triggering anaerobic respiration and shortening shelf-life. In fruit such as oranges, this can cause an increase in ethanol, acetaldehyde, and off-flavors in the final product (Ahmad et al., 1979).

Mineral oil is commonly added to dry fruit to prevent aggregation, decrease water loss, improve appearance, and inhibit development of insect eggs or larvae under the fruit surface (Kochhar and Rossell, 1982). It is used commonly on limes, lemons, cucumbers, and tomatoes. Waxes also decrease water loss in citrus.

PROCESSED AND LIGHTLY PROCESSED FOOD PRODUCTS

Enrobing food with fat, a process called "larding," was practiced in England in the sixteenth century (Kamper and Fennema, 1984a). Currently, fats are still used to coat frozen foods such as poultry. A product called Myvacet (Eastman Chemical Products, Inc.) is a distilled acetylated monoglyceride that is utilized as a protective coating for this purpose. The coating helps maintain the original fresh quality of the product during prolonged storage due to its excellent oxygen and moisture barrier properties. Applied as a dip or spray prior to freezing, the coating may be left on the product during cooking. This product has also been used to coat shrimp, nuts, dried fruits, breaded products, meat patties, sausages, and vegetables (Kruger and Igoe, 1971). Similar coatings have been used to coat nuts for use in cakes and ice cream to prevent oxidation.

The Ambrosia Chocolate Company developed a water-in-oil emulsion as a flavor carrier and protective containment for meat. The benefits of this coating included increased yield (2–6% less cook-out loss), decreased moisture loss during storage (2–17% less for coated samples), flavor improvement and innovation, production efficiency due to high speed and continuous application, and improvement in meat tenderness (Kroger and Igoe, 1971). A water-in-oil emulsion for application to frozen meat products was patented (Bauer et al., 1968) which was claimed to form a continuous substantially water vapor impervious coating. Fats used in preparation of this emulsion could be any edible fat of animal or vegetable origin, but most included corn oil, cotton seed oil, safflower seed oil, soybean oil, or fat from chicken, beef, or pork. Sometimes waxes, such as beeswax, candelilla, or carnauba were substituted to coat frozen meat pieces. Coated meats were reported to undergo extended storage periods without substantial dehydration (Daniels, 1973).

A vegetable oil blend, Durkex 500, was used to coat dried fruits which traditionally were coated with a film of mineral oil. It was found that an oiling agent maintains the "free-running" properties of the fruit and prevents aggregation into large clumps as well as further drying out by loss of moisture, which results in crystallization of sugars on the surface of the fruit. These coatings also retarded aging and degeneration of the dried fruit and added sheen (Kochhar and Rossell, 1982). Surfactants, such as ethyl esters of fatty acids (C_{10}–C_{18}), for example ethyl oleate, have been used to coat grapes and plums to increase the drying rate in the manufacture of raisins, prunes, and other waxy fruits (Loncin et al., 1984; Ponting and McBean, 1970). The use of ethyl oleate to increase the drying rate of unripe corn has also been reported (Suarez et al., 1984).

Krochta et al. (1990) prepared an edible film with a casein-lipid emulsion for application on lightly processed foods such as fruit slices. This film is reported to prevent microbial and enzymatic degradation. The emulsion was prepared by adding monoglycerides as the lipid phase and glycerol as plasticizer. The addition of calcium ascorbate to this formulation is reported to prevent some oxidation (discoloration) on the surface of the cut fruit. Solid fats such as cocoa butter and palm oils are used to cover confectionery products.

Historically, waxes were also used on foods such as cheese as peelable protection to be removed before consumption. In small amounts this type of material is edible and is used on such things as raisins and confections (Guilbert, 1986). Beeswax and vegetable oil extended the shelf-life of raisins stored with cereal (Kester and Fennema, 1986; Watters and Brekke, 1960). Waxes such as beeswax are added to chewing gum, and carnauba wax is allowed for use on hard candy and chewing gum to provide surface gloss. Shelf-life of eggs stored at low temperature (35°C) was extended with the use of paraffin wax. Coating eggs with paraffin waxes prevented weight loss and reduced the air space in the egg (Alvin et al., 1979).

Vojdani and Torres (1990) developed composite films with methylcellulose, hydroxypropyl methylcellulose and fatty acids of different chain length to prevent migration of preservatives like potassium sorbate from the surfaces of foods such as cheeses. Kamper and Fennema (1984a) prepared bilayer films with hydroxymethyl cellulose or hydroxypropyl methylcellulose and a hydrophobic component using beeswax, paraffin, hydrogenated palm oil, stearic acid, or stearic-palmitic acid. Some of these films prevented the transfer of water from high- to low-moisture foods, such as tomato paste on crackers.

Shellac and other resins are also used to coat chocolate-panned items (candy) to add a polished, glossy appearance and to protect the chocolate from humidity and air during storage (Potter, 1986).

SUMMARY AND CONCLUSIONS

This use of lipids in coatings for food products has been practiced for centuries. In the past, waxes and oils were used alone, but now they are often blended with solvents, emulsifiers, surfactants, plasticizers, resins, etc. These lipid coatings are good barriers to water and gases. In the case of fresh fruit and vegetable products, care must be taken to allow adequate respiration to occur so that ripening may proceed normally where appropriate, and off-flavors do not accumulate in the fruit tissue. The lipid coatings should, however, retard moisture loss and subsequent shriveling of fresh produce. The same reasoning applies to resinous "wax" coatings (with little or no wax in the coating formulation) for fresh fruits. Such coatings also restrict water loss, but can cause anaerobic conditions to occur at high solids content, leading to off-flavors. This is especially true when fruits come out of cold storage for marketing display, or for fruit that is chilling sensitive and therefore must be stored at relatively high temperatures.

Lipid coatings for procesed food products serve a slightly different purpose. Coatings for meats, nuts, candies, dried fruits, and confectioneries should be as impervious to water vapor and oxygen as possible to prevent moisture loss and oxidation. For some processed products, lipid and resin coatings are used to carry or encapsulate flavor components and, as with fresh produce, provide an attractive sheen.

Recently, lipid and resin materials have been combined with other film-forming agents such as polysaccharides and proteins to create composite or bilayer coatings. The goal is to combine the desirable properties of different materials to improve gloss, nutritional value, permeability characteristics, strength, flexibility, and general performance of coating formulations. Better and more stable emulsions are now available for testing in coating formulations. Lipids, which are shorter in carbon chain length, are becoming available in forms that are resistant to oxidation and may soon be incorporated into lipid coatings.

Consumer trends are leaning toward more natural products, and petroleum-based waxes, such as polyethylene and paraffin, are becoming increasingly unpopular and restricted in use. Most likely, the naturally derived waxes, such as beeswax, carnauba, etc., will be considered more acceptable in the next few years. The same reasoning applies to oils, with mineral oil being replaced by vegetable-based oils sometime in the future.

Resins are also under scrutiny and, along with polyethylene wax, are banned in some countries overseas. Wood rosin and coumarone indene resin (also a petroleum-based product) are restricted in the United States, although shellac resin has recently been granted GRAS status (not yet reflected in the Federal Code of Regulations). Nevertheless, lipid- and resin-based edible coatings will continue to play an important role in the food industry by extending the shelf- and storage-life of many products. Such coatings help prevent food losses due to decay and spoilage.

REFERENCES

Ahmad, M., Z. M. Khalid and W. A. Farooqi. 1979. "Effect of Waxing and Lining Materials on Storage Life of Some Citrus," *Proc. Florida State Hort. Soc.*, 92:237.

Akamine, E. K., H. Kitagawa, H. Subramaniyam and P. G. Long. 1975. "Packinghouse Operations," in *Postharvest Physiology, Handling and Utilization of Tropical and Subtropical Fruits and Vegetables,* E. B. Pantastico, ed., Westport, CT: AVI Publ. Co., pp. 267–283.

Albrigo, L. G. and M. A. Ismail. 1983. "Potential and Problems of Film-Wrapping Citrus in Florida," *Proc. of Florida State Hort. Soc.*, 96:329.

Allied Signal, Inc. 1986. "A-C Polyethylenes for Fruit and Vegetable Coatings," Brochure No. CTG-2, pp. 1–6.

Alvin, A. S., M. Arshad and M. Afzal. 1979. "Preservation of Shell Eggs with Different Coating Agents," *Pakistan J. Sci. Ind. Res.*, 22(6):341.

Bain, J. M. and D. M. McBean. 1967. "The Structure of Cuticular Wax of Prune Plums and Its Influence as a Water Barrier," *Aust. J. Biol. Sci.*, 20:895.

Bauer, C. D., G. L. Neuser and H. A. Pinkalla. October 15, 1968. U.S. patent 3,406,081.

Bennett, H. 1975. *Industrial Waxes, Vol. 1.* New York, NY: Chemical Publ. Co.

Brown, G. E. 1984. "Efficacy of Citrus Postharvest Fungicides Applied in Water or Resin Solution Water Wax," *Plant Disease* (May):415.

Brusewitz, G. H. and R. P. Singh. 1985. "Natural and Applied Wax Coatings on Oranges," *J. Food Proc. and Preservation,* 9:1–9.

Claypool, L. L. 1939. "The Waxing of Deciduous Fruits," *Proc. Am. Soc. Hort. Sci.*, 37:443.

Claypool, L. L. and J. R. King. 1941. "Fruit Waxing in Relation to Character of Cover," *Proc. Amer. Soc. Hort. Sci.*, 38:261–265.

Code of Federal Regulations, Title 21. 1991.

Cohen, E., Y. Shalom and I. Rosenberger. 1990. "Postharvest Ethanol Buildup of Off-Flavor in 'Murcott' Tangerine Fruits," *J. Amer. Soc. Hort. Sci.*, 115:775–778.

Dalal, V. B., W. E. Eipeson and N. S. Singh. 1971. "Wax Emulsion for Fresh Fruits and Vegetables to Extend Their Storage Life," *Indian Food Packer,* 25(5):9.

Daniels, R. 1973. *Edible Coatings and Soluble Packaging,* Park Ridge, NJ: Noyes Data Corp., p. 360.

Drake, S. R., J. K. Fellman and J. W. Nelson. 1987. "Postharvest Use of Sucrose Polymers for Extending the Shelf-Life of Stored 'Golden Delicious' Apples," *J. Food Sci.*, 52:1283.

Durand, V. J., L. Orean, U. Yanko, G. Zauberman and N. Fuchs. 1984. "Effect of Waxing on Moisture Loss and Ripening of 'Fuerte' Avocado Fruit," *Hort. Sci.,* 19(3):421.

Eastman Chemical Inc. 1984. "A Guide to the Use of Epolene Waxes under United States FDA Food Additive Regulations," Publication No. F-243D, pp. 1–6.

Eastman Chemical Inc. 1990. "Emulsification of Epolene Waxes," Publication No. F-302, pp. 1–17.

Erickson, D. R., H. D. Pryde, O. L. Brekke, T. L. Mounts and R. P. Falb. 1980. "Handbook of Soy Oil Pressing and Utilization." Champaign, IL: AOCS Publishing House.

Fox, R. C. 1958. "The Relationship of Wax Crystal Structure to the Water Vapor Transmission Rate of Wax Films," *TAPPI,* 41(6):283.

Greener, I. K. and O. R. Fennema. 1989. "Evaluation of Edible, Bilayer Films for Use as Moisture Barriers for Food," *J. Food Sci.,* 54(6):1400.

Griffin, W. C. 1979. "Emulsions," *Kirk-Othmer Encyclopedia of Chemical Technology, 3rd Edition, Vol. 8,* pp. 913–916.

Guilbert, S. 1986. "Technology and Application of Edible Protective Films," in *Food Packaging and Preservation. Theory and Practice,* M. Mathlouthi, ed., London, England: Elsevier Appl. Sci. Publ. Co.

Hagenmaier, R. D. and P. E. Shaw. 1990. "Moisture Permeability of Edible Films Made with Fatty Acid and (Hydroxypropyl)methylcellulose," *J. Agric. and Food Chem.,* 38:1799.

Hagenmaier, R. D. and P. E. Shaw. 1991. "Permeability of Shellac Coatings to Gases and Water Vapor," *J. Agric. and Food Chem.,* 39(5):825.

Hall, D. J. 1981. "Innovations in Citrus Waxing—An Overview," *Proc. Florida State Hort. Soc.,* 94:258.

Hernandez, E. and R. A. Baker. 1991. "Candelilla Wax Emulsion, Preparation and Stability," *J. Food Sci.,* 56(5):1382, 1383, 1387.

Hitz, C. W. and I. C. Haut. 1942. "Effects of Waxing and Pre-Storage Treatments upon Prolonging the Edible and Storage Qualities of Apples," *Bulletin Agr. Exp. Stat. U . of Maryland,* No. A14, pp. 1–44.

Kamper, S. L. and O. R. Fennema. 1984a. "Water Vapor Permeability of Edible Bilayer Films," *J. Food Sci.,* 49:378.

Kamper, S. L. and O. R. Fennema. 1984b. "Water Vapor Permeability of an Edible Fatty Acid, Bilayer Film," *J. Food Sci.,* 49:1482.

Kamper, S. L. and O. R. Fennema. 1988. "Use of an Edible Film to Maintain Water Vapor Gradients in Foods," *J. Food Sci.,* 50:382–384.

Kanig, J. L. and H. Goodman. 1962. "Evaluative Procedures for Film-Forming Materials Used in Pharmaceutical Applications," *J. Pharm. Sci.,* 51(1):77.

Kaplan, H. J. 1986. "Washing, Waxing and Color Adding," in *Fresh Citrus Fruits,* W. F. Wardowdki, S. Nagy and W. Grierson, eds., Westport, CT: AVI Publishing Co., p. 379.

Kester, J. J. and O. R. Fennema. 1986. "Edible Films and Coatings: A Review," *Food Tech.,* 40(12):47.

Kester, J. J. and O. R. Fennema. 1989a. "An Edible Film of Lipids and Cellulose Ethers: Barrier Properties to Moisture Vapor Transmission and Structural Evaluation," *J. Food Sci.,* 54(6):1383–1389.

Kester, J. J. and O. R. Fennema. 1989b. "Resistance of Lipid Films to Water Vapor Transmission," *J. Am. Oil Chem. Soc.,* 66(8):1139.

Kester, J. J. and O. R. Fennema. 1989c. "The Influence of Polymorphic Form on Oxygen and Water Vapor Transmission through Lipid Films," *J. Am. Oil Chem. Soc.*, 66(8):1147.

Kochhar, S. P. and J. B. Rossell. 1982. "A Vegetable Oiling Agent for Dried Fruits," *J. Food Tech.*, 17:661.

Krochta, J. M., A. E. Pavlath and N. Goodman. 1990. "Edible Films from Casein-Lipid Emulsions for Lightly Processed Fruits and Vegetables," in *Engineering and Food, Volume 2*, W. E. L. Spiess and H. Schubert, eds., New York, NY: Elsevier Applied Science.

Kroger, M. and R. S. Igoe. 1971. "Edible Containers," *Journal Series of the Penn. Agric. Exp. Sta.*, #3954.

Kumins, M. 1965. "Transport through Polymer Films," *J. Polymer Sci.*, Part C, 10:1–9.

Lakshminarayana, S., L. Sarmiento and J. I. Ortiz. 1974. "Extension of Storage Life of Citrus Fruits by Application of Candelilla Wax Emulsion and Comparison of Its Efficiency with Tag and Flavor Seal," *Proc. Florida State Hort. Soc.*, 87:325.

Lebovitz, A. 1966. "Permeability of Polymers to Gases and Liquids," *Mod. Plastics*, 43(7):139–142.

Loncin, M., T. Roth and C. Bornhardt. 1984. "Influence of Surface Active Agents on the Drying Rate of Solids," *Proceedings 4th Int. Drying Symp.*, 1:107–111.

Long, F. A. and L. J. Thompson. 1955. "Diffusion of Water Vapor in Polymers," *J. Polymer Sci.*, 15:413–426.

Martin, J. 1982. "Shellac," in *Kirk-Othmer Encyclopedia of Chemical Technology.* New York, NY: Wiley Interscience, pp. 737–747.

Martin-Polo, M., C. Mauguin and A. Voilley. 1992a. "Hydrophobic Films and Their Efficiency against Moisture Transfer. 1. Influence of the Film Preparation Technique," *J. Agric. Food Chem.*, 40:407–412.

Martin-Polo, M., A. Voilley, B. Colas, M. Mesnier and N. Floquet. 1992b. "Hydrophobic Films and Their Efficiency against Moisture Transfer. 2. Influence of the Physical State," *J. Agric. Food Chem.*, 40:413–418.

Neville Chemical Co. 1988. "Production Information, Resins, Plasticizers, Nonstaining Antioxidants and Chlorinated Paraffins," Brochure No. NCCFAL88, pp. 1–15.

Nisperos-Carriedo, M. O., P. E. Shaw and E. A. Baldwin. 1990. "Changes in Volatile Flavor Components of Pineapple Orange Juice as Influenced by the Application of Lipid and Composite Film," *J. Agric. and Food Chem.*, 38:1382–1387.

Paredes-Lopez, O., E. Camargo-Rubio and Y. Gallardo-Navarro. 1974. "Use of Coatings of Candelilla Wax for the Preservation of Limes," *J. Sci. Food and Agric.*, 25:1207.

Ponting, J. D. and D. M. McBean. 1970. "Temperature and Dipping Treatment Effects on Drying Rates and Drying Times of Grapes, Prunes and Other Waxy Fruits," *Food Tech.*, 24(12):85–88.

Potter, N. N. 1986. "Confectionary and Chocolate Products," in *Food Science, Fourth Edition.* Westport, CT: AVI Publishing Co., Inc., pp. 564–579.

Prince, L. M. 1977. "Formulation" in *Microemulsions. Theory and Practice*, L. M. Prince, ed., New York, NY: Academic Press.

Shinoda, K. and H. Kunieda. 1984. "Phase Properties of Emulsions," in *Encyclopedia of Emulsion Technology, Vol. 1*, P. Becher, ed., New York, NY: Marcel Dekker, Inc.

Suarez, C., M. Loncin and J. Chirife. 1984. "A Preliminary Study on the Effect of

Ethyl Oleate Dipping Treatment on Drying Rate of Corn," *J. Food Sci.,* 49:236–238.

Sward, G. G. 1972. "Natural Resins," *Am. Soc. Test. Mat.,* pp. 77–91.

Tulloch, A. P. 1970. "The Composition of Beeswax and Other Waxes Secreted by Insects," *Lipids,* 5(2):247–258.

Ukai, N., S. Ishibashi, T. Tsutsumi and K. Marakami. February 25, 1975. U.S. patent 3,997,674.

Vojdani, F. and J. A. Torres. 1990. "Potassium Sorbate Permeability of Methylcellulose and Hydroxypropyl Methylcellulose Coatings: Effect of Fatty Acids," *J. Food Sci.,* 55(3):841.

Watters, G. G. and J. E. Brekke. 1960. "Stabilized Raisins for Dry Cereal Products," *Food Technol.,* 14:236–238.

Wineman, R. D. 1984. "Water-Emulsion Fruit and Vegetable Coatings Based on Epolene Waxes," Publication No. F-257A, Eastman Chemical Products, Inc., pp. 1–4.

Wong, W. W., F. A. Gastineau, K. S. Gregorski, A. J. Tillin and A. E. Pavlath. 1992. "Chitosan-Lipid Films: Microstructure and Surface Energy," *J. Agric. Food Chem.,* 40:540–544.

CHAPTER 11

Edible Coatings and Films Based on Polysaccharides

MYRNA O. NISPEROS-CARRIEDO[1]

INTRODUCTION

THE application of edible films and coatings for the extension of shelf-life of fresh and processed food products is not new. Food coating processes involving wax and gelatin were developed as early as the 1800s (Guilbert, 1986). Works in the 1930s employed waxes and other oil-based materials as protective coatings to prevent desiccation and provide attractive gloss to fresh citrus fruits (Kaplan, 1986).

The development of coatings from water-soluble polysaccharides has brought a surge of new types of coatings for extending the shelf-life of fruits and vegetables because of the selective permeabilities of these polymers to O_2 and CO_2. Polysaccharide-based coatings can be utilized to modify the atmosphere, thereby reducing fruit and vegetable respiration (Nisperos-Carriedo and Baldwin, 1990; Motlagh and Quantick, 1988; Drake et al., 1987; Banks, 1984).

Water-soluble polysaccharides are long-chain polymers that dissolve or disperse in water to give a thickening or viscosity-building effect (Glicksman, 1982). These compounds serve numerous diverse roles such as providing hardness, crispness, compactness, thickening quality, viscosity, adhesiveness, gel-forming ability, and mouthfeel (Whistler and Daniel, 1990). They are of significant importance to the food industry because they are widely available, usually are of low cost, and are nontoxic (Whistler, 1991).

Polysaccharides are obtained from a variety of sources. Table 11.1 lists

[1]Food Research Institute, Dept. of Agriculture, Werribee, Victoria, Australia.

305

TABLE 11.1. Classification of Polysaccharides.

Type/Examples	21 CFR No.	Food Use
Cellulose and Derivatives:		
Carboxymethylcellulose	182.1745	multiple purpose GRAS substance
Methylcellulose	182.1480	multiple purpose GRAS substance
Hydroxypropyl cellulose	172.870	emulsifier, film-former, protective colloid, stabilizer, suspending agent, thickener
Hydroxypropyl methylcellulose	172.874	emulsifier, film-former, protective colloid, stabilizer, suspending agent, thickener
Microcrystalline cellulose	182.70	GRAS substance
Starches and Derivatives:		
Raw	182.70, 182.90	GRAS substance
Modified starch	172.892, 182.70	component of batter, multipurpose substance
Pregelatinized	172.892	multipurpose additive
Dextrin	184.1277	formulation aid, processing aid, stabilizer, thickener, surface-finishing agent
Maltodextrin	184.1444	GRAS substance
Pectins	184.1588	emulsifier, stabilizer, thickener
Seaweed Extracts:		
Agar	184.1115	drying agent, flavoring agent, stabilizer, thickener, surface-finisher, formulation aid, humectant
Alginates	184.1133, 184.1187, 184.1610, 184.1724	stabilizer, thickener, humectant, texturizer, formulation aid, firming agent, flavor adjuvant, flavor enhancer, processing aid, surface active agent

TABLE 11.1. (continued).

Type/Examples	21 CFR No.	Food Use
Carrageenans	172.620	emulsifier, stabilizer, thickener
Furcellaran	172.655	emulsifier, stabilizer, thickener
Exudate Gums:		
Gum arabic	184.1330	emulsifier, formulation aid, stabilizer, thickener, humectant, texturizer, surface-finishing agent, flavoring agent, adjuvant
Gum ghatti	184.1333	emulsifier
Gum karaya	184.1349	formulation aid, stabilizer, thickener
Gum tragacanth	184.1351	emulsifier, thickener, stabilizer, formulation aid
Seed Gums:		
Guar gum	184.1339	emulsifier, formulation aid, stabilizer, thickener, firming agent
Locust bean gum	184.1343	stabilizer, thickener
Microbial Fermentation Gums:		
Xanthan	172.695	stabilizer, emulsifier, thickener, suspending agent, bodying agent, foam enhancer
Gellan gum	172.665	stabilizer, thickener
Chitosan	not approved	

various polysaccharides and their functions in foods. The U.S. Food and Drug Administration classifies these compounds as either food additives or "generally regarded as safe (GRAS) substances" (FDA, 1991). Regulations of the specific use of these polysaccharides are outlined in Title 21 of the Code of Federal Regulations (CFR) with the corresponding part numbers listed in Table 11.1.

Because of the diverse capabilities of the water-soluble polysaccharides, often referred to as gums or hydrocolloids (Glicksman, 1982), this group of compounds has become a big business. A market survey indicated that the consumption of nonstarch hydrocolloids by five countries (U.S., France, Italy, West Germany, and the U.K.) in 1986 was 111,900 metric tons valued at $654 million (Anonymous, 1988). The projected consumption of nonstarch hydrocolloids for 1991 is 126,000 metric tons valued at $748.7 million (Dziezak, 1991).

Because of the immense scope of polysaccharide chemistry, technology and application, the reader is referred to the books written by Aspinall (1970), Glicksman (1969, 1982, 1983, 1984), Mitchell and Ledward (1986), Philips et al. (1984, 1986, 1988, 1990), Whistler and BeMiller (1973, 1992), Whistler et al. (1984), and Wurzburg (1986). This chapter focuses mainly on polysaccharides that are presently being used as edible coatings in the food industry and those that are currently being developed by private and government research agencies.

POLYSACCHARIDES USED FOR EDIBLE COATINGS

CELLULOSE AND DERIVATIVES

Cellulose is present in all land plants and is the structural material of plant cell walls. Cellulose is a polysaccharide composed of linear chains of $(1->4)$-β-D-glucopyranosyl units (see Figure 11.1). Native cellulose is insoluble in water due to the high level of intramolecular hydrogen bonding in the celulose polymer. A high degree of crystallinity is also present which reduces solubility. The crystallization energy in the unit cell is stronger than the solvation energy in water, therefore, cellulose is soluble only in solvents that can form coordination complexes such as aqueous solutions of ethylenediamine copper hydroxide. By placing along the chain substituents that interfere with the formation of the crystalline unit cell, it is possible to solvate the polymer. This is done by etherification, i.e., reacting cellulose with aqueous caustic, then with methyl chloride, propylene oxide or sodium monochloroacetate to yield methylcellulose (MC) [Figure 11.1(a)], hydroxypropyl methylcellulose (HPMC) [Figure 11.1(b)], hydryoxypropylcellulose (HPC) [Figure 11.1(c)], and sodium carboxy-

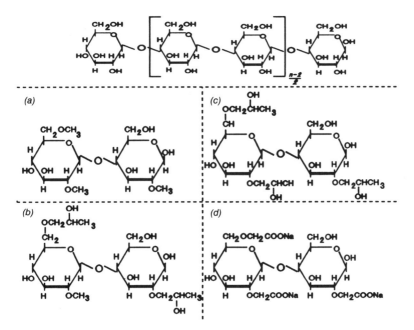

Figure 11.1 Structure of cellulose, and (a) methylcellulose, (b) hydroxypropyl methylcellulose, (c) hydroxypropylcellulose, and (d) Na⁺ salt of carboxymethylcellulose.

methylcellulose (CMC) [Figure 11.1(d)]. Changing the level of methoxyl, hydroxypropyl and carboxymethyl substitution affects a number of physical and chemical properties such as water retention properties, sensitivity to electrolytes and other solutes, dissolution temperatures, gelation properties, and solubility in nonaqueous systems. Another cellulose derivative that is of significant importance is microcrystalline cellulose (MCC), formed by controlled acid hydrolysis of native cellulose.

Nonionic Cellulose Ethers

MC, HPMC, and HPC, are nonionic, water-soluble ethers with good film-forming properties. The starting material for the manufacture of these compounds is highly purified and bleached cellulose. Each repeating unit of cellulose, the cellobiose, consists of two β-D-glucopyranosyl units which in turn have three hydroxyl groups each at positions 2, 3 and 6 that are available for etherification. Under controlled temperatures and pressures, alkali cellulose is allowed to react with methyl chloride to form MC [Figure 11.1(a)], methyl chloride and propylene oxide to form HPMC [Figure 11.1(b)], or propylene oxide to form HPC [Figure 11.1(c)]. The number of substituted hydroxyl groups per monomeric unit, expressed as degree of substitution (DS), can vary from 0 and 3. The range of DS permitted under the CFR for MC is 1.64 to 1.92 which corresponds to methoxyl substitution of 27.5% to 31.5%. The CFR for HPMC allows for a DS range of 1.11 to 2.03 for methoxyl and an MS (molar substitution or the average number of molecules of substituent which has been added per monomeric unit) range of 0.073 to 0.336. This corresponds to the range of 3.0 to 12.0% for hydroxypropyl and 19.0 to 30.0% for methoxyl substitution. For HPC, CFR allows for hydroxpropyl groups to be less than 4.6 per monomeric unit. MC, HPMC, and HPC are available commercially in powder or granular form in varying molecular weight and DS. They are soluble in cold but not hot water. Partial solubilization of MC in organic solvents is noted at a DS of 2.1, corresponding to 34% methoxyl content. Complete solubilization in organic solvents occurs at a DS of 2.6 with 40% or higher methoxyl substitution. The presence of hydroxypropyl groups in HPMC broadens the range of both water and organic solvent solubility. HPC is soluble in cold water, insoluble in hot water, and readily soluble in many hot or cold polar organic solvents.

Dissolution of nonionic cellulose ethers may be considered a two-step process: dispersion and hydration. Incomplete dissolution of particles may lead to the formation of little balls or agglomerates with a swollen outer skin of partially hydrated gum called "fish eyes." Four procedures are recommended for the preparation of lump-free solutions of these compounds. The first is by adding the powder to the vortex of well-agitated water at room temperature. The rate of addition must be slow enough to permit particles to separate in water and their surfaces to become individually wetted. Agitation is continued until all particles are dissolved and the solution is completely free of gels. Wherever possible, the MC, HPMC, or HPC should be put into solution before other soluble ingredients are added. Other dissolved materials compete for the solvent and slow the solution rate of these compounds. Second is by dry blending. Dry blending the powder with any inert or nonpolymeric soluble material that is used in the formulation is also effective. Blending separates the powder and reduces the tendency to lump. The third procedure is by dispersing the cellulose ethers in a water miscible nonsolvent such as glycerin, ethanol, or propylene glycol and then adding the slurry to water. Dispersing the powder in an edible oil is also possible. The fourth method involves the use of a stainless steel mixing device, developed by Hercules, where the powder is fed through a smooth wall funnel into a water jet ejector and is dispersed by the turbulence of water flowing at high velocity (Keller, 1984).

MC, HPMC, and HPC solutions are stable at pH 2–11. They are compatible with surfactants, other water-soluble polysaccharides, and with salts, depending on the type and concentration of the salt. Concentrations of 2% salts such as aluminum sulfate, calcium chloride, sodium acetate, sodium chloride, and sodium carbonate are generally compatible with the nonionic cellulose ether solutions. If relatively high concentrations of dissolved salts are used, these cellulose derivatives tend to "salt out" from solutions, forming finely divided and highly swollen precipitates. Solutions of these nonionic cellulose ethers can be preserved using benzoic acid and its sodium salt; sorbic acid and its potassium, sodium, and calcium salts; sodium propionate; and methyl and propyl parahydroxybenzoate.

These compounds are good film-formers. They are capable of yielding tough and flexible transparent films owing to the linear structure of the polymer backbone (Krumel and Lindsay, 1976). The films are soluble in water and resistant to fats and oils. MC, being the least hydrophilic of the cellulose ethers, would produce films that have relatively high water vapor permeability. MC films are tough and flexible, while the internal plasticizing effects of the high level of hydroxypropyl substitution in HPC results in a film of lower tensile strength and greater elongation. HPC also shows the unique property of being a water-soluble thermoplastic that is capable of injection molding and extrusion. Insolubility of MC and HPMC films in water can be obtained by cross-linking with melamine formaldehyde resins or with other multi-functional resins. With HPC, insolubility can be attained by incorporating zein, shellac, or ethyl cellulose using common solvents. Plasticity can be improved by adding polyglycols, glycerin or propylene glycol.

The film-forming characteristics of MC and HPMC upon heating provide an effective film that can reduce oil absorption in certain reformed products, notably potato, during frying (Sanderson, 1981). They are particularly suited for use in dry foods in which they create a barrier to oil absorption by the product, retard loss of natural product moisture, and improve the adhesion of batter to the product (Dziezak, 1991). Bauer et al. (1969) described a process that prevents the loss of coating composition during cooking and increases the absorption of the coating on pork and poultry pieces by MC. The high-quality film-forming characteristics of HPC have been applied to retard moisture absorption and the development of oxidative rancidity in nutmeats (Ganz, 1969), and to retard spoilage and moisture absorption in coated nuts and candies (Krumel and Lindsay, 1976). MC films, with thickness ranging from 0.2 to 1.0 mil (0.0051 to 0.025 mm), showed excellent barrier properties to lipid migration making them suitable in confectionery products provided their levels do not cause objectionable flavor (Nelson and Fennema, 1991). Bilayer films, composed of HPMC and solid lipids such as beeswax, paraffin, hydrogenated palm oil, or stearic acid yielded water vapor permeabilities that were lower than that of low density polyethylene (Kamper and Fennema, 1985). A bilayer

film consisting of stearic-palmitic acid and HPMC slowed moisture transfer from the high-moisture food (tomato paste) to the low-moisture food
(crackers) (Kamper and Fennema, 1985). Formulations involving MC,
HPMC, and HPC (later named Nature-Seal) resulted in a delay of ripening
and browning, an increase in volatile flavor components, and a higher larvae kill for quarantine treatments for some fresh commodities detailed in
Chapter 2 (Nisperos-Carriedo et al., 1992; Hallman et al., 1991; Nisperos-
Carriedo and Baldwin, 1990).

Anionic Cellulose Ether

CMC or cellulose gum, an anionic cellulose ether, is produced by reacting alkali cellulose with sodium monochloroacetate under rigidly controlled conditions. A DS of 0.4 is the minimum substitution needed to
place the gum in solution. A DS of 0.7 is the most common type utilized
in food systems. The CFR allows for a maximum substitution of 0.95
carboxymethyl groups per monomeric unit for food grade CMC. With increasing DS, increased solubility, compatibility with other ingredients,
and acid stability of CMC are noted (Keller, 1984). CMC is available in a
variety of types based on particle size, DS, viscosity, and hydration characteristics for different food applications. It is soluble in either hot or cold
water, but insoluble in organic solvents. However, the gum dissolves in
suitable mixtures of water and water-miscible solvents such as ethanol or
acetone.

The procedure for the preparation of lump-free, clear cellulose gum
solutions follows that of the nonionic cellulose ethers, except for pH conditions. CMC solutions are stable at pH 7–9. Above pH 10, a slight decrease
in viscosity can occur. Below pH 4, the less soluble free acid CMC pre-
dominates and viscosity may increase significantly. CMC is compatible
with a wide range of other food ingredients including protein, sugar,
starches and other hydrocolloids. CMC is protein reactive, forming a complex with casein that causes a large increase in viscosity (Keller, 1984).
The effect of salt varies with the type and concentration of a particular salt,
DS, and the manner in which the salt and CMC come in contact (Aqualon,
1988). In general, monovalent cations form soluble salts of CMC, divalent
cations are borderline, and trivalent cations form insoluble salts. Highly
substituted cellulose gums (DS, 0.09–1.2) have greater tolerance for most
salts. Easier dissolution can also be obtained by dissolving the cellulose
gum before adding the salt. Solutions of CMC are preserved by the addition of sodium benzoate, sodium propionate, sorbic acid and its sodium
and potassium salts.

CMC is by far the most important cellulose derivative for food applications (Sanderson, 1981). Its basic function is to bind water or impart viscosity to the aqueous phase thereby stabilizing the other ingredients or

preventing syneresis. It is seldom used to prepare free or unsupported films. However, its ability to form strong, oil-resistant films is of great importance in many applications. Plasticizers such as ethanolamine, glycerol, polyglycols, or propylene glycol have been reported to be effective with cellulose gum. Water-insoluble films and coatings can be developed by cross-linking the gum with polyfunctional resins such as melamine-formaldehyde, urea-formaldehyde and polyamide-epichlorohydrin.

CMC, when used in a dry coating process for freshly cut celery prior to canning, retained the original texture, firmness, and crispness of the product (Mason, 1969). An air-permeable coating consisting of CMC and calcium sorbate can suppress mold formation on hard cheese and hard sausage (Luck, 1968). Coating fresh fruits and vegetables with a semipermeable film composed of CMC and sucrose fatty acid esters was proposed by Lowings and Cutts (1982). These coatings reduce oxygen uptake without causing an equivalent increase in carbon dioxide level in internal atmospheres of fruit or vegetable tissues. This is advantageous in that it prevents the occurrence of anaerobic respiration. Two commercially available coatings based on CMC and sucrose esters are Semperfresh (United Agriproducts, Greeley, Colorado) and TAL Pro-long (Courtaulds Group, London). These coatings extended the shelf-life and preserved important flavor components of some fresh commodities (Nisperos-Carriedo et al., 1990; Santerre et al., 1989; Curtis, 1988; Meheriuk and Lau, 1988; Nisperos-Carriedo and Baldwin, 1988; Drake et al., 1987; Banks, 1985, 1984; Smith and Stow, 1984).

Microcrystalline Cellulose

MCC is prepared by controlled acid hydrolysis of native cellulose. Its main use has been in a number of low-calorie foods where it acts as a bulking agent and provides stabilization as a result of its ability to form a thixotropic gel-like structure. It is used to gel a variety of sugar-based products, icings, and pectin-based bakery fillings; to replace 20–25% of the amount of starch thickener; to replace oil in emulsions; to control ice crystal growth; and for suspension of particulates such as chocolate in sterilized chocolate drinks (Dziezak, 1991). Avicel, a mixture of MCC and CMC, has been reported to increase film strength, although the mixture does not have significant film-forming property (FMC, 1985).

STARCHES AND DERIVATIVES

Raw Starches

Starch, the reserve polysaccharide of most plants, occurs widely in nature, and is the most commonly used food hydrocolloid (Whistler and

Paschall, 1967, 1965). This is partly because of the wide range of functional properties it can provide in its natural and modified forms, and partly because of its low cost relative to alternatives. Starches can be derived from tubers (potato, tapioca, arrowroot, and sweet potato), stem (sago), and cereals (corn, waxy maize, wheat, and rice). Potato starch is an important food starch in Europe, while in North America, corn or maize starch is the most widely used (Sanderson, 1981). In the native state, starch exists as insoluble granules which have characteristic shape and some crystallinity. Starch granules are insoluble in cold water due to the hydrogen bonding of polymer chains. On heating, the granules gradually swell and absorb water as the hydrogen bonds are broken.

The two distinct polymers contained in starch granules are amylose [Figure 11.2(a)] and amylopectin [Figure 11.2(b)]. Amylose is the linear polymer which is composed of (1->4)-α-D-glucopyranosyl monomers ranging from 200–3000, depending on the source. Amylopectin contains the amylose backbone with side units of D-glucopyranosyl linked by α-1,6-glycosidic bonds. It is a larger molecule than amylose and is highly branched. Starches contain 18–30% amylose, except for waxy corn types which are virtually all amylopectin. Each raw starch has its own characteristic viscosity/temperature profile as a result of its particular granular composition and structure. For applications where viscosity, stability, and thickening power are desired, amylopectin is used. For film-forming and for the preparation of strong gels, amylose is required.

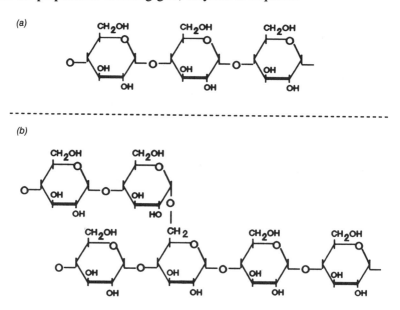

Figure 11.2 Structure of (a) amylose, and (b) amylopectin.

In unmodified forms, starches have very limited use in the food industry. Raw starches do not have the functional properties needed by modern processors (Smith, 1984). For example, unmodified waxy corn starches hydrate with ease, swell rapidly, rupture, lose viscosity, and produce weak-bodied, very stringy, and very cohesive pastes. These characteristics have limited use in foods today. Coating of a puffed cereal matrix with a starch-fat slurry produces a product similar to a nut in taste, texture, and appearance, according to Maloney et al. (1970). Starch also serves as an excellent outer coating for dry instant rice pudding because it hardens upon drying to produce a durable nontacky surface, rendering a free-flowing product (D'Ercole et al., 1969).

Films developed from amylose are described as isotropic, odorless, tasteless, colorless, nontoxic, and biologically absorbable. They exhibit physical characteristics, chemical resistance, and mechanical properties similar to those of plastic films (Wolff et al., 1951). Amylose films, obtained by coagulation with caustic solution or salt and acid, are strong, resilient, and stable (Kunz, 1962; Hiemstra and Muetgeert, 1959). The potential use of amylose for preparation of films depended greatly on the development of methods for fractionating starch, and the discovery of plant varieties with high amylose starch content. The discovery by Vineyard and Bear (1952) of the ae (amylose extender) gene in maize was followed by breeding and crop production of high amylose corn. Commercial development has resulted in the availability of two basic starches, of 55 and 70% amylose content, respectively. The 70% high amylose starch was found to give stronger, tougher, and more flexible films. The problem, however, is that this starch is difficult to disperse in water, and retrogrades or gels very rapidly once paste cooling is initiated (Jokay et al., 1967).

Processes have been developed for the production of coatings and films from amylose (Rankin et al., 1958) and amylomaize starch (Mark et al., 1966). These films have low permeability to oxygen at 5 and 25°C and at relative humidities up to 100%. The addition of plasticizers and absorption of water molecules by hydrophilic polymers can increase polymer chain mobility and generally lead to increased gas permeability (Banker, 1966). Amylose coating, when applied to potatoes and specialty potato products, was found to improve appearance, texture, taste, and stability (Murray et al., 1971). Dried raisins coated with an aqueous solution of corn amylose plasticizer poured freely, and did not clump or stick together (Moore and Robinson, 1968).

Modified Starches

Modification of native starch by disruption of hydrogen bonding through reduction of molecular weight and/or chemical substitution leads to lower gelatinization temperatures and reduced tendency for retrogradation. Modification by acid treatment shortens the chain length of the amylose fraction and converts some of the branched amylopectin into linear amylose units. The end result is to significantly lower the hot paste viscosity, which allows high-solid pastes to be prepared. On cooling, however, these sols still produce opaque rigid gels because the amylose units, even though they are degraded, can still effectively associate and retrograde. Cross-linking of starch polymers stabilizes the granules and ensures that they do not overswell or rupture, especially in cases where food processing demands stability to high shear and low pH. Chemical cross-linking reinforces the hydrogen bonds already present in the granule. When the hydrogen bonds are broken during heating or shearing, the chemical cross-links remain intact. The net result is that the granules of cross-linked starches are much more robust and retain their swollen integrity under conditions of excessive heat, shear, and reduced pH.

It is possible to react the hydroxyl groups of starch to produce esters such as acetates of phosphates, and ethers such as the hydroxypropyl ether derivative. The introduction of substituents on the hydroxyl groups weakens the hydrogen bonding in the starch, making the gelatinization temperature lower, and improving the freeze-thaw stability as well as the clarity by interfering with the ability of the linear amylose polymer to associate. The ether linkage tends to be more stable than the ester linkage, which can be important where extreme pH is encountered. Pre-gelatinized starches are prepared by instantaneously cooking/drying starch suspensions on steam-heated rollers, puffing/extruding, and then spray drying the product. The dry product swells or dissolves in cold water and, therefore, can be used by food processors who have no heating facilities, or in the formulation of instant products such as desserts.

Dextrinization refers to the reduction in size of the starch molecule to smaller fractions or dextrins. To accomplish this, dry starch, while being agitated, is sprayed with an acid such as hydrochloric or sulfuric, and subjected to controlled heating. The degree of hydrolysis is dependent upon the time, temperature, and pH conditions of the treatment. The relatively low viscosities of dextrins allow for their use at high concentrations in food processing.

Water-soluble, transparent films have been produced from hydroxypropylated (1.1%) amylomaize starch having an apparent amylose content of 71% (Roth and Mehltretter, 1967). This film exhibited very low permeability to oxygen. Hydroxypropylation reduced the dry tensile strength of amylomaize starch film, but increased bursting strength and elongation considerably. This hydroxypropyl derivative may be compounded with other ingredients in order to improve coating pliability, speed setting upon cooling and/or drying, and to control rate or resolubility (Jokay et al., 1967). Coating formulations from this type of starch have been developed for candies, prunes, raisins, dates, figs, nuts, and beans to retard the development of oxidative rancidity during storage. A laminate of an amylose ester of a fatty acid having from twelve to twenty-six carbon atoms (wherein the amylose has at least 70% of the available hydroxyl group esterified) and a fusible protein was developed to prevent oxidative degradation, mold attack, and moisture penetration in dehydrated food products (Cole, 1969). Use of an aqueous slurry of an amylosic starch ether in gelatinized form as a protective coating for foods is described by Mitan and Jokay (1969). The National Starch and Chemical Co. offers several starch formulations for use as films and coatings for baked and microwaved products (Batter Bind S, Crisp Film, Micro-Crisp), panned items (CrystalGum, K-4484), french fries (Crisp Film, Textaid A), confection (Film-Set, Crystal Gum, K-4484, Purity Gum 59, Purity Gum 40), batter mix (Batter Bind S), and fried products (Crisp Film).

Dextrins are often used as film-formers and edible adhesives to replace natural gums in products such as pan coated nuts and candy (Smith, 1984). Dextrins are also used extensively as fillers, encapsulating agents, and carriers for flavors. Their lower viscosity permits use of higher concentrations, and this requires less moisture to evaporate during filming, coating, or spraying operations. Coatings from dextrins can provide better resistance to transport of water vapor than starch films (Allen et al., 1963). A coating system composed of dextrins and other gums developed for doughnuts has a high degree of moisture resistance and stability (Thompson, 1972). Carboxylated dextrins prepared from cereal starches such as corn, wheat, waxy maize, and waxy sorghum provide efficient encapsulating agents (Evans and Herbst, 1964).

Murray and Luft (1973) used starch hydrolysates as coatings and reported that, in comparison with uncoated samples, coated almonds resulted in a more desirable texture; coated apple slices showed better texture, color, and flavor; and dried coated apricots had better flavor and higher yield. This was attributed to a reduction in the rate of moisture absorption in the case of coated almonds. Dextrin-based coatings may exhibit some resistance to oxygen transmission as shown by the reduction in the degree of browning in sliced apples (Murray and Luft, 1973). A gluten-dextrin coating was used to coat dry roasted peanuts prior to application of salt (Noznick and Bundus, 1967).

Pre-gelatinized starch was used as the essential ingredient in a coating powder for application on caramel cubes, semi-dried prunes, dates, figs, dried peaches, and an assortment of cubed fruits used in fruit cake (Scheick et al., 1970).

PECTINS

Pectins are a complex group of structural polysaccharides which occur widely in land plants (Aspinall, 1970). The major commercial sources of pectins are citrus peel and apple pomace. Pectic substances are polymers composed mainly of (1->4)-α-D-galactopyranosyluronic acid units. One aspect of difference among the pectic substances is their content of methyl esters or degree of esterification (DE), which has a vital effect on their solubility and gelation properties. The highest DE that can be achieved by extraction of natural raw material is approximately 75%. Pectins with DE from 20–70% are produced by controlled de-esterification in the manufacturing process.

The DE of 50% divides commercial pectins into high-methoxyl (HM) [Figure 11.3(a)] and low-methoxyl (LM) [Figure 11.3(b)] pectin. High-methoxyl pectins require a minimum amount of soluble solids (55–80%) and a pH within a fairly narrow range (pH 2.8–3.7) in order to form gels. Low-methoxyl pectins require the presence of a controlled amount of calcium and do not require sugar and/or acid. The degree of esterification of HM pectins controls their relative speed of gelation, as reflected by the designations of "slow set" (DE ≈ 60%) and "rapid set" (DE ≈ 75%) high-ester pectin.

The use of low-methoxyl pectinate as a coating agent for certain foods has been proposed because it is edible, and it gives an attractive, non-sticky surface to covered foods (Schultz et al., 1949). Macklay and Owens (1948), Miers et al. (1953), and Swenson et al. (1953) employed a two-step technique where an aqueous solution of low-methoxyl pectin was applied first, followed by treatment with a calcium solution to promote gelation. These coatings have high water vapor permeabilities and the only way they could prevent dehydration is by acting as sacrificial agents. Coating the

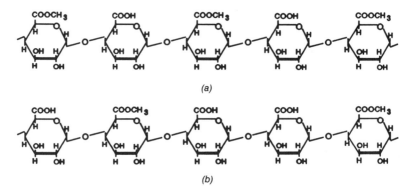

Figure 11.3 Structure of (a) high-methoxyl, and (b) low-methoxyl pectin.

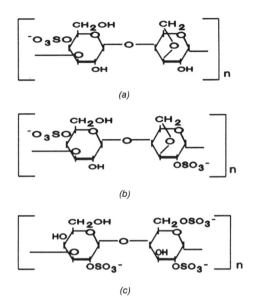

(a)

(b)

(c)

Figure 11.4 Structure of (a) $\varkappa$-carrageenan, (b) ι-carrageenan, and (c) λ-carrageenan.

original film with lipids may increase the resistance to water vapor transmission rates (Schultz et al., 1949).

SEAWEED EXTRACTS

Carrageenans

Carrageenan is extracted from several species of red seaweeds, mainly *Chondrus crispus,* with water and a small amount of alkali, then filtered and recovered by alcohol precipitation. Carrageenan is a complex mixture of several polysaccharides. The three principal carrageenan fractions, kappa ($\varkappa$), iota (ι), and lambda (λ), differ in sulfate ester and 3,6-anhydro-α-D-galactopyranosyl content (Morris et al., 1980; Glicksman, 1982) [Figures 11.4(a), (b), and (c)]. $\varkappa$-Carrageenan fractions contain the lowest number of sulfate groups and the highest concentrations of the 3,6-anhydro-α-D-galactopyranosyl units. ι-Carrageenan differs from $\varkappa$-carrageenan with an additional sulfate group at the 2 position. Gelation of $\varkappa$- and ι-carrageenans occur with both monovalent and divalent ions. In general, K^+, Rb^+, and Cs^+ cations favor gelation, whereas Li^+ and Na^+ cations are less effective at inducing gelation. Functionally, an increase in anhydride content increases the potassiumn sensitivity and gelling capacity (Glicksman, 1982), making $\varkappa$- types exhibit greater potassium sensitivity. As the amount of 2-sulfate groups increases to as high as 25 to 50%, as in the case of ι-carrageenan, calcium sensitivity becomes predominant. λ-Carrageenan differs from $\varkappa$- and ι-carrageenan by having variable amounts of sulfate groups and no 3,6-anhydro-α-D-galactopyranosyl residues. It does not form gels but can be used in combination with other gelling carrageenans to reduce brittleness and decrease the tendency for syneresis.

Commercial carrageenans are mixtures of these three fractions and skill is required in selecting the best combination for a given application. Locust bean gum is compatible with carrageenan, forming a cooperative association with the double helix structure, and increasing the elasticity of the resultant gels. Other hydrocolloids such as pectin, guar, CMC and starches can be used to modify the gel texture, but the effect is more subtle than with locust bean gum. The sulfate groups are mainly responsible for the protein reactivity of carrageenan.

In food systems, carrageenans are used mainly in gel formation, in stabilizing suspensions and emulsions, and for gelation and structural viscosity of milk-based products. Carrageenan gels can be used as food coatings (Meyer et al., 1959; Pearce and Lavers, 1949). Enhanced stability against growth of surface microorganisms in an intermediate-moisture cheese analog was obtained by enrobing it in a carrageenan-agarose gel matrix that contained sorbic acid (Torres et al., 1985). A carrageenan-based coating applied on cut grapefruit halves resulted in less shrinkage, leakage, or deterioration of taste after two weeks of storage at 40°F (Bryan, 1972).

Alginates

Alginates are salts of alginic acid. They occur naturally as the major structural polysaccharides of brown seaweeds known as Phaeophyceae. Principal seaweeds for commercial production include *Macrocystis pyrifera, Laminaria hyberborea, Laminaria digitata,* and *Ascophyllum nodosum* (Dziezak, 1991). The giant kelp, *Macrocystis pyrifera,* is harvested mechanically off the coast of California, whereas other seaweed species tend to be much smaller and are harvested by hand in various parts of the world. Alginic acid is considered to be a linear (1->4) linked polyuronic acid containing three types of block structures: poly-β-D-mannopyranosyluronic acid (M) blocks, poly-α-L-gulopyranosyluronic acid (G) blocks, and MG blocks containing both polyuronic acids [Figures 11.5(a), (b), and (c)] (Whistler and Daniel, 1990; Cottrell and Kovacs, 1980; McDowell, 1973). These block regions determine the shape of the polymer, which in turn governs how effectively the chains associate during gel formation. The block contents of the alginate are responsible for the different gel strengths of products derived from different seaweeds. Gels

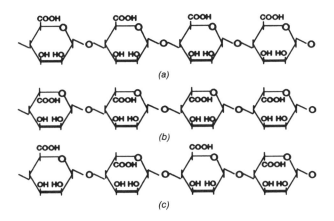

Figure 11.5 Structure of alginates consisting of (a) M block, (b) G block, and (c) alternating M and G blocks.

prepared with alginates rich in L-gulopyranosyluronic acid tend to be stronger, more brittle and less elastic than those prepared with alginates rich in D-mannopyranosyluronic acid. Alginates in solution are compatible with a wide variety of materials including other thickeners, synthetic resins, organic solvents, enzymes, surfactants, plasticizers, and alkali metal salts (Kelco, 1987).

Alginates possess good film-forming properties, which makes them particularly useful in food applications. Alginate films tend to be quite brittle when dry but may be plasticized by the inclusion of glycerol (Glicksman, 1983). Films and coatings can be made from a sodium alginate solution by rapid reaction with calcium in the cold, forming intermolecular associations involving the G-block regions. Earle (1968) describes a process of extending the shelf-life of meat or seafood products by coating with an aqueous dispersion of algin and dextrose, followed by the application of a water-soluble source of calcium ions. A similar two-step procedure for alginate coating application of various meats such as beef, cut pork, and poultry parts resulted in less dehydration than from uncoated samples (Williams et al., 1978; Mountney and Winter, 1961).

Alginate gel coating improved the adhesion between batter and product, causing reduced losses of batter from the surface of meat, fish, fruits, and vegetables (Fischer and Wong, 1972), or onion rings during frying (Smader, 1972). Reduction in dehydration is attributed mainly to the fact that the gel coating acts as a sacrificing agent; i.e., moisture in the gel evaporates prior to any significant desiccation of the enrobed food (Kester and Fennema, 1986; Allen et al., 1963). In a study involving the use of a plastic wrap and a calcium alginate coating for the control of lamb carcass shrinkage, Lazarus and co-workers (1976) indicated that although plastic film caused the least amount of shrinkage, the alginate coating was more favored due to lower surface microbial growth and faster chill rate.

A thixotropic alginate gel coating developed for ice cream eliminated drippings (Jenkinson and Williams, 1973). Flavor-Tex, a multifunctional calcium alginate coating (D. H. McKee, Inc., 1989), is used to reduce shrinkage, oxidative rancidity, moisture migration, oil absorption, and seal in volatile flavors of various food products including meat, poultry, seafood, dough products, vegetables, extruded foods, and cheese. Alginate coatings can also provide protection against lipid oxidation of food ingredients as in the case of pre-cooked, ground pork patties. Coated samples did not develop warmed-over flavor (an indication of rancidity) and exhibited better texture when compared to uncoated samples.

Agar and Furcellaran

Agar is extracted commercially from a number of marine algae of the class *Rhodophyceae,* called "red seaweeds." Agar, which is made up of β-D-galactopyranosyl linked (1->4) to a 3,6-anhydro-α-L-galactopyranosyl unit, and partially esterified with sulphuric acid, produces perceptible gelation at concentrations as low as 0.04%. Agar is best known as a culture medium and is not used to a great extent in foods.

Furcellaran is extracted from the seaweed *Furcellaria fastigiata* which is found in the waters surrounding Denmark. Furcellaran is similar to x-carrageenan in functional properties and has found substantial use in food applications. It differs from x-carrageenan mainly in the amount of ester sulfate present. It is mainly used in Europe in jams and jellies, fruit juices, confectionery, milk puddings, chocolate milk, and beer.

EXUDATE GUMS

Gum Arabic

Exudate gums are structurally complex heteropolysaccharides obtained from certain shrubs and smalls trees which grow predominantly in Africa and Asia (Whistler and BeMiller, 1973; Smith and Montgomery, 1959). The term "exudate" refers to the fact that these gums are secreted or exuded in response to injury to the plant tissue, and upon exposure to the atmosphere, they form extremely hard, glossy nodules or flakes which are harvested by hand.

Gum arabic or acacia is the dried, gummy exudate from the stems or branches of *Acacia senegal* and related species of *Acacia.* The trees which produce commercial grades of gum arabic grow primarily in the African Sahel (the "gum belt") (Balke, 1984).

Gum arabic is a neutral or slightly acidic salt of a complex polysaccharide containing calcium, magnesium and potassium ions. The gum consists of D-galactopyranosyl, L-rhamnopyranosyl, L-arabinopyranosyl, L-arabinofuranosyl, and D-glucopyranosyluronic acid units (Prakash et al., 1990; Whistler and Daniel, 1990), and a small amount of protein (about 2%). The molecular weight is about 600,000.

Gum arabic dissolves readily in hot or cold water. It is the least viscous and the most soluble of the hydrocolloids. A 40% solution produces an excellent mucilage for adhesiveness. Aqueous solutions of over 50% concentrations may be prepared. Gum arabic is insoluble in alcohol and most organic solvents. It is compatible with other plant hydrocolloids, proteins and starches. It produces stable emulsions with most oils over a wide pH range. Electrolytes generally reduce the viscosity of gum arabic solutions. Many salts, particularly trivalent metal salts, give precipitates. Gum arabic solutions are subject to bacterial attack and can be preserved using 0.1% concentrations of sodium benzoate or benzoic acid.

The uses of gum arabic are based upon its action as a protective colloid or stabilizer and the adhesiveness of its water solutions. More than one-half of the world's supply is used in confections where the gum acts to retard sugar crystallization, and to thicken candies, jellies, glazes, and chewing gums (Anonymous, 1985; Fogarty, 1988). The flavor industry uses the gum as a flavor-fixative, protecting the flavor from evaporation, oxidation, and absorption of moisture from the air. It is also used as a foam stabilizer and agent to promote adhesion of foam to glass. Gum arabic has been used for coating pecan nut halves to eliminate moist, oily appearance and at the same time providing a low-calorie product (Arnold, 1963). Two water-soluble adhesive film-forming polymers called "Spraygum" and "Sealgum" (Colloides Naturels, Inc., 1988) are based on acacia gum, and acacia gum with gelatin, respectively. This coating can be used as a protective film on oily and nonoily food materials including chocolates, nuts, cheese, as well as pharmaceutical tablets. These two edible coatings, in combination with calcium chloride, showed considerable promise as inhibitors of after-cooking darkening of potatoes (Mazza and Qi, 1991).

Gum Tragacanth

Gum tragacanth is the dried gum exuded by the stems of *Astragalus gummifer* and other Asiatic species of *Astragalus*. The plants grow wild in certain sections of Asia Minor and in the arid and mountainous regions of Iran, Syria, and Turkey.

Gum tragacanth has a complex structure, and on hydrolysis yields D-galacturonic acid, L-fucose, D-galactose, D-xylose, and L-arabinose

(Meir et al., 1973). It consists of a mixture of a water-insoluble component, tragacanthic acid, which constitutes about 60–70% of the gum, and a water-soluble component, tragacanthin or arabinogalactan, in which L-arabinose is the preponderant sugar (Glicksman, 1983).

Gum tragacanth swells rapidly in either cold or hot water to form highly viscous colloidal sols or semi-gels, which act as protective colloids and stabilizing agents. It is insoluble in alcohol and other organic solvents, but is compatible with other plant hydrocolloids as well as carbohydrates and proteins. Viscosity is most stable at pH 4 to 8. It is compatible with relatively high salt concentrations and with most other natural and synthetic gums. Gum tragacanth jellies can be preserved with 0.1% sodium benzoate.

The gum is mainly used as a thickener and stabilizer in salad dressings, sauces, bakery emulsion, toppings, ice cream, and confectionery. Its film-forming properties are useful in nonfood systems such as hair lotions, hand lotions, and creams.

Gum Ghatti and Gum Karaya

Gum ghatti, also known as Indian Gum, is an amorphous translucent exudate of the *Anogeissus latifolia* tree of the family Combretaceae which is native to India (Glicksman, 1983). The gum is a water-soluble, complex polysaccharide consisting of L-arabinofuranosyl, D-galactopyranosyl, D-mannopyranosyl, D-xylopyranosyl, and D-glucopyranosyluronic acid units. It is made up of a soluble fraction and an insoluble but swellable "gelling" fraction. The gum, as a whole, is nongelling but can be dispersed in hot or cold water to give a collodial sol that develops due to the soluble fraction. Maximum viscosity develops at about pH 5 to 7. The gum closely resembles the viscosity and emulsifying properties of gum arabic. It has been used effectively in food systems as an emulsifier and stabilizer. Films formed from ghatti dispersions are relatively soluble, brittle, and are not considered very useful. Gum ghatti is seldom used today because of its limited supply (Dziezak, 1991).

Gum karaya is the dried gummy exudate of the *Sterculia* tree which grows in central and northern India. It occurs naturally as a complex, partially acetylated branched polysaccharide containing about 37% uronic acid residues and approximately 8% acetyl groups. Karaya contains a central chain of D-galactopyranosyl, L-rhamnopyranosyl and D-galactopyranosyluronic acid units, with some side chains containing D-glucopyranosyluronic acid (Glicksman, 1983). Gum karaya is water-swellable rather than water-soluble and absorbs water very rapidly to form viscous colloidal dispersions at low concentrations. In the past, the gum has been used as an emulsifier, stabilizer, or binder in the frozen desserts, dairy products, salad dressings, or meat products, but today it has been replaced by better stabilizers. Practical applications of karaya make use of its thickening and adhesive properties. Gum karaya forms smooth films that can be plasticized with compounds such as glycols to reduce brittleness.

SEED GUMS

Locust Bean Gum

The galactomannan gums—locust bean gum and guar gum—come from seeds produced by certain plants of the leguminose *Ceratonia* and *Cyamopsis* genera. Seed gums have been used as food for humans and animals since ancient times and they are still used extensively today as food additives (Whistler and BeMiller, 1973; Glicksman, 1969).

Locust bean or carob gum is processed from the seeds of the evergreen tree, *Ceratonia siliqua* which is indigenous to Mediterranean countries. It is a neutral polysaccharide consisting of β-D-mannopyranosyl and α-D-galactopyranosyl units in a ratio of 4:1 [Figure 11.6(a)].

Locust bean gum is insoluble in cold water and must be heated to be dissolved. The highest viscosity is obtained by dispersing the gum at 95°C and then cooling. The gum is insoluble in most organic solvents. It is compatible with other hydrocolloids, as well as carbohydrates and proteins. It

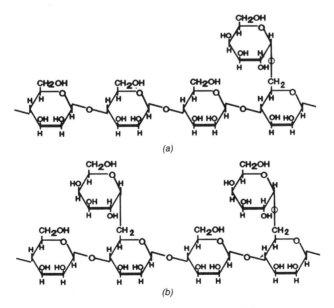

(a)

(b)

Figure 11.6 Structure of (a) locust bean gum, and (b) guar gum.

forms gels with carrageenan, agar, or xanthan gum. It is stable at a pH range of 2–10.

For food applications, locust bean gum is used as a thickener or viscosity modifier, binder of free water, suspending agent, or stabilizer in cheeses, frozen confections, bakery products, pie fillings, meat, sauces, and salad dressings. Its film-forming property is utilized in the textile industry, making it ideal as a sizing and finishing agent, as well as in the pharmaceutical and cosmetics industry for the production of lotions and creams.

Guar Gum

Guar gum is derived from the ground endosperm of the guar plant, *Cyamopsis tetragonolobus*. The plant is cultivated commercially in India and Pakistan for human and animal consumption, and to a limited extent, in the semi-arid southwestern U.S. Guar gum, like locust bean gum, is a polysaccharide having a straight chain of β-D-mannopyranosyl with single α-D-galactopyranosyl units attached as side chains in the ratio of 1:2 [Figure 11.6(b)].

Guar gum will disperse and swell almost completely in cold or hot water to form a highly viscous sol. It is insoluble in organic solvents but is compatible with most other plant hydrocolloids, chemically modified starches, or cellulose, and water-soluble proteins. Guar sols are very stable in the pH range of 4 to 10.5.

Guar gums are based primarily on thickening aqueous solutions and controlling the mobility of dispersed or solubilized materials in the water. Food applications include dairy, bakery, and meat products, as well as beverages and salad dressings. Its film-forming property is used in the textile industry for finishing agents.

MICROBIAL FERMENTATION GUMS

Xanthan Gum

Xanthan gum is produced from the organism *Xanthomonas campestris* by controlled fermentation. Each xanthan gum unit contains five sugar residues: two β-D-glucopyranosyl, two β-D-mannopyranosyl, and one β-glucopyranosyluronic acid units (Jansson et al., 1975; Melton et al., 1976) (Figure 11.7). The molecule has a cellulosic backbone which is rendered water soluble by the presence of short side chains attached to every second glucose residue in the main chain. As a consequence of this covalent or primary structure, the molecule exists in solution as a rigid rod sta-

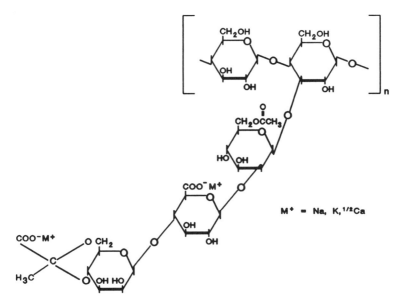

Figure 11.7 Structure of xanthan gum.

bilized by noncovalent interactions between the backbone and the side chains (Morris et al., 1977). This ordered conformation gives rise to the unique solution properties of xanthan and viscosity stability over a wide temperature. Its stability in varying pH changes is exceptional, therefore it is regarded as the most acid-stable gum. Xanthan gum is compatible with salts even at high concentrations (e.g., 20% sodium chloride). Aqueous solutions of xanthan gum will tolerate as much as 50–60% of water-miscible solvents such as methanol, ethanol, isopropanol, or acetone. It has a synergistic effect with the galactomannans guar and locust bean gum. The synergism takes the form of increased viscosity with guar, although the interaction with locust bean gum is much stronger. The interaction of xanthan gum with galactomannans works well in protein systems.

Xanthan gum is readily dispersed in hot or cold water. It shows remarkably high viscosity at low shear rates as a result of weak intermolecular associations, even at very low gum concentrations. Therefore, xanthan gum is functional in foods at levels of 0.05–0.5%.

Xanthan gum is used for its thickening, suspending, and stabilizing effects in salad dressings, dry mix products, icings and frostings, confectionery, dairy products, fruit gels, sauces, gravies, pie fillings, syrups, toppings, and baked goods. It can be used to provide uniform coating, good cling, and improved adhesion in wet batters, and to prevent moisture migration during frying.

Gellan Gum

Gellan gum, a relatively new multifunctional hydrocolloid developed and patented by Kelco (1990), is produced by fermentation of a pure culture of *Pseudomonas elodea*. The gum has a linear tetrasaccharide as repeating unit, consisting of (1->3)-β-D-glucopyranosyl, (1->4)-β-D-glucopyranosyluronic acid, (1->4)-β-D-glucopyranosyl, and (1->4)-α-L-rhamnopyranosyl units. It is functional at very low levels, forming gels as low as 0.05%. Gels are formed by dispersing the gellan gum in water, heating, adding cations, and then cooling to set. It can be used with other hydrocolloids to improve product stability or vary its texture. Gel texture can be modified by blending with other gums, especially the gelling gums. It has good stability over a wide pH range (3.5–8.0).

Present uses of gellan gum are as stabilizers and thickeners in glazes, icings, frostings, jams, and jellies. Future applications are expected to parallel those of other gelling stabilizers already in use, or perhaps to replace or blend with other heating/cooling gel systems such as carrageenan, agar, starch, or pectin.

CHITOSAN

Chitin is a β-1,4 linked linear polymer of 2-acetamido-2-deoxy-D-glucopyranosyl residues (BeMiller, 1965). It occurs as the major organic skeletal substance of invertebrates and as a cell wall constituent of fungi and green algae. Chitin does not exist alone, but is found in close association with calcium carbonate and/or protein and other organic substances, making the compound stable to most reagents. Fusion of chitin with alkalies gives the product chitosan, a heterogeneous substance in various stages of deacetylation and depolymerization.

Chitosan is nontoxic (Arai et al., 1968). Chitosan applications include coatings, flocculating agents, and ingredients for foods and feeds. In the U.S., however, chitosan is not approved as a food ingredient. Arai et al. (1968) have recommended chitosan as a food ingredient based on dosage results indicating the same order of lethality as sugar and salt.

Chitosan can form a semi-permeable coating which can modify the internal atmosphere, thereby delaying ripening and decreasing transpiration rates in fruits and vegetables (Arai et al., 1968). Theoretically, it should be an ideal preservative coating material based on the numerous positive effects of chitosan. It can inhibit the growth of fungi and phytopathogens (El Ghaouth et al., 1991; Hirano and Nagao, 1989; Kendra and Hadwiger, 1984; Stossel and Leuba, 1984; Hadwiger and Beckman, 1980; Allan and Hadwiger, 1979), and can induce chitinase, a defense enzyme (Mauch et al., 1984).

Carolan and co-workers (1991) have developed a method for preparing a chitosan derivative with a wide variety of agricultural and industrial applications. *N,O*-carboxymethyl chitosan (NOCC) was prepared from the reaction of chitosan with monochloroacetic acid under alkaline conditions. NOCC is water-soluble, biodegradable, and forms selectively permeable nontoxic films. Nutri-Save, an NOCC-based coating (Nova Chem, Halifax, Nova Scotia), was reported to have some success as a postharvest edible coating for fresh fruits (El Ghaouth et al., 1991; Meheriuk, 1990; Meheriuk and Lau, 1988; Elson, 1985). Chitosan–acetic acid complex membrane showed high permeabilities for O_2 and CO_2 and synthetic polymers can modify the permeation behavior of the chitosan membrane for these two gases (Bai et al., 1988).

POTENTIAL ADVANTAGES OF POLYSACCHARIDE-BASED COATINGS

The ability of some water-soluble polysaccharides to form thermally induced gelatinous coatings has found wide application in breadings and batters for the reduction of oil absorption during frying. The attractive, nongreasy, and low-calorie polysaccharide films and coatings can be used to extend the shelf-life of fruit, vegetables, seafood, or meat products by preventing dehydration, oxidative rancidity, and surface browning. Their application in the agricultural produce field has become popular during the last few years due to their ability to modify internal atmospheres. Fresh fruit and vegetable coatings, based on waxes or oils, have been restricted to applications seeking reduced moisture loss or cosmetic effects. Numerous studies, however, have reported the occurrence of off-flavor as a result of over-modification of the internal atmospheres in wax- or oil-coated fruits and vegetables. The ability of water-soluble polysaccharides to reduce O_2 and increase CO_2 levels in internal atmospheres reduces respiration rates, thereby prolonging the shelf-life of fruits and vegetables in a manner analogous to controlled atmospheres.

Sufficient evidence is available to establish the beneficial effects of coatings on agricultural produce, among which are (1) improved retention of flavor, acids, sugars, texture and color, (2) increased stability during shipping and storage, (3) improved appearance, and (4) reduced spoilage. The ease with which the permeability properties can be manipulated reduces the chances that anaerobic conditions will occur. This results in improvement or retention of the original fresh flavor and aroma of the stored coated fruit or vegetable. The inability of these coatings to provide sufficient gloss or prevent moisture loss can be improved by the incorporation of functional food ingredients. Food additives such as resins and rosins, plasticizers, surfactants, oils, waxes, and emulsifiers can be added provided the coating is not rendered impermeable. "Tailoring" the coating to specific fruits or vegetables to allow proper permeability, while providing gloss and reducing moisture loss, remains a major challenge to the food or polymer scientist.

REFERENCES

Allan, C. R. and L. A. Hadwiger. 1979. "The Fungicidal Effect of Chitosan on Fungi of Varying Cell Wall Composition," *Exp. Mycol.*, 3:285.

Allen, L., A. I. Nelson, M. P. Steinberg and J. N. McGill. 1963. "Edible Corn-Carbohydrate Food Coatings. I. Development and Physical Testing of Starch-Algin Coating," *Food Technol.*, 17:1437.

Anonymous. 1964. "Edible Water-Soluble Starch Film," *Food Proc.*, 25(5):112.

Anonymous. 1985. "The Natural Solution," *Food, Flavor, Ingred. Proc. Pack.*, 7(11):33.

Anonymous. 1988. "Market Survey of Food Thickeners and Stabilizers in the USA and Western Europe," *Carbohyd. Polymers*, 8:63.

Aqualon. 1988. "Aqualon Gum Product Bulletin," Wilmington, DE: Aqualon Co.

Arai, L., Y. Kinumaki and T. Fujita. 1968. "Toxicity of Chitosan," *Bull. Tokai Reg. Fish. Res. Lab.*, 56:89.

Arnold, F. W. May 14, 1963. U.S. patent 3,383,220.

Aspinall, G. O. 1970. *Polysaccharides*. Oxford: Pergamon Press.

Bai, R., M. Huang and Y. Jiang. 1988. "Selective Permeabilities of Chitosan-Acetic Acid Complex Membrane and Chitosan-Polymer Complex Membranes for Oxygen and Carbon Dioxide," *Polymer Bull.*, 20:83–88.

Balke, W. J. 1984. "The Use of Natural Hydrocolloids as Thickeners, Stabilizers and Emulsifiers," in *Gum Starch and Technology. 18th Annual Symposium*, D. L. Downing, ed., New York: Cornell University, Institute of Food Sci., pp. 31–35.

Banker, G. S., A. Y. Gore and J. Swarbrick. 1966. "Water Vapour Transmission Properties of Free Polymer Films," *J. Pharm.*, 18:487.

Banks, N. H. 1984. "Some Effects of TAL Pro-long Coating on Ripening Bananas," *J. Exper. Bot.*, 35(150):127–137.

Banks, N. H. 1985. "Internal Atmosphere Modification in Pro-long Coated Apples," *Acta Hortic.*, 157:105.

Bauer, C. D., G. L. Neuser and H. A. Pinkalla. December 9, 1969. U.S. patent 3,483,004.

BeMiller, J. N. 1965. "Chitosan," in *Methods in Carbohydrate Chemistry. General Polysaccharides, Vol. V*, R. L. Whistler, ed., New York: Academic Press, pp. 103–106.

Bryan, D. S. December 26, 1972. U.S. patent 3,707,383.

Carolan, C., H. S. Blair and S. J. Allen. 1991. "Chitosan-Derivative Keeps Apples Fresh," *Postharvest News and Info.*, 2(2):75.

Cole, M. S. November 18, 1969. U.S. patent 3,479,191.

Colloides Naturels Inc. 1988. "Sealgum: Something New in Films." Bridgewater, NJ: Colloides Naturels Inc., Product Bulletin.

Cottrell, I. W. and P. Kovacs. 1980. "Alginates," in *Handbook of Water-Soluble Gums and Resins*, R. Davidson, ed., New York: McGraw-Hill, p. 18.

Curtis, G. J. 1988. "Some Experiments with Edible Coatings on the Long-Term Storage of Citrus Fruits," *Proc. Sixth Inter. Citrus Congress,* March 6–11, 1988, Tel-Aviv, Israel, pp. 1515–1520.

D'Ercole, A., A. Krias and R. Tewey. September 16, 1969. U.S. patent 3,467,528.

D. H. McKee, Inc. 1989. "Flavor-Tex Food Coating System." Tampa, FL: Food Research, Inc., Product Bulletin.

Drake, S. R., J. K. Fellman and J. W. Nelson. 1987. "Postharvest Use of Sucrose Polyesters for Extending the Shelf-Life of Stored 'Golden Delicious' Apples," *J. Food Sci.,* 52:685–690.

Dziezak, J. D. 1991. "Special Report: A Focus on Gums," *Food Tech.,* 45(3):116–132.

Earle, R. D. July 30, 1968. U.S. patent 3,395,024.

El Ghaouth, A., J. Arul, R. Ponnampalan and M. Boulet. 1991. "Chitosan Coating Effect on Storability and Quality of Fresh Strawberries," *J. Food Sci.,* 56:1618.

Elson, C. W., E. R. Hayes and P. D. Lidster. 1985. "Development of the Differentially Permeable Fruit Coatings 'Nutri-Save' for the Modified Atmosphere Storage of Fruit," in *Proceedings of the Fourth National Controlled Atmosphere Research Conf.,* S. M. Blankenship, ed., Raleigh, NC: Dept. of Hort. Sci., North Carolina State Univ., Rpt. 126:248.

Evans, R. B. and W. Herbst. December 1, 1964. U.S. patent 3,159,585.

FDA. 1991. *Title 21—Food and Drugs.* Washington, DC: Federal Register, U.S. Food and Drug Administration, Code of Federal Regulations.

Fisher, L. G. and P. Wong. July 11, 1972. U.S. patent 3,676,158.

FMC. 1985. "Avicel Cellulose Gel." Philadelphia, PA: FMC Corp., Product Bulletin.

Fogarty, D. 1988. "No Longer a Raw Deal," *Food, Flavor, Ingred. Proc. Pack.,* 10(7):25.

Ganz, A. J. 1969. "CMC and Hydroxypropylcellulose—Versatile Gums for Food Use," *Food Prod. Dev.,* 3(6):65.

Glicksman, M. 1969. *Gum Technology in the Food Industry.* New York: Academic Press.

Glicksman, M. 1984. *Food Hydrocolloids. Vol. III;* 1983, *Vol. II;* 1982, *Vol. I.* Boca Raton, FL: CRC Press.

Guilbert, S. 1986. "Technology and Application of Edible Protective Films," in *Food Packaging and Preservation. Theory and Practice,* M. Mathlouthi, ed., London, UK: Elsevier Applied Science Publishing Co., p. 371.

Hadwiger, L. A. and J. M. Beckman. 1980. "Chitosan as a Component of Pea-*Fusarium solani* Interactions," *Plant Physiol.,* 66:205–211.

Hallman, G. J., M. O. Nisperos-Carriedo, E. A. Baldwin and C. A. Campbell. 1994. "Mortality of Caribbean Fruit Fly (Diptera: Tephritidae) Immatures in Coated Fruits," *J. Econom. Ent.* (in press).

Hiemstra, P. and J. Muetgeert. September 1, 1959. U.S. patent 2,902,336.

Hirano, S. and N. Nagao. 1989. "Effect of Chitosan, Pectic Acid, Lysozyme and Chitinase on the Growth of Several Phytopathogens," *Agric. Biol. Chem.,* 53:3065.

Jansson, P. E., L. Kenne and B. Lindberg. 1975. "Structure of the Extracellular Polysaccharide from *Xanthomonas campestris,*" *Carbohyd. Res.,* 45:275–282.

Jenkinson, T. J. and T. P. Williams. August 14, 1973. U.S. patent 3,752,678.

Jokay, L., G. E. Nelson and E. L. Powell. 1967. "Development of Edible Amylaceous Coatings for Foods," *Food Technol.,* 21:12–14.

Kamper, S. L. and O. R. Fennema. 1984. "Water Vapor Permeability of Edible Bilayer Films," *J. Food Sci.*, 49:1478–1485.

Kamper, S. L. and O. R. Fennema. 1985. "Use of Edible Film to Maintain Water Vapor Gradients in Foods," *J. Food Sci.*, 50:382–384.

Kaplan, H. J. 1986. "Washing, Waxing and Color-Grading," in *Fresh Citrus Fruits*, W. F. Wardowski, S. Nagy and W. Grierson, eds., Westport, CT: AVI, p. 379.

Kelco. 1987. "Alginate Products for Scientific Water Control." San Diego, CA: Kelco, Product Bulletin.

Kelco. 1990. "Gellan Gum—Multifunctional Gelling Agent." Rahway, NJ: Merck & Co., Technical Bulletin.

Keller, J. 1984. "Sodium Carboxymethylcellulose (CMC)," in *Gum Starch and Technology. 18th Annual Symposium*, D. L. Downing, ed., New York: Cornell University, Institute of Food Science, pp. 9–19.

Kendra, D. F. and L. A. Hadwiger. 1984. "Characterization of the Smallest Chitosan Oligomer That Is Maximally Antifungal to *Fusarium solani* and Elicits Pisatin Formation in *Pisum sativum*," *Exp. Mycol.*, 8:276.

Kester, J. J. and O. R. Fennema. 1986. "Evaluation of an Edible, Heat-Sensitive Cellulose Ether-Lipid Film as a Barrier to Moisture Transmission," paper presented at the *46th Annual Meeting, Institute of Food Technologists*, June 15–16, 1986, Dallas, TX.

Krumel, K. L. and T. A. Lindsay. 1976. "Nonionic Cellulose Ethers," *Food Technol.*, 30(4):36–43.

Kunz, W. B. April 24, 1962. U.S. patent 3,030,667.

Lazarus, C. R., R. L. West, J. L. Oblinger and A. Z. Palmer. 1976. "Evaluation of a Calcium Alginate Coating and a Protective Plastic Wrapping for the Control of Lamb Carcass Shrinkage," *J. Food Sci.*, 41:639–641.

Lowings, P. H. and D. F. Cutts. 1982. "The Preservation of Fresh Fruits and Vegetables," *Proc. Inst. Food Sci. Tech. Ann. Symp.*, July, 1981, Nottingham, UK, p. 52.

Luck, E. July 2, 1968. U.S. patent 3,391,008.

Maclay, W. D. and H. S. Owens. 1948. "Pectinate Films," *Mod. Packag.*, 21:157.

Maloney, J. F., E. H. Sander and A. Wilhelm. April 7, 1970. U.S. patent 3,505,076.

Mark, A. M., W. B. Roth and C. E. Rist. 1966. "Oxygen Permeability of Amylomaize Starch Films," *Food Technol.*, 20(1):75.

Mason, D. F. October 14, 1969. U.S. patent 3,472,662.

Mauch, F., L. A. Hadwiger and T. Boller. 1984. "Ethylene: Symptom, Not Signal for the Induction of Chitinase and β-1,3-Glucanase in Pea Pods by Pathogens and Elicitors," *Plant Physiol.*, 76:607.

Mazza, G. and H. Qi. 1991. "Control of After-Cooking Darkening in Potatoes with Edible Film-Forming Products and Calcium Chloride," *J. Agric. Food Chem.*, 39:2163–2166.

McDowell, R. H. 1973. "Properties of Alginates," London, UK: Alginate Industrial Limited.

Meheriuk, M. 1990. "Skin Color in 'Newton' Apples Treated with Calcium Nitrate, Urea, Diphenylamine, and a Film Coating," *HortScience*, 25(7):775–776.

Meheriuk, M. and O. L. Lau. 1988. "Effect of Two Polymeric Coatings on Fruit Quality of 'Bartlett' and 'd'Anjou' Pears," *J. Amer. Soc. Hort. Sci.*, 113(2):222–226.

Meir, G., W. A. Meer and T. Guard. 1973. "Gum Tragacanth," in *Industrial Gums*,

Polysaccharides and Their Derivatives, R. L. Whistler and J. N. BeMiller, eds., New York: Academic Press, pp. 289–299.

Melton, L. D., L. Mindt, D. A. Rees and G. R. Sanderson. 1976. "Covalent Structure of the Polysaccharide from *Xanthomonas campestris*: Evidence from Partial Hydrolysis Studies," *Carbohyd. Res.,* 46(2):245–257.

Meyer, R. C., A. R. Winter and H. H. Weiser. 1959. "Edible Protective Coatings for Extending the Shelf Life of Poultry," *Food Technol.,* 13:146.

Miers, J. C., H. A. Swenson, T. H. Schultz and H. S. Owens. 1953. "Pectinate and Pectate Coatings. I. General Requirements and Procedures," *Food Technol.,* 7:229–231.

Mitan, F. J. and L. Jokay. February 18, 1969. U.S. patent 3,427,951.

Mitchell, J. D. and D. A. Ledward. 1986. *Functional Properties of Food Macromolecules.* London, UK: Elsevier Applied Science Publishers.

Moore, C. O. and J. W. Robinson. February, 1968. U.S. patent 3,368,909.

Morris, E. R., D. A. Rees and G. Robinson. 1980. "Cation-Specific Aggregation of Carregeenan Helices: Domain Model of Polymer Gel Structure," *J. Mol. Biol.,* 138:349.

Morris, E. R., D. A. Rees, G. Young, M. D. Walkinshaw and A. Darke. 1977. "Order-Disorder Transition for a Bacterial Polysaccharide in Solution: A Role for Polysaccharide Conformation in Recognition between *Xanthomonas* Pathogen and its Plant Host," *J. Mol. Biol.,* 110(1):1–6.

Motlagh, H. F. and P. Quantick. 1988. "Effect of Permeable Coatings on the Storage Life of Fruits. I. Pro-long Treatment of Limes (*Citrus aurantifolia* cv. Persian)," *Int. J. Food Sci. Tech,* 23:99–105.

Mountney, G. J. and A. R. Winter. 1961. "The Use of Calcium Alginate Film for Coating Cut-Up Poultry," *Poultry Sci.,* 40:28.

Murray, D. G. and L. R. Luft. 1973. "Low-D.E. Corn Starch Hydrolysates," *Food Technol.,* 27(3):32–40.

Murray, D. G., N. G. Marotta and R. M. Boettger. August 3, 1971. U.S. patent 3,597,227.

National Starch and Chemical Corporation. 1990. "How to Choose. A Professional Guide to Food Starches and Gums." Bridgewater, NJ: Food Products Division.

Nelson, K. L. and O. R. Fennema. 191. "Methylcellulose Films to Prevent Lipid Migration in Confectionery Products," *J. Food Sci.,* 56(2):504–509.

Nisperos-Carriedo, M.O. and E. A. Baldwin. 1988. "Effect of Two Types of Edible Films on Tomato Fruit Ripening," *Proc. Florida State Hort. Soc.,* 101:217–220.

Nisperos-Carriedo, M. O. and E. A. Baldwin. 1990. "Edible Coatings for Fresh Fruits and Vegetables," *1990 Subtropical Technology Conference Proceedings,* Oct. 18, 1990, Lake Alfred, FL.

Nisperos-Carriedo, M. O., E. A. Baldwin and P. E. Shaw. 1992. "Development of Edible Coating for Extending Postharvest Life of Selected Fruits and Vegetables," *Proc. Florida State Hort. Soc.,* 104:122–125.

Nisperos-Carriedo, M. O., P. E. Shaw and E. A. Baldwin. 1990. "Changes in Volatile Flavor Components of Pineapple Orange Juice as Influenced by the Application of Lipid and Composite Coatings," *J. Agric. Food Chem.,* 38(6):1382–1387.

Pearce, J. A. and C. G. Lavers. 1949. "Frozen Storage of Poultry. V. Effects of Some Processing Factors on Quality," *Can. J. Res.,* 27:253.

Philips, G. O., D. J. Wedlock and P. A. Williams. 1990. *Gums and Stabilizers for the Food Industry, Vol. 5;* 1988, *Vol. 4;* 1986, *Vol. 3;* 1984, *Vol. 2.* London. UK: Elsevier Applied Science Publishers.

Prakash, A., M. Joseph and M. E. Mangino. 1990. "The Effects of Added Proteins on the Functionality of Gum Arabic in Soft Drink Emulsion Systems," *Food Hydrocolloids,* 4(3):177.

Rankin, J. C., I. A. Wolff, H. A. Davis and C. E. Rist. 1958. "Permeability of Amylose Film to Moisture Vapor, Selected Organic Vapors and the Common Gases," *Ind. Eng. Chem., Chem. Eng. Data Ser.,* 3:120–123.

Roth, W. B. and C. L. Mehltretter. 1967. "Some Properties of Hydroxypropylated Amylomaize Starch Films," *Food Technol.,* 21:72–74.

Sanderson, G. R. 1981. "Polysaccharides in Foods," *Food Technol.,* 35(7):50–57, 83.

Sanford, P. A. and J. Baird. 1983. "Industrial Utilization of Polysaccharides," in *The Polysaccharides, Vol. 2,* G. O. Aspinal, ed., New York: Academic Press, pp. 411–490.

Santerre, C. R., T. F. Leach and J. N. Cash. 1989. "The Influence of the Sucrose Polyester, Semperfresh, on the Storage of Michigan Grown 'McIntosh' and 'Golden Delicious' Apples," *J. Food Proc. Preserv.,* 13:293–305.

Scheick, K. A., L. Jokay and G. E. Nelson. September 8, 1970. U.S. patent 3,527,646.

Schultz, T. H., J. C. Miers, H. S. Owens and W. D. Maclay. 1949. "Permeability of Pectinate Films to Water Vapor," *J. Phys. Colloid Chem.,* 53:1320.

Smadar, Y. March 21, 1972. U.S. patent 3,650,766.

Smith, F. and R. Montgomery. 1959. *The Chemistry of Plant Gums and Mucilages and Some Related Polysaccharides.* ACS Monograph Series 41. New York: Reinhold Pub. Corp.

Smith, P. S. 1984. "Food Starches and Their Uses," in *Gum Starch and Technology. 18th Annual Symposium,* D. L. Downing, ed., New York: Cornell University, Institute of Food Science, pp. 34–42.

Smith, S. M. and J. R. Stow. 1984. "The Potential of Sucrose Ester Coating Material for Improving the Storage and Shelf-Life Qualities of Cox's Orange Pippin Apples," *Ann. Appl. Biol.,* 104:383.

Stossel, P. and J. L. Leuba. 1984. "Effect of Chitosan, Chitin and Some Amino Sugars on Growth of Various Soilborne Phytopathogenic Fungi," *Phytopath. Z.,* 111:82.

Swenson, H. A., J. C. Miers, T. H. Schultz and H. S. Owens. 1953. "Pectinate and Pectate Coatings. II. Application to Nuts and Fruit Products," *Food Technol.,* 7:232–235.

Thompson, H. J. June 13, 1972. U.S. patent 3,669,688.

Torres, J. A., J. O. Bouzas and M. Karel. 1985. "Microbial Stabilization of Intermediate Moisture Food Surfaces. II. Control of Surface pH," *J. Food Proc. Preserv.,* 9:93.

Vineyard, M. L. and R. P. Bear. 1952. "Amylose Content," *Maize Genetics Coop. Newsletter,* 26:5.

Whistler, R. L. 1991. "Introduction to Industrial Short Course," *18th AACC Short Course on Gum Chemistry and Technology,* November 6–8, 1991, Chicago, IL.

Whistler, R. L. and J. N. BeMiller. 1973. *Industrial Gums, Polysaccharides and Their Derivatives, Second Edition.* New York: Academic Press.

Whistler, R. L. and J. N. BeMiller. 1992. *Industrial Gums, Third Edition.* New York: Academic Press.

Whistler, R. L. and J. R. Daniel. 1990. "Functions of Polysaccharides in Foods," in *Food Additives,* A. L. Branen, P. M. Davidson and S. Salminen, eds., New York: Marcel Dekker, Inc., pp. 395–424.

Whistler, R. L. and E. F. Paschall. 1967. *Starch: Chemistry and Technology, Vol. 2;* 1965, *Vol. 1.* New York: Academic Press.

Whistler, R. L., J. N. BeMiller and H. F. Paschall. 1984. *Starch: Chemistry and Technology.* New York: Academic Press.

Williams, R. B. H., J. L. Oblinger and R. L. West. 1978. "Evaluation of a Calcium Alginate Film for Use on Beef Cuts," *J. Food Sci.,* 43:292.

Wolff, I. A., H. A. Davis, J. E. Cluskey, L. J. Gundrum and C. E. Rist. 1951. "Preparation of Films from Amylose," *Ind. Chem. Eng.,* 915–919.

Wurzburg, O. B. 1986. *Modified Starches: Properties and Uses.* Boca Raton, FL: CRC Press.

CHAPTER 12

Mathematical Modeling of Moisture Transfer in Food Systems with Edible Coatings

TOM R. RUMSEY[1]
JOHN M. KROCHTA[2]

INTRODUCTION

THE effectiveness of edible coatings is primarily defined by their mass transfer properties. Mathematical models discussed in this chapter will deal with prediction of water vapor transfer through edible films applied as coatings on foods. The analyses rely on theory of diffusion in solids as outlined in the texts by Crank (1975) and Geankoplis (1983).

Biquet and Labuza (1988a) presented one of the first mathematical models to predict the moisture gain in a system using an edible coating. Their analysis is based on earlier models that predict moisture gain or loss in packaged food products using nonedible packaging materials (Labuza et al., 1972). The model can be used to calculate the average moisture content of the food material as a function of time given the relative humidity of the surrounding air. The moisture gradient within the food was assumed to be uniform, and the moisture flux through the coating was modeled as the permeance times the difference between water vapor pressure at the coating-food interface and the coating environment. Film and food properties required to evaluate the model are (1) film thickness, permeability, surface area, and (2) food weight and a linear equation for modeling the food isotherm. Hong et al. (1991) relaxed the assumption of uniform moisture in the food material in developing a model to examine moisture transfer from a slab of food through an edible coating. The diffusion equation was

[1]Department of Biological and Agricultural Engineering, University of California, Davis, CA 95616.
[2]Departments of Food Science and Technology, and Biological and Agricultural Engineering, University of California, Davis, CA 95616.

used to model moisture transfer in the food, while the moisture transfer through the coating used the same model described above. They also used the L number, a dimensionless number that describes the ratio of the moisture permeance of a food substance to that of a coating material, to predict the effectiveness of edible coating materials.

The remaining sections of this chapter describe the use of simple diffusion theory to model moisture transport in edible films, and processed foods with edible films formed as coatings. Specifically, "Analytical Solutions of Diffusion Equation for Edible Films" makes use of several analytical solutions of the one-dimensional diffusion equation to analyze moisture movement through a free-standing film. These solutions are then used to provide an explanation for the anomalous behavior of chocolate films reported by Biquet and Labuza (1988a). In "Numerical Solution of Diffusion Equation for an Edible Film" a detailed development of finite difference solution of the one-dimensional diffusion equation is given. "Numerical Solution of Moisture Diffusion through a Coating Formed on a Food Product," develops a moisture diffusion model for a slab of food with an edible coating. The model differs from that by Hong et al. (1991) in two respects: (1) the coating is allowed to accumulate moisture, and (2) the moisture isotherm for the coating is included in the analysis. A finite difference solution for the model is presented, and results predcted by the model are compared to experimental data from Biquet and Labuza (1988a) for a freeze-dried gel covered with a chocolate coating. In "Simplified Model of Moisture Diffusion through a Coating Formed on a Food Product," the equations are simplified by neglecting the accumulation of moisture in the coating, and results predicted by this model are compared to those in "Numerical Solution of Moisture Diffusion through a Coating Formed on a Food Product."

ANALYTICAL SOLUTIONS OF DIFFUSION EQUATION FOR EDIBLE FILMS

Mass transfer through edible films can be estimated using Fick's second law for the special case of one-dimensional diffusion in a medium bounded by two parallel planes (Crank, 1975). The diffusion equation in one dimension is

$$\frac{\partial c}{\partial t} = D \frac{\partial^2 c}{\partial x^2} \tag{12.1}$$

The film is subject to known concentrations at the boundaries ($x = 0$, $x = L$) and an initial distribution within the boundaries. Crank (1975) presents a number of analytical solutions to this equation for a variety of initial and boundary conditions. We will examine the case of uniform initial distribution (c_o) and constant surface concentrations (c_a, c_b).

Initial and boundary conditions are thus:

$$c(x,0) = c_o \tag{12.2}$$

$$c(0,t) = c_a \tag{12.3a}$$

$$c(L,t) = c_b \tag{12.3b}$$

where t is time.

The analytical solution of Equation (12.1) subject to Equations (12.2) and (12.3) is given by Crank (1975) as:

$$c(x,t) = c_a + (c_b - c_a)\frac{x}{L}$$

$$+ \frac{2}{\pi}\sum_{n=1}^{\infty}\frac{c_b \cos n\pi - c_a}{n} \sin \frac{n\pi x}{L} \exp(-Dn^2\pi^2 t/L^2)$$

$$+ \frac{4c_o}{\pi}\sum_{m=0}^{\infty}\frac{1}{2m + 1} \sin \frac{(2m + 1)\pi x}{L}$$

$$\times \exp(-D(2m + 1)^2\pi^2 t/L^2) \tag{12.4}$$

It is instructive to utilize this equation to calculate moisture concentrations in dark chocolate edible films as tested by Biquet and Labuza (1988a). Boundary conditions were chosen to approximate the permeability test configuration 2 in their Table 3. The chocolate film was sealed on top of a cup inside a desiccator with ambient conditions inside the desiccator kept dry, and a salt solution placed inside the cup to maintain a high water activity. Relative humidity of the air next to the top of the film was 0.0, and the bottom side of the film was exposed to air with a relative humidity of 80.8%. A coordinate system is designed such that $x = 0$ corresponds to the side of the film exposed to the atmosphere inside the cup, and $x = L$ is the outer surface exposed to ambient conditions inside the desiccator cabinet.

To find the surface concentrations, c_a and c_b, it is necessary to know the relation between relative humidity of the surrounding air and moisture concentration in the chocolate. Biquet and Labuza (1988a) experimentally determined moisture isotherms for dark chocolate and fitted the Guggenheim-Anderson-DeBoer (GAB) equation to the data:

$$m = \frac{m_o CKa_w}{(1 - Ka_w)(1 - Ka_w + CKa_w)} \tag{12.5}$$

where m is the equilibrium moisture, percent dry basis and a_w is the water activity (which equals the relative humidity divided by 100). They found for absorption tests over the range of water activity 0.01 to 0.808 that $m_o = 0.545$ g water/100 g solids, $K = 1.024$ and $C = 103.857$. Using a chocolate density of 1750 kg/m³ and an equilibrium moisture calculated from the GAB equation, surface concentrations for configuration 2 were calculated to be $c_a = 53$ kg/m³ and $c_b = 0$.

Biquet and Labuza (1988a) determined an effective diffusion coefficient for dark chocolate of $D = 1.08 \times 10^{-13}$ m²/s for absorption when the relative humidity was 75%. Film thickness was 0.61 mm, and initial moisture concentration was 12 kg/m³.

A FORTRAN program was written on a personal computer to evaluate Equation (12.4). Predicted concentration profiles at selected times are shown in Figure 12.1 using the experimental D from Biquet and Labuza (1988a). Note that steady state has not been reached after 200 hours, as the profile is still slightly curved.

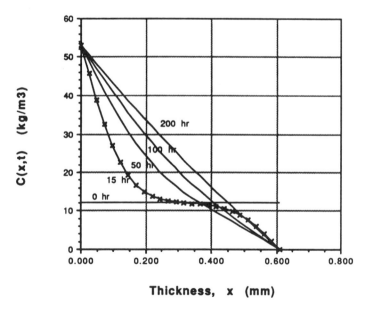

Figure 12.1 Moisture concentration profiles in chocolate film, configuration 2.

Unsteady state diffusion theory can be used to explain some of the results Biquet and Labuza (1988a) reported for water vapor permeability of chocolate films. While evaluating test procedures using the ASTM cup method, they reported anomalous behavior for the chocolate films: the permeability appeared to depend on whether the low humidity side was inside or outside the test cup. They tried two procedures: in configuration 1 a salt solution was placed outside the cup and a desiccant inside; in configuration 2, the positions of the salt solution and desiccant were reversed. Theoretically, the water vapor transmission rate (WVTR) at steady state should be the same for both configurations; however, they reported widely different values depending on the test configuration. For a 0.61 mm chocolate film subject to 0–80.8% relative humidity differential, they measured a WVTR 3.21 g/m²·day for configuration 1, and a WVTR of 0.67 g/m²·day for configuration 2.

WVTR is determined experimentally by measuring Q_t, the total amount of substance per unit film area that diffuses across the film. For the ASTM cup method, this is simply the change in weight of the cup, film and its contents. At steady state, Q_t is a linear function of time, and WVTR is calculated from the slope of the Q_t versus time curve. It reportedly took 15 hours to reach steady state for configuration 1 and 75 hours for configuration 2.

Unsteady state diffusion theory can be used to predict Q_t as a function of time, given values for D and the initial and surface concentrations. An expression for Q_t is obtained by integrating the flux at the outer surface of the film over time:

$$Q_t(t) = \int_0^t \left[-D \frac{\partial c}{\partial x}(L,t) \right] dt \qquad (12.6)$$

The result of this integration is given by Crank (1975) for uniform initial and constant boundary conditions:

$$Q_t(t) = D(c_a - c_b)\frac{t}{L} + \frac{2L}{\pi^2} \sum_{n=1}^{\infty} \frac{c_a \cos n\pi - c_b}{n^2} \{1 - \exp(-Dn^2\pi^2 t/L^2)\}$$

$$+ \frac{4c_o L}{\pi^2} \sum_{m=0}^{\infty} \frac{1}{(2m+1)^2}\{1 - \exp(-D(2m+1)^2\pi^2 t/L^2)\} \qquad (12.7)$$

This equation forms the basis of a method for determining the diffusion constant, the permeability constant, and the solubility of a gas in a film (Crank, 1975).

Evaluation of Equation (12.7) using Biquet and Labuza's (1988a) experimental value for diffusion coefficient demonstrates that the films were not at steady state after 15 or even 75 hours. Plots of Q_t versus time predicted by the equation for the two test configurations are shown in Figure 12.2. Boundary and initial concentration values for configuration 1 were $c_a = 0$ kg/m³, $c_b = 53$ kg/m³, $c_o = 12$ kg/m³, and for configuration 2 were $c_a = 53$ kg/m³, $c_b = 0$ kg/m³, $c_o = 12$ kg/m³. The slopes of the two curves do not become constant (and equal) until nearly 400 hours.

Biquet and Labuza (1988a) assumed steady state to be reached when the change in weight of the test cup over time became constant as determined by plotting the data and visually determining when the slope became constant. A plot of Q_t for the first 80 hours in configuration 2 is replotted in Figure 12.3. Because of the slowly changing nature of Q_t, it is easily concluded that steady state has been achieved after 75 hours; a straight line fits the last portion of the curve quite well.

Instantaneous values of WVTR into (or out of) the cup can be calculated by differentiating Equation (12.4) with respect to x, multiplying the result by D, and evaluating the resulting expression at $x = L$. Predicted instantaneous values of the WVTR for the configurations are plotted in Figure 12.4; note the transmission rates do not become equal until nearly 400 hours.

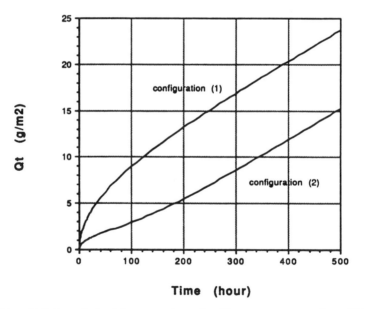

Figure 12.2 Amount of moisture per unit area that diffuses across outer surface of film.

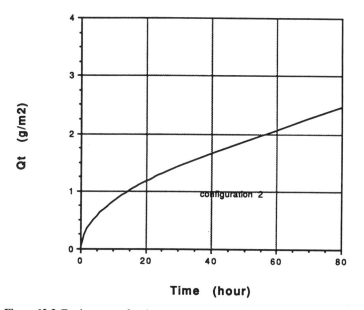

Figure 12.3 Total amount of moisture across outer surface of film, configuration 2.

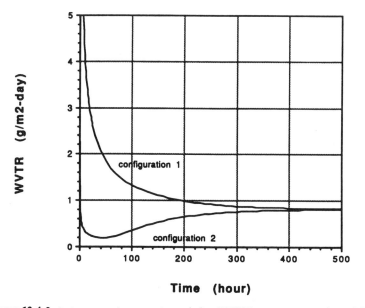

Figure 12.4 Instantaneous water vapor transmission (WVTR) across outer surface of film.

Theoretical values for WVTR from the model are 2.83 g/m²·day at 15 hours for configuration 1 and 0.47 g/m²·day at 75 hours for configuration 2, while Biquet and Labuza (1988a) measured values of 3.21 g/m²·day and 0.67 g/m²·day respectively. The model predicts a steady state value of 0.81 g/m²·day for both configurations.

The reason for the difference in initial values of WVTR for the two configurations is the difference in initial driving forces for moisture transfer at the outer surface of the film. Figure 12.1 shows the profiles for configuration 2, with $x = 0$ corresponding to the inner surface of the film, and $x = 0.61$ mm the outer surface. A plot of concentrations for configuration 1 would be mirror image of Figure 12.1, with the highest concentrations at the right hand side of the graph. Since the outer surface concentration is 53 kg/m³ for configuration 1 and 0 kg/m³ for configuration 2, the difference between the outer surface concentration and the initial film concentration of 12 kg/m³ is greater for configuration 1, resulting in a larger initial driving force. Once steady state reached, the driving forces are equal, since the profiles are linear.

NUMERICAL SOLUTION OF DIFFUSION EQUATION FOR AN EDIBLE FILM

Numerical solutions of diffusion equations are useful in studying the migration of moisture through films and food materials. Hong et al. (1991) employed the Crank-Nicolson finite difference method to solve a moisture diffusion equation that simulated the movement of moisture through a coated banana chip.

The Crank-Nicolson finite difference method (Smith, 1978) will be used to solve Equation (12.1) for a single material, and then used to solve the problem of a film coating a food material. The method was chosen because it is unconditionally stable; that is, predicted results do not grow exponentially due to the choice of time step. As illustrated in Figure 12.5, the film is divided up into a number of equally spaced segments of width Δx, and the concentration is solved at the edge of each segment at discrete intervals of time, Δt. The concentration at position $i\Delta x$ and time $n\Delta t$ is denoted by

$$c((i - 1)\Delta x, n\Delta t) = c_i^n \qquad (i = 1, 2, \ldots, \quad n = 0, 1, 2, \ldots) \quad (12.8)$$

Finite difference approximations are substituted for the derivatives in Equation (12.1) at each of the interior points $(i = 2, 3, \ldots, nf - 1)$ for the time $(n + 0.5)\Delta t$.

$$\frac{c_i^{n+1} - c_i^n}{\Delta t} = \frac{1}{2} \left[\frac{c_{i+1}^{n+1} + c_{i-1}^{n+1} - 2c_i^{n+1}}{\Delta x^2} + \frac{c_{i+1}^n + c_{i-1}^n - 2c_i^n}{\Delta x^2} \right] \quad (12.9)$$

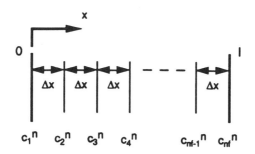

Figure 12.5 Division of film into discrete segments.

This results in one algebraic equation for each of the unknown position concentrations in the film interior. The boundary conditions at $x = 0$ and L are written as prescribed in Equations (12.3). The resulting algebraic equations for the concentrations at each interior and boundary point at time $(n + 1)\Delta t$ can be put into matrix form as shown below. Only nonzero entries are shown in the coefficient matrix multiplying the column vector of unknown concentrations.

$$
\begin{bmatrix}
1 & & & & \\
-r & (1+2r) & -r & & \\
 & \cdot & \cdot & \cdot & \\
 & & \cdot & \cdot & \cdot \\
 & & -r & (1+2r) & -r \\
 & & & & 1
\end{bmatrix}
\begin{bmatrix}
c_1^{n+1} \\
c_2^{n+1} \\
\cdot \\
\cdot \\
c_{nf-1}^{n+1} \\
c_{nf}^{n+1}
\end{bmatrix}
$$

$$
=
\begin{bmatrix}
c_a^{n+1} \\
rc_1^n + (1 - 2r)c_2^n + rc_3^n \\
\cdot \\
\cdot \\
rc_{nf-2}^n + (1 - 2r)c_{nf-1}^n + rc_{nf}^n \\
c_b^{n+1}
\end{bmatrix}
\qquad (12.10)
$$

where

$$
r \equiv \frac{D\Delta t}{2\Delta x^2} \qquad (12.11)
$$

The calculational procedure is to start with the initial conditions on the right hand side of Equation (12.10) and solve for the concentrations at the next time step. These are then substituted into the right hand side and the procedure is repeated again. A FORTRAN program was written on a personal computer to evaluate the numerical solution. The program has a structure similar to the Crank-Nicolson algorithm 11.3 listed in Burden and Faires (1985). It uses the Thomas algorithm to take advantage of the tridiagonal nature of the coefficient matrix multiplying the unknown concentrations (Holland and Liapis, 1983). A plot of concentrations predicted by this method (denoted by the small x symbols) is overlaid on the analytical solution for $t = 15$ hours in Figure 12.1. The values for Δx and Δt used in the numerical calculations were $\Delta x = 0.0244$ mm and $\Delta t = 0.0833$ hour, and it can be seen that the numerical results are in excellent agreement with the analytical solution.

NUMERICAL SOLUTION OF MOISTURE DIFFUSION THROUGH A COATING FORMED ON A FOOD PRODUCT

Feasible mathematical models for prediction of mass transfer through coated food systems are based on Fickian models developed for mass transfer in uncoated foods or foods packaged with polymeric films.

The problem analyzed is that of a film of thickness L_2 placed on a food product of thickness $2L_1$ in order to prevent moisture from entering or leaving the product. A transient, one-dimensional problem will be considered, and the diffusion equation [Equation (12.1)] is assumed to hold in each material. The origin of the x-axis is placed at the center of the food, so that because of symmetry, we only need to analyze the concentrations for the positive values of x. The diffusion equations for the food and film are:

$$\frac{\partial c}{\partial t} = D_1 \frac{\partial^2 c}{\partial x^2} \qquad (0 < x < L_1) \tag{12.12}$$

$$\frac{\partial c}{\partial t} = D_2 \frac{\partial^2 c}{\partial x^2} \qquad (L_1 < x < L_1 + L_2) \tag{12.13}$$

Four boundary conditions are needed—two for each partial differential equation. The four conditions used in this study are:

(1) The symmetry at the center line of the product is modeled by

$$D_1 \frac{\partial c}{\partial x}(0,t) = 0 \tag{12.14}$$

(2) The flux of moisture out of the product, $f(L_1,t)$, equals the flux into the film

$$f(L_1,t) = -D_1 \frac{\partial c}{\partial x}(L_1^-,t) = -D_2 \frac{\partial c}{\partial x}(L_1^+,t) \tag{12.15}$$

where the + and − superscripts indicate positions on the food and film sides of L_1.

(3) There is a unique relationship between the concentration of moisture in the product and film at the interface. The moisture isotherms for the two materials are plotted together, and a series of straight line segments are curved to fit over the water activity range of interest. The interface boundary condition is written as

$$c(L_1^-,t) = Pc(L_1^+,t) + Q \qquad (12.16)$$

The relationship is nonlinear, as the coefficients P and Q are, in general, functions of concentration just inside the film.

(4) Prescribed concentration at the outer film boundary is a known function of time

$$c(L_1 + L_2,t) = c_b(t) \qquad (12.17)$$

The initial conditions in the two materials are

$$c(x,0) = c_{o1} \ (0 < x < L_1) \qquad (12.18a)$$

$$c(x,0) = c_{o2} \ (L_1 < x < L_1 + L_2) \qquad (12.18b)$$

Again, the Crank-Nicolson finite difference method was used to approximate the partial differential equations [Equations (12.12) and (12.13)] subject to the boundary and initial conditions [Equations (12.14) through (12.18)]. The food and film were divided into segments, as shown in Figure 12.6. Note that there are provisions for two concentrations at the food-film interface, one for the food and one for the film.

The partial differential equations [Equations (12.12 and 12.13)] and the boundary conditions are approximated using finite differences, and the resulting algebraic equations can be put into matrix form as before, with

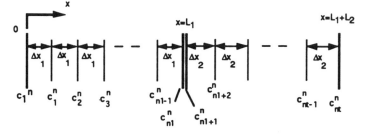

Figure 12.6 Division of food and film into discrete segments.

the coefficient matrix being tridiagonal in form. Only the nonzero elements of the matrix are shown.

$$
\begin{bmatrix}
(1+2r_1) & -2r_1 \\
-r_1 & (1+2r_1) & -r_1 \\
& \cdot & \cdot & \cdot \\
& & -r_1 & (1+2r_1) & -r_1 \\
& & & -2r_1 & (1+2r_1) & 2\dfrac{r_1\Delta x_1}{D_1} \\
& & & & 1 & 0 & -P \\
& & & & & -2\dfrac{r_2\Delta x_2}{D_2} & (1+2r_2) & -2r_2 \\
& & & & & & -r_2 & (1+2r_2) & -r_2 \\
& & & & & & & \cdot & \cdot & \cdot \\
& & & & & & & & -r_2 & (1+2r_2) & -r_2 \\
& & & & & & & & & 0 & 1
\end{bmatrix}
\begin{bmatrix}
c_1^{n+1} \\
c_2^{n+1} \\
\cdot \\
\cdot \\
c_{n1}^{n+1} \\
f^{n+1} \\
c_{n1+1}^{n+1} \\
\cdot \\
\cdot \\
c_{nt-1}^{n+1} \\
c_{nt}^{n+1}
\end{bmatrix}
$$

$$
=
\begin{bmatrix}
(1-2r_1)c_1^n + 2r_1 c_2^n \\
r_1 c_1^n + (1-2r_1)c_2^n + r_1 c_3^n \\
\cdot \\
r_1 c_{n1-1}^n + (1-2r_1)c_{n1}^n + r_1 c_{n1+1}^n \\
2r_1 c_{n1-1}^n + (1-2r_1)c_{n1}^n - 2r_1\Delta x_1 f^n \\
Q \\
2r_2\Delta x_2\dfrac{D_1}{D_2}f^n + (1-2r_2)c_{n1+1}^n + 2r_2 c_{n1+2}^n \\
r_2 c_{n1+1}^n + (1-2r_1)c_{n1+2}^n + r_1 c_{n1+3}^n \\
\cdot \\
r_2 c_{nt-2}^n + (1-2r_1)c_{nt-1}^n + r_1 c_{nt}^n \\
C_b^{n+1}
\end{bmatrix}
\tag{12.19}
$$

where

$$
r_1 \equiv \frac{D_1\Delta t}{2\Delta x_1^2} \qquad r_2 \equiv \frac{D_2\Delta t}{2\Delta x_2^2}
\tag{12.20}
$$

Several of the equations are nonlinear, because Q and P are functions of concentration at the interface. This was resolved by using the values for concentrations from the previous time step to calculate P and Q. The equations were solved on a personal computer using a program written in FORTRAN. The numerical solution was first verified by considering the case where the properties of the food and film were the same, and comparing predicted results to the previous model.

Next, to verify the theory, the simulation was used to predict the behavior of the model system studied by Biquet and Labuza (1988a). The food was a dried gel (Biquet and Labuza, 1988b), while the film was dark chocolate.

A plot of the moisture isotherms for the gel and dark chocolate calculated from the GAB equation is shown in Figure 12.7. GAB model parameters for the gel were taken from Biquet and Labuza (1988b) and were equal to $m_o = 7.789$, $K = 0.706$ and $C = 10.062$. For each water activity, there are unique concentration values in the gel and film. These values are plotted against one another in Figure 12.8.

The resulting curve was approximated by three straight line segments, as shown in Figure 12.9. A linear regression feature in the plotting program was used to do the curve fitting. The P values are the coefficients of the x term and the Qs are the intercepts.

The numerical values for P and Q along with the range of chocolate concentration over which they are valid are:

$$\left.\begin{array}{l} P = 0.553 \\ \\ Q = -2.401 \end{array}\right\} \quad 5.0 < C_{\text{choc}} < 19.8$$

$$\left.\begin{array}{l} P = 0.188 \\ \\ Q = 4.328 \end{array}\right\} \quad 19.8 < C_{\text{choc}} < 34.8$$

$$\left.\begin{array}{l} P = 0.0564 \\ \\ Q = 9.432 \end{array}\right\} \quad 34.8 < C_{\text{choc}} < 101.6$$

A summary of the remaining physical and numerical simulation parameters is presented in Table 12.1.

Average moistures predicted by the model are plotted in Figure 12.10,

TABLE 12.1. Parameter Values Used in Coated Gel Analysis.

Parameter	Freeze Dried Gel	Chocolate Coating
Diffusivity, D (m²/s)	1.04×10^{-9}	1.08×10^{-13}
Density, ϱ (kg/m³)	75.7	1750.
Width, L (m)	9.70×10^{-2}	6.04×10^{-4}
Initial Conc., c_o (kg/m³)	5.58	5.50
Surface Conc., c_b (kg/m³)	—	53.0
Distance Step, Δx (m)	9.70×10^{-4}	6.04×10^{-5}
Time Step, Δt (s)	864.	864.

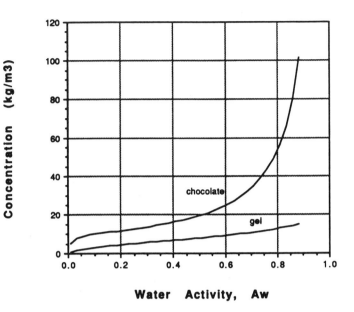

Figure 12.7 Moisture isotherms for gel and chocolate.

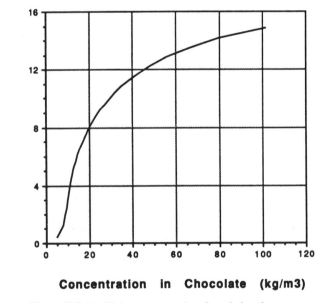

Figure 12.8 Equilibrium concentrations for gel-chocolate.

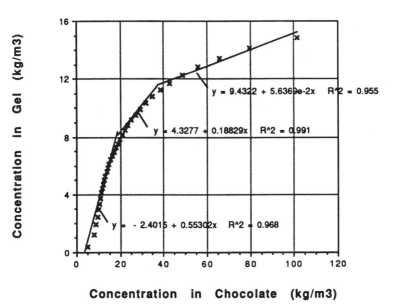

Figure 12.9 Piece-wise approximation of gel-chocolate equilibrium curve.

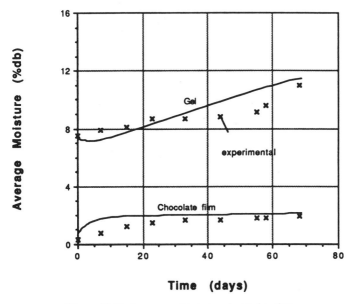

Figure 12.10 Average moisture in gel and chocolate.

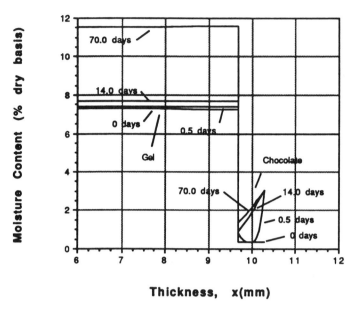

Figure 12.11 Predicted moisture profiles in gel and chocolate.

along with data taken from Figure 4 of Biquet and Labuza (1988a). This model predicts the experimental data as closely as the simplified model developed in their work. They assumed no accumulation of moisture in the film, and a uniform moisture distribution in the gel. Moisture profiles in the gel and chocolate for several times are shown in Figure 12.11.

SIMPLIFIED MODEL OF MOISTURE DIFFUSION THROUGH A COATING FORMED ON A FOOD PRODUCT

The time derivative term in Equation (12.13) may be neglected in cases when the rate of change of moisture in the coating is rapid in comparison to that in the food.

$$\frac{\partial^2 c}{\partial x^2} = 0 \qquad (L_1 < x < L_1 + L_2) \qquad (12.21)$$

The coating acts as if it is in steady state and there is a linear concentration gradient. This assumption is similar to that used by Hong et al. (1991); by characterizing the film covering foods with a permeability model, the accumulation of diffusing substance in the film is neglected. Equation (12.21) can be integrated once with respect to x to give

$$\frac{\partial c}{\partial x} = k_1 + k_2(t) \tag{12.22}$$

where k_1 and $k_2(t)$ are to be determined. Integrating Equation (12.22) gives

$$c(x,t) = (k_1 + k_2(t))x + k_3 + k_4(t) \tag{12.23}$$

where k_3 and $k_4(t)$ are also to be determined. By examining Equation (12.23), it is seen that the concentration in the film is a linear function of x, with a slope [k_1 and $k_2(t)$] and intercept [k_3 and $k_4(t)$] that vary with time. The slope can be determined in terms of the concentrations just inside both edges of the film. Writing the edge concentrations using Equation (12.23):

$$c(L_1^+,t) = (k_1 + k_2(t))L_1^+ + k_3 + k_4(t) \tag{12.24}$$

$$c(L_1 + L_2^-,t) = (k_1 + k_2(t))(L_1 + L_2^-) + k_3 + k_4(t) \tag{12.25}$$

Subtracting Equation (12.24) from Equation (12.25) and solving for the slope:

$$k_1 + k_2(t) = \frac{c(L_1 + L_2^-,t) - c(L_1^+,t)}{L_2} \tag{12.26}$$

This can be substituted into Equation (12.22) to give the gradient of the concentration in the film:

$$\frac{\partial c}{\partial x} = \frac{c(L_1 + L) - c(L_1^+,t)}{L_2} \tag{12.27}$$

At the interface, the flux boundary condition [Equation (12.15)] is then

$$-D_1 \frac{\partial c}{\partial x}(L_1^-,t) = -D_2 \frac{\partial c}{\partial x}(L_1^+,t) = -D_2 \frac{c(L_1 + L_2^-,t) - c(L_1^+,t)}{L_2} \tag{12.28}$$

Solving for the gradient in the food:

$$\frac{\partial c}{\partial x}(L_1^-,t) = \frac{D_2}{D_1} \frac{c(L_1 + L_2^-,t) - c(L_1^+,t)}{L_2} \tag{12.29}$$

The equilibrium relation, Equation (12.16), can be solved for the concentration in the film:

$$c(L_1^+,t) = \frac{1}{P}(c(L_1^-,t) - Q)$$ (12.30)

which is substituted into Equation (12.29)

$$\frac{\partial c}{\partial x}(L_1^-,t) = \frac{D_2}{D_1 L_2 P}(Q - c(L_1^-,t) + Pc(L_1 + L_2^-,t))$$ (12.31)

The model for the concentration in the food is then the diffusion Equation (12.12) subject to boundary conditions [Equations (12.14) and (12.31)].

Thus, the problem has been greatly simplified, as it is not necesary to solve the diffusion in the film. The effect of the film shows up in the food's outer surface boundary condition, Equation (12.31). The Crank-Nicholson method was used to solve the equations. Average moisture in the gel predicted by this simplified model and the full model are plotted in Figure 12.12. There is no lag in the uptake of moisture by the gel as was found for the full model. The model gives conservative results for designing; i.e., it

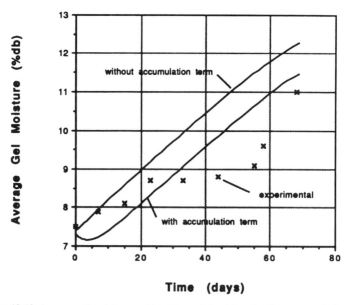

Figure 12.12 Average gel moisture predicted by models with and without accumulation term in film diffusion equation.

underestimates the time that a given film would allow the food to reach a given average moisture.

NOMENCLATURE

a_w water activity, (decimal)
C parameter in GAB moisture isotherm Equation (12.5)
C_{choc} concentration of chocolate in piecewise isotherm model (kg/m^3)
c_a moisture concentration at $x = 0$ (kg/m^3)
c_b moisture concentration at $x = L$ (kg/m^3)
c_o initial moisture concentration (kg/m^3)
$c(x,t)$ moisture concentration at position x and time t (kg/m^3)
D diffusion coefficient (m^2/s)
$f(x,t)$ flux of moisture at position x and time t (kg/m^2s)
K parameter in GAB moisture isotherm Equation (12.5)
$k_1, k_2(t)$ parameters in solution for concentration gradient Equation (12.22)
$k_3, k_4(t)$ parameters in solution for concentration Equation (12.23)
L film thickness (m)
m equilibrium moisture (% dry basis)
m_o parameter in GAB moisture isotherm Equation (12.5)
$P(c)$ parameter in interface boundary Equation (12.16)
$Q(c)$ parameter in interface boundary Equation (12.16)
$Q_t(t)$ total amount of water per unit area that diffuses across film (kg/m^2)
r non-dimensional ratio $= D\Delta t/(2\Delta x^2)$
t time (s)
x position (m)
Δt width of time step in finite difference scheme (s)
Δx width of distance step in finite difference scheme (m)

SUBSCRIPTS

1   food material
2   coating material

REFERENCES

Biquet, B. and T. P. Labuza. 1988a. "Evaluation of the Moisture Permeability Characteristics of Chocolate Films as an Edible Moisture Barrier," *J. of Food Sci.*, 53:989–998.

Biquet, B. and T. P. Labuza. 1988b. "New Model Gel System for Studying Water Activity of Foods," *J. of Food Proc. & Pres.,* 12:151–161.

Burden, R. L. and J. D. Faires. 1985. *Numerical Analysis, Third Edition,* Chapter 11. Boston, MA: PWS Publishers.

Crank, J. 1975. *The Mathematics of Diffusion, Second Edition,* Chapter 4. Oxford, England: Clarendon Press.

Geankoplis, C. J. 1983. *Transport Processes and Unit Operations, Second Edition,* Chapters 6, 7. Englewood Cliffs, NJ: Prentice Hall.

Holland, C. D. and A. I. Liapis. 1983. *Computer Methods for Solving Dynamic Separation Problems,* Chapter 4. New York: McGraw-Hill Book Co., Equations 4.23, 4.24, p. 132.

Hong, Y. C., C. M. Koelsch and T. P. Labuza. 1991. "Using the L Number to Predict the Efficacy of Moisture Barrier Properties of Edible Food Coating Materials," *J. of Food Proc. and Pres.,* 15:45–62.

Labuza, T. P., S. Mizrahi and M. Karel. 1972. *Mathematical Models for Optimization of Flexible Film Packaging of Foods for Storage, Vol. 15.* Trans. ASAE, p. 150.

Smith, G. D. 1978. *Numerical Solution of Partial Differential Equations: Finite Difference Methods, Second Ed.* Oxford, England: Clarendon Press.

Index